WITHDRAWN

AF699127

TECHNOLOGY OF ELECTRICAL MEASUREMENTS

WILEY SERIES IN MEASUREMENT SCIENCE AND TECHNOLOGY

Chief Editor

Peter H. Sydenham
University of
South Australia

Instruments and Experiences: Papers on Measurement and Instrument Design
R. V. Jones

Handbook of Measurement Science, Volume 1
Edited by P. H. Sydenham

Handbook of Measurement Science, Volume 2
Edited by P. H. Sydenham

Handbook of Measurement Science, Volume 3
Edited by P. H. Sydenham and R. Thorn

Introduction to Measurement Science & Engineering
P. H. Sydenham, N. H. Hancock and R. Thorn

Temperature Measurement
L. Michalski, K. Eckersdorf and J. McGhee

Technology of Electrical Measurements
Edited by L. Schnell

TECHNOLOGY OF ELECTRICAL MEASUREMENTS

Edited by

L. Schnell
Department of Measurement and Instrument Engineering
Technical University of Budapest
Hungary

Authors

I. Kollár **P. Pataki**
Gy. Korányi **L. Schnell**
P. Osváth **I. Zoltán**

JOHN WILEY & SONS
Chichester · New York · Brisbane · Toronto · Singapore

Copyright © 1993 by John Wiley & Sons Ltd.
Baffins Lane, Chichester
West Sussex PO19 1UD, England

All rights reserved.

No part of this book may be reproduced by any means,
or transmitted, or translated into a machine language
without the written permission of the publisher.

Other Wiley Editorial Offices

John Wiley & Sons, Inc., 605 Third Avenue,
New York, NY 10158-0012, USA

Jacaranda Wiley Ltd, G.P.O. Box 859, Brisbane,
Queensland 4001, Australia

John Wiley & Sons (Canada) Ltd, 22 Worcester Road,
Rexdale, Ontario M9W 1L1, Canada

John Wiley & Sons (SEA) Pte Ltd, 37 Jalan Pemimpin #05-04,
Block B, Union Industrial Building, Singapore 2057

TK
7878
.T43
1993

British Library Cataloguing in Publication Data

A catalogue record for this book is available
from the British Library

ISBN 0 471 93435 6

Typeset in T_EX by I. Kollár, J. Horváth and D. Miklós
Illustrations by T. Mányal
Printed and bound in Great Britain by Bookcraft (Bath) Ltd

''But do I thus measure, O my God, and know not what I measure?''

St. Augustine: *Confessions*

Contents

Preface

Measurement plays a fundamental role in every field of science and technology. While the methods and techniques of measurement may be different in various application domains, there are common principles and solutions which can be treated in a unified manner.

Measurement is a much more complex procedure than just the reading of the display of an instrument. The necessary preparations include the collection of *a priori knowledge* about the investigated phenomenon, and the definition of the measurement goal. In accordance with the goal, a model of the measured object or phenomenon has to be established. This *modelling step* is a crucial point in every measurement: the quality of the model is usually decisive for the whole measurement process. It must also be emphasized that modelling includes subjective components, therefore the quality of measurement strongly depends on the skill and experience of the person who designs the whole procedure.

Having defined the model, the means and methods of information gathering and processing have to be designed, and the *measurement devices* have to be chosen. The key design criterion is the *measurement error.* The measured quantities, the necessary measurement time, and the processing algorithms have to be selected so as to satisfy the required level of accuracy. If necessary, the model may also have to be refined to provide a more precise description of the object.

While the physical nature of the investigated object is taken into account primarily in modelling, the same general principles of signal and system theory and signal processing can be applied to very different physical problems. Noisy measurements are usually evaluated using the tools of estimation and decision theory; measurement results are usually considered and processed as signals rather than scalar quantities.

However, the measurement of *electrical quantities* has outstanding importance in present day engineering. In the current technology electrical signals are the primary media of information transfer. Measurement of nonelectrical quantities is usually performed by converting them into electrically measurable forms; signal processing and information processing are generally realized by means of electrical or electronic systems. Quality assurance, testing, process monitoring and control all rely on electrical equipment.

This book has been written keeping the complex nature of measurement in mind, and recognizing the utmost importance of electrical quantities. The key concepts are *modelling, measurement devices and methods,* and *measurement error.* The aim is to present the basic principles with their advantages and disadvantages. The application field covers the frequency range up to about 50 MHz, at low and medium voltages (below 1000 V).

There is a fairly rich assortment of instruments on the market for the measurement of electrical signals and quantities, from highly complex signal analyzers to automatic bridges, multimeters etc. However, with the increase of processing power, correction capability and increasing precision of the instruments, the task of the user becomes very complex. Interpretation of measurement results is not simple any more. For example, in high precision capacitance measurements one should be aware of the frequency and waveform of the signals used, and of the model of the impedance applied in the instrument. Or, using a signal analyzer, the variance of the estimate of some functions may result in strange forms that can easily be misinterpreted by an unsuspecting user. In general, the user should be familiar with at least the basics of the applied measurement and signal processing method, otherwise he may be misled using his expensive arsenal.

There is an abundance of technical literature available on specific measurement problems. However, an overview of electrical measurements is quite often difficult to attain. There are various books which orient themselves on the problems of basic applications like measurement of electrical machines etc, or on the apparatus applied to perform specific electrical measurements. The approach of this book is different. It concentrates on the measurement methods classified according to the electrical quantities: voltage, current, impedance etc.

Although signal analysis could be treated in a self-contained form, we found on the basis of our research and educational work that quite a lot of electrical measurements rely on signal processing methods and signal modelling concepts. Thus a common treatment of the two fields has many advantages and therefore has been used in this book.

The book consists of 7 chapters. The first one deals with the modelling procedure, gives a very brief overview of signal models, and discusses the estimation, treatment and presentation of measurement errors.

Chapters 2–5 deal with the principles, devices and methods of the measurement of the basic electrical quantities: voltage, current, power, energy, impedances, time and frequency. The presented devices are discussed at block diagram level, and the attention is focused on the working principles and the estimation of the errors. Wherever it is important, the investigation covers the effects of the load of the measuring device, of the measuring leads, of the stray impedances and couplings, and of external noise, and deals with the usual models of the measured object, too.

Chapter 6 gives an overview of the next step of the general measurement procedure: it deals with the methods of signal analysis, and the available devices for signal processing. Proofs and complicated mathematics are avoided; instead, explanations based on simple models and illustrations are given. A speciality of the treatment is that the errors of basic signal processing methods are analyzed not only for random signals with continuous spectrum, but also for sine waves with fixed or random phase.

Chapter 7 briefly summarizes the metrological background of electrical measurements: the definition, realization and accuracy of the base units (metre, kilogram, second, ampere), and practical realizations of the standards.

The authors are all present or former staff members of the Department of Measurement and Instrument Engineering, Technical University of Budapest. The book partly relies on a more comprehensive book published in 1985 in Hungarian (Schnell, L., ed., Measurement Technology of Signals and Systems, Műszaki Könyvkiadó, Budapest), which contains in addition a large amount of material on modelling, signal and system theory, estimation theory, structure and functional units of measurement systems. However, making use of the experience obtained through the use of the original Hungarian version in education, significant parts of the relevant chapters have been re-written to improve intelligibility and to incorporate some more recent achievements in the field.

Though focusing on electrical measurements, the treatment is such that it is not necessary to deal with details of electrical circuitry. Thus readers who are not electrical engineers will hopefully be able to follow the train of thought, understand the physical principles and modelling background of individual methods, and get acquainted with error estimation procedures.

The material described in the book has been successfully used in the teaching of measurement technology at the Technical University of Budapest. The first five chapters cover the material of a one-semester course, while Chapter 6 may serve as a basis for a course in the second semester. Chapters 1 and 6 may also be used for a stand-alone introduction to modelling and signal analysis. We also believe that the book is useful for practicing engineers and scientists, since it presents the latest developments along with the classical solutions using a modern, comprehensive approach to measurement.

For the use of the book, only some basic knowledge is supposed. The reader should be familiar with the basics of electrical circuits (Ohm's law, Kirchoff's laws, operational impedances etc). For Chapters 1 and 6 the knowledge of some basic concepts of signal theory is assumed (Fourier series expansion and Fourier transform), and furthermore, some basics of probability calculus (probability density function, expected value, standard deviation, variance).

The authors gratefully acknowledge the very useful remarks and suggestions of several colleagues who read some parts of the manuscript. Especially valuable help was provided by Professor János Sztipánovits (Vanderbilt University, Nashville, TN), Dr. Reiner Thomä (Techniche Hochschule Ilmenau, Germany), Professor Johan Schoukens (Vrije Universiteit Brussel, Belgium) and Dr. István Novák (Technical University of Budapest).

Budapest, January 1993.

László Schnell

Glossary of Symbols

B	bandwidth, magnetic induction
$\mathrm{cov}\{x, y\}$	covariance
$C(\tau)$	covariance function
c_k	complex Fourier coefficient
C	capacitance
D or $\tan\delta$	dissipation factor or loss tangent
$\exp(\cdot)$	exponential function, also $e^{(\cdot)}$
$E(f)$	energy density spectrum
$\mathcal{E}\{x\}$	expected value (mean value)
f	frequency
f_s	sampling frequency, sampling rate
f_0	centre frequency
f_1	first harmonic
$f(x)$	probability density function
$\mathcal{F}\{\cdot\}$	Fourier transform: $\int_{-\infty}^{\infty} x(t)e^{-j2\pi ft}\,\mathrm{d}t$
$F(x)$	probability distribution function
G	conductance
$h(t)$	impulse response
$H(f)$	frequency response, transfer function
$i(t)$	time function of current
I	effective value of current
$\mathrm{Im}\{\cdot\}$	imaginary part
j	$\sqrt{-1}$
$\lg(\cdot)$	base 10 logarithm
$\ln(\cdot)$	natural logarithm (base e)
L	inductance
$\mathcal{L}\{\cdot\}$	Laplace transform
M	mutual inductance
$n(t)$	noise
N	number of samples, number of turns of a coil
$N(\mu, \sigma)$	normal distribution
p_i	probability

P	power
$\mathrm{P}\{\cdot\}$	probability of an event
q	quantum size in quantization
Q	quality factor or weighting coefficient
$r(t)$	normalized covariance
R	resistance
$\mathrm{Re}\{\cdot\}$	real part
$R(\tau)$	correlation function
s	Laplace variable, empirical standard deviation
s^*	corrected empirical standard deviation
$S(f)$	power spectral density
$S^c(f)$	covariance power spectral density
$\mathrm{sgn}(x)$	sign function
$\mathrm{sinc}(t)$	$\sin(x)/x$
T	time constant, integration time
T_m	measurement time
T_p	period length
T_r	record length
$\tan\delta$ or D	loss tangent or dissipation factor
u	standard normal random variable
$u(t)$	time function of voltage
U	effective value of voltage
U_p	peak value
U_pp	peak-to-peak value
$\mathrm{var}\{x\}$	variance
$w(t)$	window function in the time domain
$W(f)$	window function in the frequency domain
$W_x(\alpha)$	characteristic function
x, $x(\xi)$	random variable
$x(t)$	measured time function
X	reactance, general signal amplitude
$X(f)$	Fourier transform of $x(t)$
$X(f, T)$	finite Fourier transform of $x(t)$
Z	impedance
δ	angle error
Δf	frequency increment
Δt	time increment (sampling interval)
ε_0	permittivity of the vacuum
ε_r	relative permittivity
ϵ	error
ϵ_c	width of confidence interval
ϵ_r	relative error
$\in$	element of set, value within given interval
φ	phase angle
$\gamma(f)$	coherence function
μ	mean value
σ	standard deviation
Σ	covariance matrix
$\star$	convolution
$\widehat{\ }$	e. g. $\widehat{x}$, estimated value

Symbol	Meaning	Symbol	Meaning
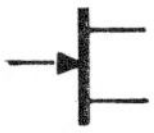	FET	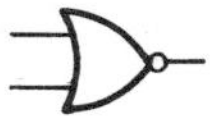	NOR gate
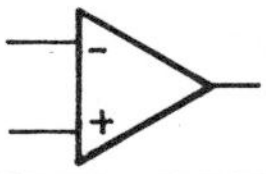	Operational amplifier		Input/output port
	Ground (general)	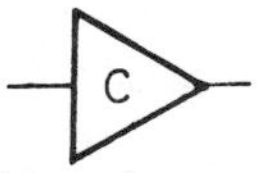	Comparator
	Ground (earth)	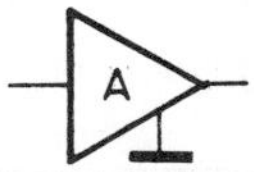	Amplifier
	Source ground	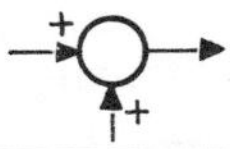	Summer
	Power supply ground	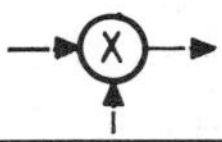	Multiplier
	Junction		Multiplying DAC
	Shielded lead		Lowpass filter
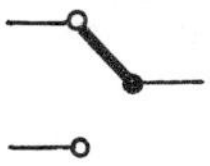	Alternate switch		Bandpass filter
	Switch	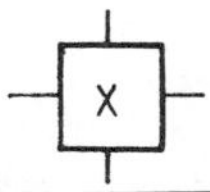	Hall generator
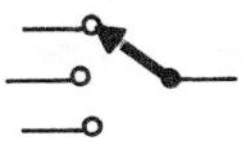	Multiple switch		Thermocouple
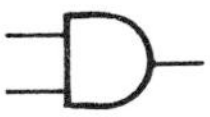	AND gate	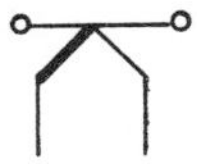	Thermal converter

Symbol	Description	Symbol	Description
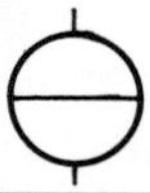	Current source (general)	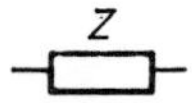	Impedance
	Voltage source (general)	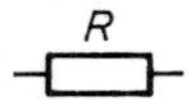	Resistor
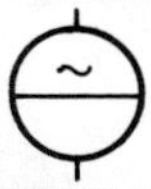	AC current source		Adjustable resistor
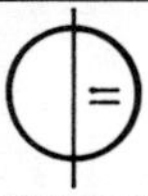	DC voltage source	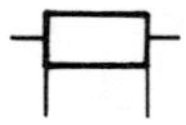	Shunt, Four-terminal resistor
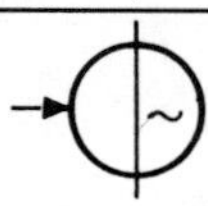	Controlled AC voltage source		Capacitor
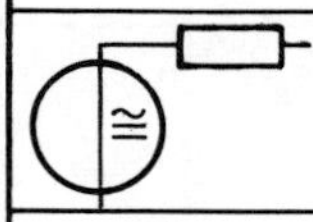	Non-ideal AC/DC voltage source		Adjustable capacitor
	Battery		Inductor
	Measuring instrument	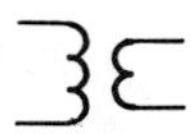	Mutual inductor
	Voltmeter	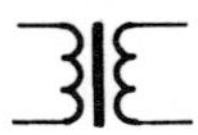	Transformer with iron core
	Ampermeter		Diode
	Null detector		Esaki (Zener) diode
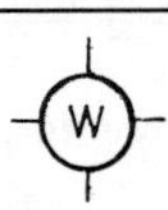	Wattmeter	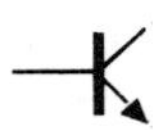	Transistor

Acronyms and Abbreviations

ac	alternating current
A/D	analog-to-digital
ADC	analog-to-digital converter
AF	audio frequency (20 Hz – 20 kHz)
AFC	automatic frequency control
AGC	automatic gain control
ALU	arithmetic and logic unit
AM	amplitude modulation
BW	bandwidth
CCD	charge coupled device
CMR	common-mode rejection
CMRR	common-mode rejection ratio
CRT	cathode ray oscilloscope
CT	current transformer (*not* computer tomograph in this book)
D/A	digital-to-analog
DAC	digital-to-analog converter
dBV	decibels relative to 1 V
dBm	decibels relative to 1 mW
dc	direct current
DFM	digital frequency meter
DFT	discrete Fourier transform
DMM	digital multimeter
DRFT	direct Fourier transform
DSP	digital signal processing (digital signal processor)
DUT	device under test
DVM	digital voltmeter
ECL	emitter coupled logic
emf	electromagnetic force
FET	field effect transistor
FFT	fast Fourier transform

FM	frequency modulation
HF	high frequency (3–30 MHz)
HP	highpass (sometimes: Hewlett-Packard)
HV	high voltage
IA	instrumentation amplifier
IC	integrated circuit
IF	intermediate frequency
I/O	input/output
LC	inductance-capacitance (circuit)
LF	low frequency (30–300 kHz)
LP	lowpass (*not* long play in this book)
LS	least squares
LSB	least significant bit
LSI	large scale integration
MDAC	multiplying digital-to-analog converter
MF	medium frequency (0.3–3 MHz)
MSB	most significant bit
MUX	multiplexer
NMR	normal-mode rejection
PA	power amplifier
pdf	probability density function
PF	power factor
PLL	phase-locked loop
PSD	power spectral density (function)
PWM	pulse-width modulation
RAM	random access memory
RC	resistance-capacitance (circuit)
RF	radio frequency
rms	root mean square
ROM	read only memory
S/H	sample-hold
SI	Système International (d'Unités): International System of Units
SNR	signal-to-noise ratio
TC	temperature coefficient
TDM	time division multiplier
U/D	up/down
UHF	ultra high frequency (300 MHz – 3 GHz)
ULF	ultra low frequency (0–300 Hz)
VAR	volt-ampere reactive
VCO	voltage controlled oscillator
V/F	voltage-to-frequency
VFC	voltage-to-frequency converter
VHF	very high frequency (30–300 MHz)
WFTA	Winograd Fourier transform algorithm
XO	crystal oscillator

1

Basic Concepts of Modelling and Measurement

L. Schnell

Editorial introduction

Measurement in a general sense is the procedure of obtaining information about the world by technical means. However, the surrounding world is extremely complex, and it is not possible to deal with every detail. Therefore, science uses simplified equivalents of real phenomena: it neglects unimportant aspects in order to catch the essential features. This simplification is called *modelling.*

Every measurement is subject to errors due to theoretical and practical reasons. These *measurement errors* may originate from modelling, from the circumstances and methods of measurement, from the properties of the measuring devices and from external disturbances.

Measurement errors can be analyzed and measurement procedures can be properly designed if the basic concepts of measurement and modelling are clear. The aim of this chapter is to provide such a background.

1.1 MODEL AND MODELLING

Measurements can be properly designed, performed and interpreted only when it is exactly defined, what is to be measured. The desire to obtain information about a certain phenomenon is still no specification. Specification means a careful description of the essential characteristic properties of the phenomenon under test, and the selection of the parameters of interest. The choice of characteristic properties and parameters, and the description of them by some formalism is the task of modelling.

A physical object may have a lot of different models. The type and the complexity of the model to be used strongly depend on the purpose of modelling, and on the important aspects of the reality. By the use of the model

Technology of Electrical Measurements, Edited by L. Schnell
© 1993 John Wiley & Sons Ltd.

- a part of the reality can be described, eventually emphasising some of its aspects;
- the phenomenon can be simplified, without effecting its essence, in order to make the quantitative analysis of the phenomenon simpler, or possible at all;
- the level of our knowledge, concerning the real phenomenon, is clearly stated and expressed.

For example, a *resistance* is the model of a resistor. The property that a current, flowing through it, will cause a voltage proportional to the current, is emphasized. It is neglected that thermal noise is produced, that each physical object has a self-inductance and a stray capacitance, and that the resistance changes with temperature, because for the given application these are unimportant.

1.1.1 Types of Model

There are three main types of model: conceptional, physical, and mathematical models. In engineering practice, mathematical models are the most important ones.

The knowledge about the phenomenon under test, represented by the mathematical model, can be classified as follows.

- Fundamental physical laws, determining the general form of the equations that describe the mathematical model.
- The structure, which describes the internal parts of the phenomenon and their relations.
- Parameters, i. e. the values of the coefficients in the equations.
- State variables, fully describing the process of a phenomenon together with the influencing external effects.

In the case of the resistor, this knowledge is summarized by Ohm's law: $U = IR$, the structure is a two-pole, the parameter is R. There are no external effects taken into account.

1.1.2 The Basic Concepts of Modelling

There are some universal principles to be applied in any modelling procedure. These are as follows (Courtois, 1985; Eykhoff, 1974):

Separation

The objects of the physical world are in mutual interaction, influencing each other's behaviour to a smaller or greater extent. The formulation of a modelling task requires the ability to separate one part of the world from the rest. This part is called the *system* or phenomenon to be modelled. Concepts, connected to the separation, are the system border, through which the system is interfacing its environment, and the elements (i. e. the subsystems) of the entire system, which are in mutual interaction.

Selection

A number of interaction types exist within a phenomenon, and also between it and the environment. However, the goal of an examination is the analysis of a phenomenon from the point of view of a given interaction, which means that, during the model creation process, the important interactions must be selected.

Economy

Due to the principles of separation and selection, a model is always simplified, thus, it is in some aspects an erroneous image of reality. Although this is true, separation and selection are inevitable, since overcomplex models cannot be handled.

The principle of economy means that the model, while corresponding to the goal of the test, should be the simplest of all. The simplicity of the model means the simplicity of the structure, and the minimum number of parameters and state variables.

The information necessary for the derivation of the model may come from two sources:

- *a priori knowledge:* all the knowledge, concerning the phenomenon to be modelled, at the beginning of the investigation;
- *a posteriori knowledge:* all the knowledge at the end of the modelling procedure. This is greater than the a priori knowledge by the information gained during the observation.

The first phase of the modelling procedure is the collection and systematic treatment of the a priori knowledge. The source of the a priori knowledge about the given phenomenon is a preceding analysis of it, comprising any possible connection to other phenomena, and physical laws. The a priori knowledge is of key importance in modelling. Unfortunately, its amount is always limited, since the attention is concentrated upon the description of a phenomenon, while its environment is usually less well known. Furthermore, the knowledge about the phenomenon itself is never complete, since considering all of its aspects is impossible. Even if the general rules are known, the specific relationships and quantities are usually unknown.

The a priori knowledge and the goal of the modelling determine the final type of model, the accuracy requirement, the type of specific modelling procedure, the complexity of the model, and the method and cost of its realization.

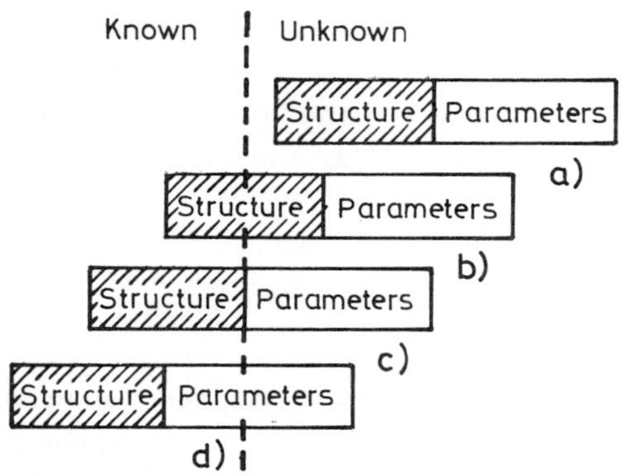

Fig. 1.1 Missing a priori knowledge

As to the missing information, four cases can be distinguished, as illustrated in Fig. 1.1.

The missing a priori knowledge can be substituted by its possible variations, by which a class of models is generated. The task is now to select an element of this class (i. e. a model), which corresponds best to the goal of the modelling task.

Among the four cases, illustrated by Fig. 1.1, there are substantial differences. While in the last two cases one single model is accepted (the structure is known) and it is only the parameter values that are not completely given, in the first two cases the structure itself is to be determined. In cases *c)* and *d)* the values of the parameters are sought, i. e. parameter estimation (or parameter identification) is to be carried out. The solutions of this problem are rather elaborate and standardized. At the same time, there are no similar methods for the selection of the structure; here engineering intuition and trial play a significant role.

An example for the above cases is the measurement of the transfer function of a given filter. It may be possible that practically nothing is known about the internal structure (e. g. a special integrated circuit is applied in it), which corresponds to case *a)*. When the circuit diagram is at hand, but it is uncertain how the operational amplifiers are to be modelled, case *b)* is faced. If the filter is constructed of discrete components, the structure is usually completely known however, because of the tolerances, the parameters are uncertain (case *c)*). It may also happen that a part of the circuit is realized by using a well-specified block whose parameters are known; this corresponds to case *d)*.

1.1.3 Methods of Modelling

Before an overview of the steps of model creation, the two extreme cases of model creating methods will be briefly summarized: the deductive one, using a priori knowledge only, and the inductive one, using experimental data only.

Deductive model creation

By deductive model creation, the modelling procedure is based on general laws of nature. During the theoretical analysis of the problem, the boundaries of the system under test are defined; its elements — subsystems — are separated, and, applying conservation and continuity laws, the boundary conditions and the interactions among the subsystems are determined. The gained a priori knowledge is obtained in the form of a certain structure, with its parameters. When both the physical laws and the knowledge, expressed by a structure and its parameters, are available, the generated mathematical model unambiguously describes the phenomenon under test (within the limitations of the uncertainties, caused by selection and separation). Since the internal structure of the phenomenon is known, or it can be accessed, the system is 'transparent' for the model maker, therefore this case is called 'white box'.

Deductive model creation is applied e. g. when the transfer function of a filter is calculated from the circuit diagram.

Inductive model creation

The observations during the experiments provide information about the interactions between the phenomenon and its environment, i. e. about the input and output signals of the system. During the experimental analysis, a description — or model — of the phenomenon is sought, which can imitate it; that is, it is able to reproduce the output signals. In a pure inductive case, nothing is known about the internal structure, the system is 'nontransparent', it is a so-called 'black box'. The phenomenon can usually be imitated by several models of different structure: inductive modelling is ambiguous. This is the case e. g. when the transfer function of a dynamic system is sought on the basis of measurements only, with very limited a priori knowledge, that is, even the order is not known.

Experimental data can be obtained in passive or active ways.

In the case of a *passive* experiment, the phenomenon can be observed under so-called normal operating conditions. The model maker in this case is not able, or not willing, to force artificial input signals upon the system. Observations originating from the system and its environment, not depending on the experimenter, have to be accepted.

In the case of an *active* experiment, the phenomenon under test is influenced by so-called test signals. Thus, the model maker directly intervenes in the interactions of the system and its environment, he influences the input signals partly or entirely, and observes the output signals as the response of the system.

The pure deductive or inductive methods are idealizations; they are extreme cases of the modelling processes, realized in practice. The models are usually somewhere between these extremes ('grey box').

1.1.4 The Steps of Model Creation

The modelling process generally is a very complex one, consisting of several consecutive steps. The consecutive models are based on each other, they combine the new information with the previous model. The knowledge, comprised by a model, can serve as a priori knowledge for another modelling problem. Therefore, an illustration of the modelling process is only possible with strong simplifications (see Fig. 1.2).

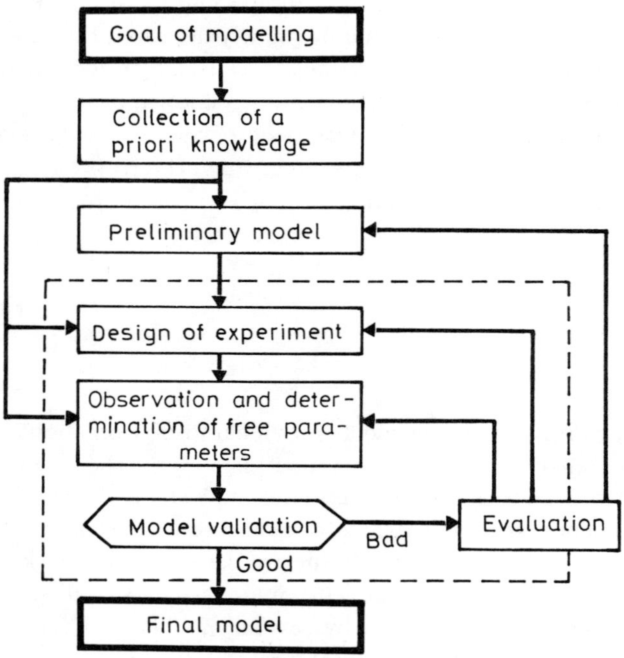

Fig. 1.2 Steps of the modelling process

The process starts by setting the *goal of modelling,* which determines the type of the model and gives aspects for the determination of the necessary accuracy.

The next step is the *collection of the a priori knowledge,* to be used for the construction of the model. It consists of all the physical laws and supplementary knowledge applicable in the investigation of the phenomenon, and of available structural and parametric information. On the basis of the a priori knowledge, the deductive phase of the model creation can be carried out, resulting in the *preliminary model.*

Using the preliminary model and the a priori knowledge, the design of the *experiment* can be prepared. In the case of a passive experiment, the design of the experiment means a selection of the quantities to be observed, and the determination of the time and duration of observation. In the case of an active experiment, this is extended by the selection of excitation points and the determination of the characteristics of the excitation signals.

After the design of the experiment, the *observation* of the phenomenon and the *recording* of the observed data follow, along with the *determination of the free parameters* of the preliminary model, reflecting the a priori knowledge.

The created model should be *validated,* thus the similarity between the phenomenon and the model must be evaluated, using some similarity criteria. If the result of the validation is that the model corresponds to the goal of modelling, the modelling process is finished, and the *final model* has been produced. The choice of the final model, as already mentioned, is ambiguous. It can be said that the less a priori knowledge is available, the more final models can be produced, satisfying the goal of modelling.

Another possible outcome of validation may be that the model is not satisfactory. The *evaluation* of the circumstances can lead to a number of different interventions. In the simplest case, the experiment should be continued, and the accuracy of the model can be improved by the information obtained through the new observations. If this does not work, intervention is necessary at the previous level, that is, the circumstances of the experiment (e. g. the placement of the observation points or the eventual test signals) must be modified. In an extreme case, it can happen that the hypothesis, made during the specification of the preliminary model, proves to be invalid, and significant structural or parametric modification of the preliminary model must take place.

As can be seen from the above discussion, model creation is a process of iterative character, consisting of a number of steps that should be followed until the model of the phenomenon is obtained.

Having surveyed the essential concepts of model and modelling, the fundamentals of measurement will now be discussed.

1.2 FUNDAMENTAL CONCEPTS OF MEASUREMENT

Measurement is a very general term. Therefore, numerous concepts may define the basic ideas, usually in a heuristic manner. Although there have been attempts to give the axiomatic formulation of measurement (Pfanzangl, 1968), and the abstract model of measurement can also be given, using the apparatus of the set theory (Finkelstein, 1975), here a heuristic discussion of the concepts of measurement will be presented, in order to avoid theoretical formulations which are difficult to understand.

1.2.1 Relation between Modelling and Measurement

The modelling procedure includes empirical stages, where observations are used for determination of the free parameters of the preliminary model. During these measurements, either the desired parameters are obtained, or a decision is reached that the preliminary model is not suitable for the required accuracy. Therefore, measurement is always embedded into the modelling procedure. There is no measurement without modelling: the design of a measurement postulates a preliminary model of the phenomenon. It can even be said that measurement is aimed at the determination of parameters of this model, and the connection of measurement and the true world is only established via a model.

The usability of the measurement result depends on the *modelling and measurement errors.* Clearly, there is a conceptional difference between these two errors. The modelling error is due to incorrect deduction and/or inaccurate a priori parameters, in addition to the necessarily simplified image of the actual phenomenon, yielding a presentation which has finite accuracy only. The sources of the measurement error are the measurement uncertainties and the inaccuracies in the evaluation and processing of the measured data.

The modelling error is in many cases the reasonable limit in specifying the measurement accuracy. Evidently, it would be meaningless to determine with high accuracy (and thus expensively) the parameters of a rough and inaccurate model.

1.2.2 General Definition of Measurement

According to the conventional definition, measurement is defined as the characterization of a physical or chemical quantity by a numerical value, expressed in a chosen measurement unit. However, this definition can be applied for only a part of all measurement tasks. For example, in the measurement of the internal status of a digital controller, the result is given as the status of the controller storage elements, expressed by suitable codes. Evidently, the conventional definition cannot be applied to this case because we do not measure the magnitude of a quantity, the result is not necessarily a number, and the measurement unit cannot be defined conventionally.

According to the generalized definition, a symbol, adapted to the model type, is assigned to the measurement result or, in other words, the set of measurement results is mapped to a set of symbols. This means that a measurement may not be directly expressed by a numerical value, and the elements of the symbol set may be arbitrary (colour, letter, shape, status etc).

The scale information includes all conventions and agreements related to the measurement. For physical and chemical measurements, the scale information is specified by an internationally accepted system of units (see Chapter 7). It thus follows that the measurement result is expressed by a symbol *and* the scale information.

Finally, the measurement error is defined as the distance between the real and ideal measurement result, defined by the distance measure of the symbol set. This generalization means that the distance concept cannot be restricted to difference building.

Note 1.1

Practical measurement methods can be classified in various ways. One can speak, for example, of measurement based on pointer deflection, and measurement based on the null method.

When a measurement is based on pointer deflection, the physical quantity to be measured is transformed into another physical quantity. For instance, the effect of the electrical current is opposed by a spring force when using an electromechanical instrument. The energy needed for the measurement is taken from the system to be measured, resulting in a systematic error which has to be corrected.

By a measurement based on the null method, the physical quantity to be measured is compared to the same physical quantity with a known adjustable magnitude. The two quantities are made equal by adjusting the latter, resulting in zero difference between them. The unknown signal magnitude can then be determined from the circuit diagram. Typical circuits based on the null method are the bridge circuit and the potentiometer circuit (Fig. 1.3).

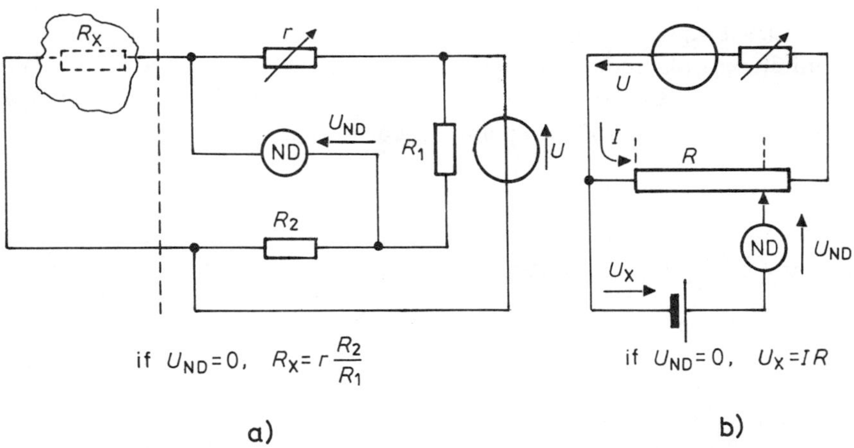

Fig. 1.3 Null method *a)* bridge *b)* potentiometer

Considering the bridge circuit, shown in Fig. 1.3a, the adjustable resistor r can be regarded as the preliminary model of resistor R_X to be measured.

1.2.3 Basic Characteristics of a Measurement Process

When solving a measurement task, a good model is sought, that is, a model which is as similar as possible to the ideal model of the system to be measured. Let us assume that within a class of models there is a best one. This model is certainly unknown, and the goal is to find this model, or a model close to it. If a criterion (cost function) is defined, which describes the deviation of the actual and ideal models in some sense, and the measurement procedure minimizes this criterion, the measurement process is called *optimal*, and the obtained parameter value is the optimal estimate of the parameter.

The measurement process can be characterized by the relation between its input and output. The observations, which serve as the input to the measurement process contain the information about the quantities to be measured in some hidden form, disturbed by influencing effects. However, the output of the measurement process should contain nothing but the useful information. This can be expressed by saying that the measurement process must be *selective* for the information concerning the quantities to be measured. This can be realized only if the preliminary model contains a usable conceptional or physical model of the object to be measured. This implies that measurement is aimed at quantification of a model, that is, in any measurement the model parameters and not directly the quantities of the true world are measured.

In the design of the measurement process, estimation theory can be applied if the desired quantity is of continuous nature, and decision theory can be used if it has discrete nature. Important topics of estimation theory are maximum likelihood, Bayes and least squares error estimations, while those of the decision theory are Bayes, minimax, maximum likelihood, and Neyman-Pearson decisions. These are described in the literature (Eykhoff, 1974; Bendat and Piersol, 1971; Sage and Melsa, 1971; Van Trees, 1968; Papoulis, 1965; 1977).

1.2.4 Types of Measurement Procedure

The main types of measurement procedure are distinguished according to the type of model applied in the definition of the measurement procedure, the method of realization, and the type of model applied for the measured object.

Deterministic measurement procedures

As to the definition of the measurement procedure, the adjective 'deterministic' means that when defining the criterion of similarity and the condition for optimum, probability theory is not used. This can be done by considering the a priori knowledge covering the object to be measured and the situation of observation.

The deterministic measurement procedure does not mean an error-free measurement. It refers to the use of deterministic models, in the deduction of the procedure only.

Common measurements like the measurement of a voltage or an impedance are the simplest examples of deterministic procedures.

Stochastic measurement procedures

In the derivation of stochastic measurement procedures, probabilistic models are used. The typical locations of these models within the measurement procedure are the input and the output signal of the object to be measured, the noisy signal transmission channel, and the expression of the a priori knowledge concerning the quantities to be measured.

An example is the use of estimation theory to determine unknown quantities from noisy measurements.

Analog measurement procedures

In analog measurement procedures, the quantities in the realization of the measurement procedure and in the information processing are analog. Analog realization is mainly applied in the case of simpler measurement procedures, where the measuring system can be synthesised from a small number of elements and the required accuracy is not too high.

Averaging by the use of an integrator or a lowpass filter (Subsection 6.1.4) are typical examples.

Discrete measurement procedures

Nowadays most measurement procedures are realized with the help of a computer. These work with discretized data, therefore the realized measurement procedures must be described with a proper apparatus. Sophisticated measurement systems have in most cases digital structure, e. g. transient analyzers or Fourier analyzers.

Parametric measurement procedures

The two basic types of model for the object and the associated quantity to be measured are parametric and nonparametric models.

When using a *parametric* approach, the measurement task is to determine a few parameters of a given model. Regarding the complexity of the measurement, the parametric models are more favourable, although for their definition some a priori knowledge about the structure is needed, and this is not always available.

An example is the measurement of a sine wave: in this case the aim is to determine the parameters (X, f_1, φ).

Nonparametric measurement procedures

When using nonparametric measurement procedures, the quantity to be measured is a function of one or more continuous variables. Well known examples are the frequency response of linear networks, or the autocorrelation function of a stochastic process. Nonparametric models are typically used for the description of systems with input and output, when the object to be measured has a 'black box' character.

Having surveyed the different types of measurement procedures, the discussion of the general problems of measurement and modelling will be stopped here. As a conclusion, the most important statement of this chapter is repeatedly emphasised: *the measurement is always a part of a model creating process,* and although it has its own features and special problems, it is only possible to understand the basic concept of measurement by a profound analysis of its relation to modelling. For further details, see e. g. Bendat and Piersol (1971), Eykhoff (1974), Goodwin and Payne (1977), Norton (1986).

Note 1.2

In *analog* measurements the information is represented by analog signals. Analog devices (e. g. amplifiers) produce noise, therefore the relationship between the output and the input of a measuring instrument is both deterministic and stochastic. In a chain, consisting of cascaded analog elements, the output of the ith element — with growing i — is less and less determined by the input of the chain, and it is more and more determined by the noises. Thus, in analog systems, an increase of the number of elements, that is, the increase the complexity of the system, results in a more and more noisy transfer. It is the noise effect which sets a limit to the reasonable extension of analog systems.

Analog measurement procedures are important, because on the one hand the characteristics of the objects to be measured are often represented by analog signals, on the other hand the analog output (the position of a pointer, or the appearance of a chart) is easy and comfortable to survey for the user. This is true especially in the case of a large number of displays (meter panels, facia boards) to be surveyed. Analog instruments are more suitable for rapid visual monitoring than digital ones.

In *digital systems* the measurement information is represented at every stage of the system by *digital* signals. Strictly speaking, the input signal of a digital measuring system is also digital (e. g. the number of bottles in a bottle-filling machine). In a wider sense, those systems are also considered as digital ones, where the input signal is analog, but an analog to digital (A/D) conversion is made at the input, or where just before the output a digital to analog (D/A) conversion occurs. A/D and the D/A conversions are always more or less noisy, and cause a loss of information. Digital technique theoretically ensures noise-free transfer, i. e. the input unambiguously determines the output. (This is not always true, e. g. because of rounding etc). Noise-free transfer means a transfer without distortion of the information, therefore an arbitrarily long chain can be established — the limitation is set by the reliability alone.

In *hybrid* systems the measurement information is represented by both analog and digital signals. In such measuring systems, beyond the usual digital functions, analog signal transferring and processing units (e. g. programmable amplifiers, filters etc.) also play an important role.

Hybrid systems unify the advantages of the analog and digital systems, they can realize the functions of the measurement procedure with the most favourable signal processing techniques.

Example 1.1

The relation between modelling and measurement is illustrated by the simple example of measuring the mains voltage at a wall socket. The a priori experience of the person who measures, if available, should be used for determining the measurement steps. In the case of such a trivial task, previous experience will normally result in the omission of several steps and the solution will be reached immediately. However, in the following, let us proceed step-by-step, using the basic a priori information.

First level

Mains: location of the phenomenon, external electromagnetic disturbances may be present (separation).
Voltage: one of the main characteristics, the rest of them excluded (selection).

It is thus seen that the definition of the mains and the voltage is based on a priori knowledge. Let this separation and selection be justified without further analysis, allowing the use of the mathematical voltage model $u(t)$ as a general time function. The measurement then requires the determination of the voltage function $u(t)$ in a given time interval $t_1 < t < t_2$.

Second level

Based on further a priori knowledge, the task may be specified in more detail. According to Fig. 1.4, the mains voltage is taken as a noisy periodic function given by

$$u(t) = \sum_{k=0}^{\infty} u_k e^{jk\omega_1 t} + n(t, \xi) .$$

This means that the signal to be measured consists of

- the fundamental component ($k=1$, $f_1 = \frac{\omega_1}{2\pi} = 50$ Hz),
- the sinusoidal harmonic components ($k=2, 3, \ldots$)
- the noise $n(t, \xi)$.

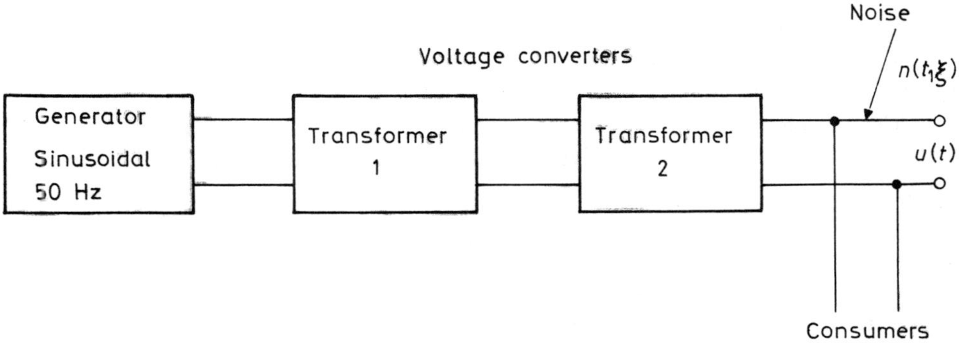

Fig. 1.4 Network model

According to our knowledge, the mains power is determined essentially by the fundamental component, and the amplitudes of the harmonics are normally much lower than the amplitude of the fundamental component. The effect of the noise source which may be either within or outside the network can generally be neglected unless the noise component itself has to be measured (economy).

Based on the above a priori knowledge, the final task is the measurement of an approximately sinusoidal periodic signal which has a fundamental frequency of 50 Hz.

Third level

Let us look for further a priori knowledge related to the method of measurement. An oscilloscope could be used to display the time function of the signal but, in spite of showing all the information related to the signal, this would not be a suitable method: a sinusoidal signal can be well characterized by a single number, condensing all information (economy). This is, e. g., the root-mean-square value or the mean absolute value of the signal.

It is known that the root-mean-square value (i. e. the effective value) is the most meaningful one so an instrument is selected, which is suitable for measuring this value. There are different kinds of rms instruments: the true rms instrument actually measures the true rms value, while other instruments measure the mean absolute value or the peak value, but they are also calibrated in rms values. The latter display the correct rms value for sinusoidal signals only, while the former has no such restriction.

We further know that the rms value of the mains voltage is around 220 V. Assume now that the voltage to be measured comprises up to 20 harmonics which should all be included in the measurement without distortion. The selected instrument will then be characterized as follows.

voltage range:	200 to 250 V
frequency range:	at least to 1000 Hz
method of measurement:	true rms

This kind of instrument could be, for example, an electrodynamic voltmeter. A medium accuracy class of 0.5 to 1 could be chosen as no accuracy requirement was specified.

1.3 SIGNAL AND SYSTEM MODELS

When designing measurement procedures, mathematical models of signals and systems are used. The theory of signals and systems deals with the synthesis, analysis, and modification of these models.

In this section, a short survey of the considerations resulting in the creation of the different model classes will be given, and the main characteristics of the most important models will be briefly listed. More detailed discussions can be found in Bendat and Piersol (1971), Eykhoff (1974), Oppenheim et al (1983), Urkowitz (1983).

When making the mathematical model, the signals or the systems are identified by giving their characteristics, that is, the relationships of their characteristic variables. The model gives abstract mathematical relationships among abstract mathematical objects. In the case of physical systems, these relationships express mutual interactions, determined by physical laws, in mathematical form. For example, the voltage and the current of an impedance must satisfy Ohm's law, or the internal relations of an electrical network can be described by Kirchhoff's laws, etc. When creating the model according to signal and system theory, these laws are applied.

1.3.1 Signal Models

Signal models can be divided into two groups: deterministic and stochastic.

a) Deterministic signals

A signal model is deterministic if it contains no random components. For example, the signal in Eq. (1.1) is a deterministic model, if X, f_1, φ are fixed (though maybe unknown) parameters.

A deterministic signal model is used if the 'outcome' of the interactions, generating the signal, seem to be unambiguously determined, and also the character of the measurement task requires the use of a deterministic model. In some cases deterministic signal models are also used when the output of the interactions does not seem to be definitely determined, but the probability relations are not known, and using the deterministic model, a better result can be expected than with the use of some assumed probability distribution (see Subsection 1.6.3).

The deterministic signals can be subdivided into a number of groups. The classification relies on the mathematical apparatus used when applying these models.

Continuous (analog) signals. Analog signals are continuous both in time and amplitude or in general terms: both the independent and dependent variables are continuous.

Applying a continuous signal model, it is ensured that characterization of the interaction, generating the signal, will be possible at any moment within a given time interval. The model is the natural one for most physical phenomena. For continuous signal models an effective mathematical apparatus is available, but a disadvantage is that they do not directly fit the widely applied digital signal processing.

Discrete signals. In discrete signal models, the signal is characterized at certain points of the space and/or at selected time instants only, and often also the amplitudes are discretized. The most important application of such a model is that if some, in practice not severe, conditions (e. g. the sampling laws) are fulfilled, this model gives a description equivalent to the continuous case. In the case of uniform distribution of the space points and time instants, there are mathematical means for the description of the discrete signal models at least as effective as those in the continuous case. These methods are suited to digital signal processing.

In the following pages the basic types of deterministic and stochastic signals will be surveyed based on the classification of Bendat and Piersol (1971).

Deterministic signal models are classified as shown in Fig. 1.5.

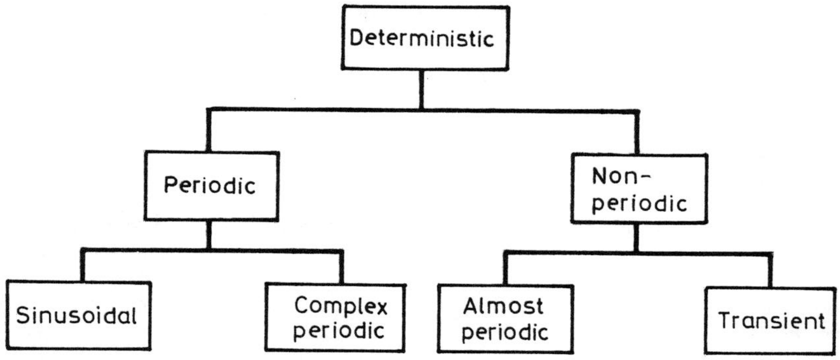

Fig. 1.5 Classification of deterministic signals

- *Sinusoidal signal* (Fig. 1.6):

$$x(t) = X\cos(2\pi f_1 t + \varphi)\,, \tag{1.1}$$

where

$x(t)$	is the instantaneous value at time t,
f_1	is the frequency,
X	is the signal amplitude,
φ	is the initial phase angle,
T_p	is the period time.

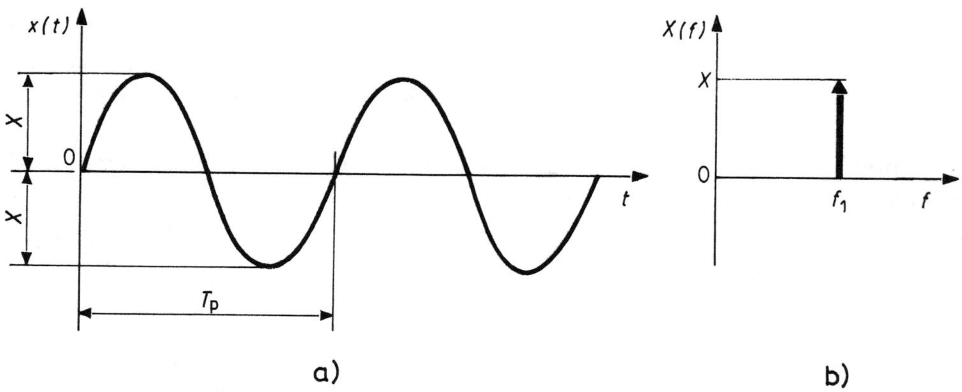

Fig. 1.6 Sinusoidal signal *a)* in the time domain *b)* in the frequency domain

- *Complex periodic signal*

The condition for periodicity is given by

$$x(t) = x(t \pm kT_p)\,, \qquad k = 1,\, 2,\, \ldots$$

In this case, $x(t)$ may generally be written as

$$x(t) = X_0 + \sum_{k=1}^{\infty} X_k \cos(2\pi k f_1 t + \varphi_k) = \sum_{k=-\infty}^{\infty} c_k e^{j2\pi k f_1 t}, \tag{1.2}$$

where

$$X_k = 2|c_k| \quad \text{and} \quad \varphi_k = \arg c_k .$$

It can be seen that the periodic signal consists of a constant component X_0 and a number of sinusoidal components, called harmonics, as shown in Fig. 1.7. The harmonic frequency values are integer multiples of f_1. Complex periodic signals can be characterized by their Fourier series. The power, associated with the frequency f_k, is $P_k = |c_k|^2$, for $k = \pm 1, 2, \ldots$. The so-called *power spectrum* is the function which is equal to P_k at f_k, otherwise it is zero.

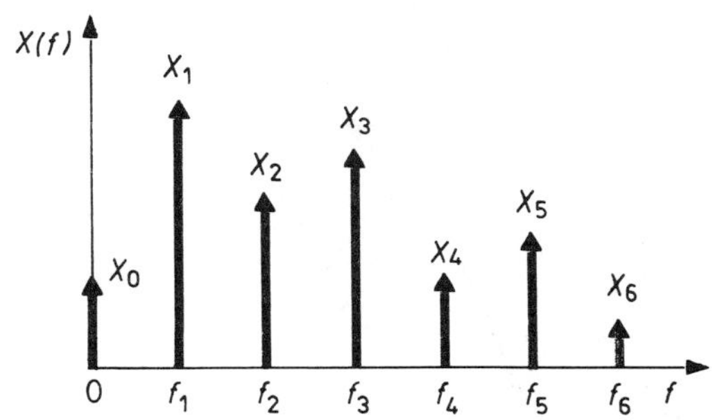

Fig. 1.7 Spectrum of a complex periodic signal

- *Almost periodic signal*

  The addition of two or more sinusoidal signals with different frequencies results in a periodic signal if the ratio of any two component frequencies can be expressed by a rational number. However incommensurate frequencies cannot be measured, since for this *no measurement error* would be allowed, thus the above definition is more adequate for practical purposes.

  This model is applied for investigating the sum or product of two independent signal sources. Due to frequency fluctuations, the resulting signal is not periodic, in spite of the approximately periodic components, having line spectra. A good example is the vibration response in a multiple engine propeller airplane when the engines are out of synchronization (Bendat and Piersol, 1971).

- *Transient signal*

  Transient data include all data previously not discussed, which can be described by some suitable time-varying function. An example is given in Fig. 1.8. An important characteristic, as opposed to periodic and almost periodic data, is that a discrete spectral representation is not possible. A continuous spectral representation can be obtained in most cases by the Fourier transform. The Fourier spectrum $X(f)$ is generally complex, and it can be written in complex polar form as

$$X(f) = |X(f)|e^{j\Theta(f)}, \tag{1.3}$$

where $|X(f)|$ is the magnitude of $X(f)$ and $\Theta(f)$ is its argument. The absolute square of the amplitude spectrum,

$$E(f) = |X(f)|^2,$$

is the so-called energy density spectrum (see also Subsection 6.7.1).

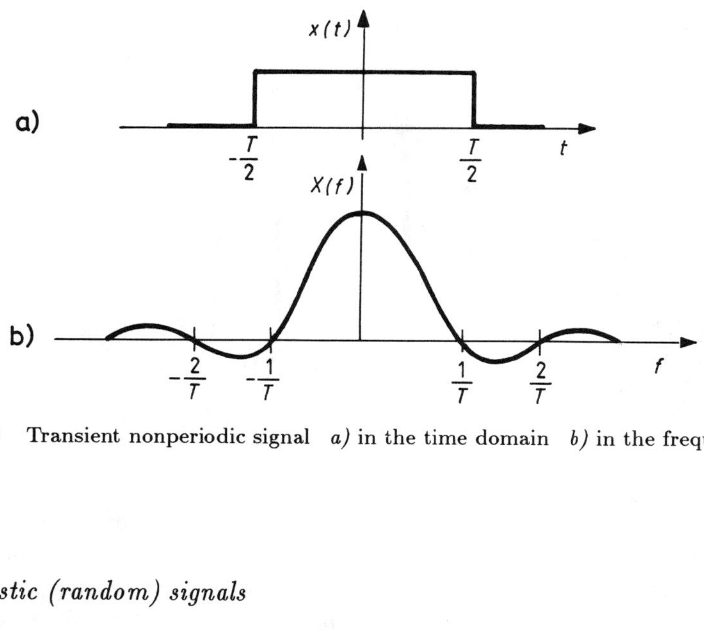

Fig. 1.8 Transient nonperiodic signal *a)* in the time domain *b)* in the frequency domain

b) Stochastic (random) signals

The model of random signals, the *stochastic process* (or *random process*) is a rather involved mathematical concept. Therefore, before the definition, a heuristic introduction will be given.

Some physical phenomena cannot be described by explicit mathematical equations: future signal values cannot be uniquely predicted from the previous time history. As an example, let us investigate the output of several thermal noise sources (e. g. resistors), generating different voltage-time functions (Fig. 1.9). One of these functions within a finite time interval can be taken as a sample record, selected from an infinite number of sample functions. A random process incorporates the ensemble of all possible sample functions $\{x(t)\}$.

If the stochastic process is discussed in the frame of probability theory, it is quite obvious to consider each sample function as a single event. The event space

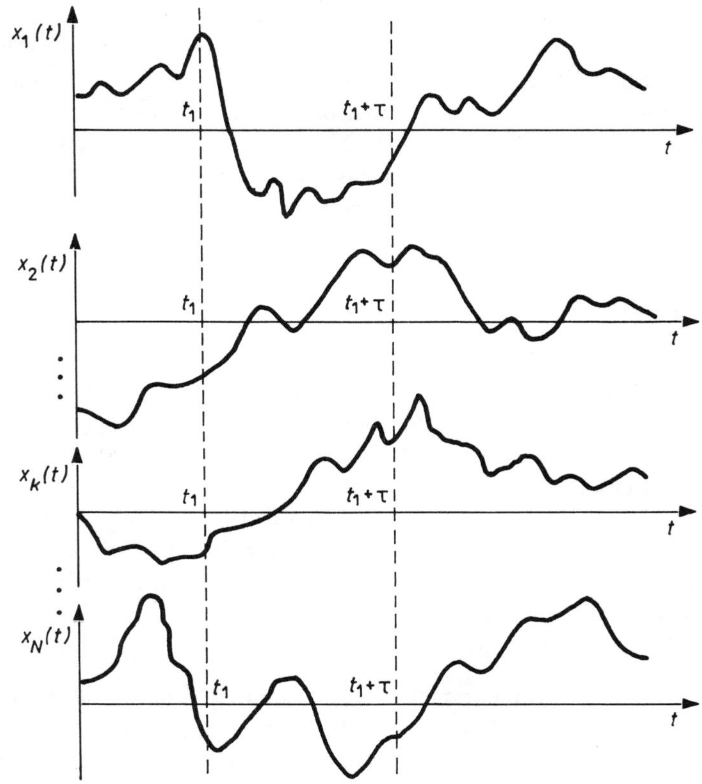

Fig. 1.9 Sample functions of a stochastic process

is then defined as the set of possible sample functions, and a probability measure is defined on this event space. This is the definition of a stochastic process.

This definition is rather abstract. Therefore a more visual way of description will be introduced, based on Kolmogorov's theorem (Gihman and Skorohod, 1979): the values $x_k(t_1)$ can be considered as sample values of the random variable $x(t_1)$, and the whole stochastic process will be considered as a set of random variables, dependent on parameter t: $\{x(t), -\infty < t < \infty\}$. Such a set of random variables can be fully characterized by all the joint probability density functions of the random variables $x(t_1)$, $x(t_2)$ ... $x(t_n)$, where $t_1, \ldots, t_n$ are arbitrary time instants, and $1 \le n \le \infty$.

It is easy to see that these joint probability density functions are still too complex to be used in practice. Therefore, simpler quantities, like first- and second-order moments (see later) are usually used to characterize a stochastic process. But before discussing these possibilities, two basic concepts, stationarity and ergodicity, will be introduced.

Stationary stochastic processes. In general, the statistical properties of a stochastic process may change with time. This may make modelling and analysis very difficult.

Fortunately, most natural phenomena do not significantly change their statistical properties during the observation time. In these cases the statistical descriptors of the model (the simple and joint probability density functions) may be chosen time-independent, and the stochastic process will be called *stationary in strict sense.*

Strict stationarity is difficult to test. Moreover, often first- and second-order moments are measured only. In such a case it is not necessary to consider the probability density functions or the higher-order moments. If the first- and the second-order moments:

$$\mu_x(t_1) = \mathcal{E}\{x(t)\} \quad \text{and} \quad R(t_1, \tau) = \mathcal{E}\{x(t_1)x(t_1+\tau)\} \tag{1.4}$$

are independent of t_1, the process is said to be *stationary in weak sense.*

The stationary process model may be applied if the properties of the physical phenomenon to be modelled seem to be independent of time.

The usability of weak stationarity can be further justified considering that since many phenomena are influenced by several, more or less independent impacts of small individual weights the distributions are in many cases Gaussian (central limit theorem). If every simple and joint probability density function is Gaussian, the stochastic process is called a Gaussian process. Since the Gaussian distribution is fully determined by the mean values and covariances, for Gaussian processes weak stationarity implies strict stationarity.

Ergodic processes. Even if a phenomenon can be modelled by a stationary process, its properties are difficult to evaluate, since (1.4) can be approximated by measuring a lot of sample records and performing averaging. Often it is even impossible to obtain several records; e. g. when studying the movement of a satellite, just one time record is available. In such a case there is no other possibility than to try to average along the time axis. In a lot of cases this will give correct results. If this is true, the associated process model is called *ergodic.* Similarly to stationarity, ergodicity can also be defined in strict sense and in weak sense. Ergodicity in weak sense means that

$$\mu_x = \mathcal{E}\{x(t)\} = \lim_{T\to\infty} \frac{1}{T}\int_0^T x_k(t)\,\mathrm{d}t = \mu_{x,k}\,, \tag{1.5}$$

and

$$R_x(\tau) = \mathcal{E}\{x(t)x(t+\tau)\} = \lim_{T\to\infty} \frac{1}{T}\int_0^T x_k(t)x_k(t+\tau)\,\mathrm{d}t = R_x = R_{x,k}(\tau)\,. \tag{1.6}$$

This means that for an ergodic process, the ensemble average and autocorrelation function yield the same result as the time average of the kth sample. In this case, a single sample of an ergodic process is sufficient for the calculation of the process characteristics. This would be true for a sample record with 'infinite' length, but

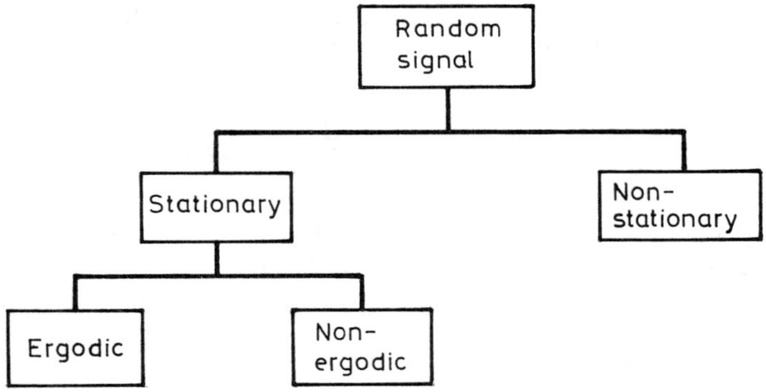

Fig. 1.10 Classification of random signals

errors due to the truncation, resulting from the finite sample, will therefore be introduced.

Stochastic processes may be classified according to Fig. 1.10.

Note 1.3 Possibilities for the description of *ergodic signals:*

- *Mean value of* $x(t)$:

$$\mu_x = \mathcal{E}\{x(t)\} = \lim_{T\to\infty} \frac{1}{T} \int_0^T x(t)\,\mathrm{d}t\,. \tag{1.7}$$

This is the static (dc) component of $x(t)$.

- *Mean square value of* $x(t)$:

$$\psi^2 = \mathcal{E}\{x^2(t)\} = \lim_{T\to\infty} \frac{1}{T} \int_0^T x^2(t)\,\mathrm{d}t\,. \tag{1.8}$$

Should $x(t)$ stand for voltage or current, $x^2(t)$ is proportional to the power dissipated by a fixed resistor. Therefore, ψ_x^2 is also called power in signal theory. The positive square root of this quantity is the rms value.

- *Variance of* $x(t)$:

$$\sigma_x^2 = \mathcal{E}\{\left(x(t)-\mu_x\right)^2\} = \lim_{T\to\infty} \frac{1}{T} \int_0^T \left(x(t)-\mu_x\right)^2 \mathrm{d}t\,. \tag{1.9}$$

This is the rms value of the signal superimposed on the static component. The positive square root of the variance is the standard deviation, denoted by σ_x. It can be shown that

$$\sigma_x^2 = \psi^2 - \mu_x^2\,. \tag{1.10}$$

- *Autocorrelation function.* Given a sample time record $x(t)$, the autocorrelation function $R_x(\tau)$ is as follows:

$$R_x(\tau) = \mathcal{E}\{x(t)x(t+\tau)\} = \lim_{T\to\infty} \frac{1}{T} \int_0^T x(t)x(t+\tau)\,\mathrm{d}t\,. \tag{1.11}$$

The autocorrelation function of random data describes the general interdependence of the values of the data at time t and at another time $t+\tau$.

The autocovariance function is a quantity, closely related to $R_x(\tau)$:

$$C(\tau) = \mathcal{E}\big\{\big(x(t) - \mu_x\big)\big(x(t+\tau) - \mu_x\big)\big\}$$

$$= \lim_{T\to\infty} \frac{1}{T} \int_0^T \big(x(t) - \mu_x\big)\big(x(t+\tau) - \mu_x\big)\,\mathrm{d}t = R_x(\tau) - \mu_x^2\,. \tag{1.12}$$

- *Power spectral density function (Power density spectrum).* This is defined as the Fourier transform of the autocorrelation function:

$$S_x(f) = \int_{-\infty}^{\infty} R_x(\tau)e^{-j2\pi f\tau}\,\mathrm{d}\tau\,. \tag{1.13}$$

Consequently:

$$R_x(\tau) = \int_{-\infty}^{\infty} S_x(f)e^{j2\pi f\tau}\,\mathrm{d}f\,. \tag{1.14}$$

Equations (1.13) and (1.14) are the Wiener-Khintchine relations.

From (1.13) it may not be clear why $S_x(f)$ is called the power spectral density function. As will be shown in Subsection 6.7.1, $S_x(f_0)\Delta f$ approximately gives the power in the band $f_0 \pm \Delta f/2$.

Sometimes the one-sided power spectral density function is also defined:

$$G_x(f) = 2\int_{-\infty}^{\infty} R_x(\tau)e^{-j2\pi f\tau}\,\mathrm{d}\tau\,, \qquad f \geq 0\,. \tag{1.15}$$

Stochastic process with periodic realizations. The structure of a stochastic process can well be characterized by the autocorrelation function or the power spectral density function. However, there are special processes which can be described by some parameters only. The most important class of such processes has periodic realizations. If the Fourier series consists of n harmonic components, it can be characterized by n triplets:

$$(X_k,\, f_k,\, \varphi_k)\,, \qquad k = 1, 2, \ldots, n\,. \tag{1.16}$$

These parameters may be fixed or may be random variables. The most often used model is the randomly timed signal, used for the description of unsynchronized measurements of periodic signals: X_k and f_k are fixed, and $\varphi_k = 2\pi f_k \tau$, where τ is uniformly distributed in $[0, 1/f_1)$. This means that each harmonic is a random phase sine wave. Such a process has special properties.

- It can be fully characterized by the above triplets, thus the measurements are reduced to the estimation of these parameters.
- The autocorrelation function is

$$R(\tau) = \sum_{k=1}^{n} \frac{X_k^2}{2} \cos 2\pi f_k \tau \,. \tag{1.17}$$

  This function does not converge to a constant value for $\tau \to \infty$.
- The power spectral density function consists of Dirac delta functions:

$$S(f) = \sum_{k=1}^{n} \frac{X_k^2}{4} \Big(\delta(f - f_k) + \delta(f + f_k) \Big) \,. \tag{1.18}$$

  Therefore, spectrum analyzers cannot properly measure $S(f)$, and the measured quantities must be carefully analyzed for the determination of the above parameters.

Because of the above properties, the performance of signal processing procedures has to be separately analyzed for these signals.

Here the question may arise, whether there are further special process types which have to be considered separately. According to an important theorem (see e. g. Loève, 1977), a distribution function (e. g. the spectral distribution function) can always be decomposed into three components:

- the first one is everywhere differentiable, this corresponds to 'regular' stochastic processes with 'regular' power spectral density functions,
- the second one is stepwise constant, which corresponds to the Dirac delta functions in $S(f)$, and thus to the randomly timed sine waves,
- the third part is almost everywhere zero, thus it has no practical importance.

Thus, the first two components have to be considered. Chapter 6 will deal with these two kinds of signal when analyzing the individual signal processing procedures.

1.3.2 System Models

Signals are produced and transformed by systems. Therefore, it is reasonable to overview here a few important aspects of system modelling. Deterministic systems will be discussed only, since first of all these models will be used in this book. A deterministic system model is used if the internal relationships of the system (its structure and parameters) seem to be unambiguously determined.

Continuous systems. In a continuous system model the internal relationships of the system can be characterized theoretically at each point of the space under test and at any moment. These relationships are results of mathematical abstraction. When 'measuring' them, the same limitations are met as with continuous signals. By refining the resolution, i. e. by the application of more precise measuring arrangements, these limitations usually lose their importance.

An example for a continuous system is a dynamic system, described by the s–domain transfer function, or the propagation of electromagnetic waves, described by partial differential equations (Maxwell equations).

Discrete systems. In the case of a discrete system model, the characterization of the system is directly possible at selected points of the space and at selected instants only. The construction of such a model may be necessary if the objects under test are concentrated at certain parts of the space, and the interactions allow them to be characterized at selected instants only. However since digital computers can more easily handle discrete models, these models are often used even if the 'natural' solution would be to use a continuous model.

An example is ARMA modelling of dynamic systems.

Distributed parameter systems. Distributed parameter system models are used if the internal relationships of the system are functions of the spatial coordinates. Examples are wave propagation in transmission lines, or heat transfer in large objects.

Lumped parameter systems. A lumped (or concentrated) parameter system model can be used if the system under test and the interactions can be well described by a few simple elements, connected to each other. Most natural phenomena are of distributed nature; however, they can often be well approximated by lumped parameter systems. This type of model is mostly used due to its relative simplicity and the elaborated mathematical apparatus. In this book the frames of the lumped parameter system models will not be exceeded. Example: modelling the measuring lead by equivalent capacitance, inductance and resistance.

Time-variant systems. A time-variant system model is used if the relationships of the system change with time, and, because of the measurement task, it is reasonable to take this into account. For example, after having been switched on, an electrical circuit is usually time-variant until it reaches thermal equilibrium.

Time-invariant systems. A time-invariant system model can be used, if from the point of view of the investigations the relationships of the system can be regarded as time-independent ones. Time-invariance is usually justified by short-time investigations. This book deals mostly with time-invariant systems.

Nonlinear systems. A nonlinear system model is necessary if the internal relationships of the system depend on the intensity of the interactions. This dependence can be of different kinds, therefore a large number of nonlinear system types exist. The handling of nonlinear system models is rather difficult; it needs relatively complicated apparatus. However, its use cannot be always avoided, e. g. the effect of a diode is strongly nonlinear if the signal crosses the opening level of it.

Linear systems. In the case of the application of a linear system model, it is assumed that the internal relationships are regarded as independent of the intensity relations of the interactions. In this case the concept of superposition is applicable, which is very favourable for the mathematical treatment. Though most natural phenomena are nonlinear, usually linear approximations are used, see e. g. the small displacement investigation of mechanical systems.

1.4 MEASUREMENT ERRORS

Measurement error is a crucial term of measurement theory and practice.

All measurements are subject to errors, for theoretical and practical reasons. These measurement errors, i. e. the deviations from the ideal measurement result (which is often considered equal to the true value), may originate from following sources:

- modelling errors,
- circumstances and methods of the observation and the measurement,
- properties of the measuring device,
- external disturbances,
- etc.

This section discusses the most important questions arising in connection with the measurement errors.

1.4.1 Types of Measurement Error

When measuring a quantity, it is assumed that the quantity to be measured has an actual value x_{t}, called the *true value.*

As the result of the measurement, a value x_{m} is produced, called the *measured value.* Because of the previously mentioned causes, this is not equal to the true value:

$$x_{\mathrm{m}} \neq x_{\mathrm{t}} \, , \tag{1.19}$$

hence

$$x_{\mathrm{m}} - x_{\mathrm{t}} = \text{error of measurement} \, . \tag{1.20}$$

This means that the value x_{m}, produced by the measurement, more or less deviates from the value x_{t}. The error of the measurement is characteristic for the quality of the measurement, therefore it is important to analyze the origin of the measurement errors and the possibilities of their reduction.

Measurement errors can be classified into two basic groups: systematic errors and random errors. In the following sections they will be considered in more detail.

The measurement errors and their treatment are illustrated by Fig. 1.11.

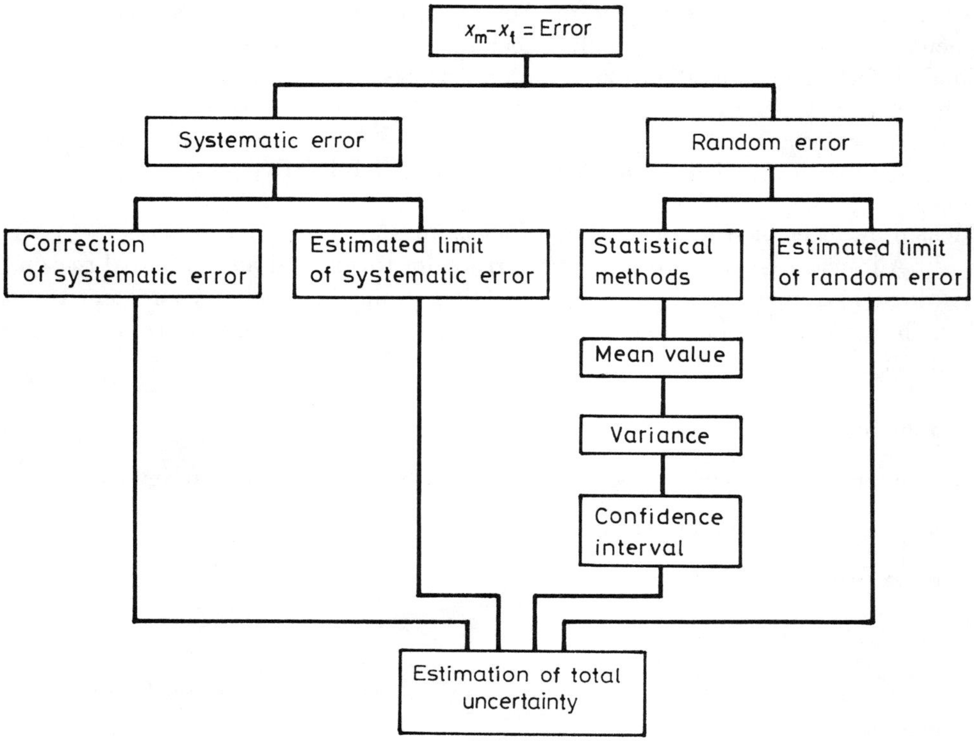

Fig. 1.11 Classification of measurement errors

1.5 SYSTEMATIC ERROR

A characteristic feature of the *systematic error* or *bias* is that this error has the same magnitude and sign in every repeated measurement. If the magnitude and sign of this error are really known, correction can be applied and the error can be eliminated.

Some examples for the sources of the systematic error: internal resistance of an instrument; capacitance or inductance of connecting wires; external magnetic field or temperature, different from the assumed value; etc.

In most cases, systematic errors cannot be easily discovered. As will be discussed later, random errors can be detected by the analysis of repeated measurements. However, if the measurements are all corrupted by the same bias, the only way of detecting this bias is to measure the same parameter by another instrument, using another method of measurement, and by analyzing the measurement results and the measurement method. The discovery of systematic errors requires careful measurement design, due foresight and critical attitude, and the evaluation of the error model. An undetected systematic error results in an erroneous measurement value.

1.5.1 Elimination of the Systematic Error by Correction

If e. g. the connection of an electrodynamic power meter according to Fig. 3.1 is analyzed, it can be stated that the power, indicated by the instrument, is greater than the accurate value, by the consumption of either the voltage or the current coil, depending on the applied connection. Knowing the coil resistance, its power consumption can be calculated, and the systematic error can be corrected.

If the data necessary for the correction are available, and the correction is not negligible compared to the random errors or to the required accuracy, the correction should be made.

1.5.2 Estimated Limit of the Systematic Error

As an example, let us consider a universal current and voltage measuring instrument with 20 different measurement ranges. Systematic errors may occur in every range, because of the inaccurate adjustment of the series and shunt resistances, or because of hysteresis etc. Theoretically a correction table could be assigned to each range, but the use of the instrument would become intolerably complicated. Even in a microprocessor controlled instrument, the bias may change with time or temperature, thus regular self-calibration would be necessary.

Another example where the correction cannot be easily performed, is non-parametric transfer function measurement. Here the input and output signals are Fourier transformed, and the output complex amplitudes, divided by the input amplitudes, give the estimate. It is easy to see that if the input is corrupted by Gaussian noise, the transfer function estimate will be biased, and the bias depends on the signal-to-noise ratio (see Subsection 6.8.2).

If not all the data are available for proper correction, or the correction would be too complicated, it is practical to choose another solution. Using a priori knowledge, an interval can often be given around the indicated value, which — according to our judgement — will comprise the systematic errors in each range. This interval will be regarded thereafter as a random error and correction will not be used. However, it should not be forgotten, that averaging of repeated measurements will not eliminate this type of error.

1.6 RANDOM ERROR

If the measured value, x_{m}, shows such random-like changes in repeated measurements, which cannot be explained by changes of the true value, x_{t}, or by changes of an influencing quantity, like temperature, magnetic field, etc, this error is modelled by a random variable, and it is called random error. On the basis of the measurements, corrupted by random errors, an estimate of the true value used to be given, using estimation theory (Sage and Melsa, 1971; Bendat and Piersol 1971; van Trees, 1968; etc).

The estimate is usually marked by a hat ($\widehat{\ }$) over the actual symbol.

The optimal estimate cannot always be determined, sometimes even the information is not sufficient for this purpose. In such cases simple and straightforward solutions can be chosen or special estimates can be taken.

In the following subsections the most often used quantities, the arithmetic mean and the sample variance, will be treated, and the concept of the confidence interval will be introduced.

1.6.1 Arithmetic Mean Value and Sample Variance

Let us consider an experiment where a quantity will be measured several times in succession, under unchanged external circumstances. Let the measurements be corrupted by a random error in each measurement, independent of the result of other measurements.

Making N subsequent measurements, a sample of N elements is generated:

$$x_1, x_2, \ldots x_k, \ldots x_N , \tag{1.21}$$

where x is the random variable. This sample is called the measurement series.

a) Mean value

In the processing of the measured data, the arithmetic mean value, as an estimate of the expected value, has an important role.

In the continuous case the expected value of the random variable x is defined by the relationship:

$$\mu_x = \mathcal{E}\{x\} = \int_{-\infty}^{+\infty} x f(x) \, \mathrm{d}x , \tag{1.22}$$

and in the discrete case:

$$\mu_x = \mathcal{E}\{x\} = \sum_r x_r p_r , \tag{1.23}$$

where $f(x)$ is the probability density function, and p_r is the probability of the value x_r.

The expression

$$\widehat{\mu}_x = \overline{x} = \frac{1}{N} \sum_{k=1}^{N} x_k \tag{1.24}$$

is called the arithmetic mean value or empirical expected value, and it is generally denoted by $\overline{x}$. The sample mean value is an unbiased estimator of the expected value:

$$\mathcal{E}\{\widehat{\mu}_x\} = \mathcal{E}\{\overline{x}\} = \mu_x \,. \tag{1.25}$$

b) Sample variance

It is important to know how wide the measurement results are scattered around the expected value. This scattering can be characterized by the standard deviation σ_x, or its square, the variance.

In the continuous case:

$$\mathrm{var}\{x\} = \sigma_x^2 = \mathcal{E}\{(x-\mu_x)^2\} = \int_{-\infty}^{+\infty} (x-\mu_x)^2 f(x)\,\mathrm{d}x \,. \tag{1.26}$$

In the discrete case:

$$\mathrm{var}\{x\} = \sigma_x^2 = \sum_r (x_r - \mu_x)^2 p_r \,. \tag{1.27}$$

The variance is usually not known. An estimate of it, using the sample of N elements of independent measurement results, is the empirical variance, denoted by s^2:

$$s^2 = \frac{1}{N}\sum_{k=1}^{N} (x_k - \overline{x})^2 \,. \tag{1.28}$$

However, this is not an unbiased estimator of σ_x^2, since

$$\mathcal{E}\{s^2\} = \frac{N-1}{N}\sigma_x^2 \,. \tag{1.29}$$

To overcome this problem, let us introduce the following modification:

$$s^{*2} = \frac{N}{N-1}s^2 = \frac{1}{N-1}\sum_{k=1}^{N} (x_k - \overline{x})^2 \,, \tag{1.30}$$

for which:

$$\mathcal{E}\{s^{*2}\} = \sigma_x^2 \,. \tag{1.31}$$

s^{*2} is the corrected empirical variance, and it is an unbiased estimator of σ_x^2.

From Eqs. (1.28) and (1.30) it can be seen that the bias is significant only if N is small. With greater N values the bias disappears: the estimate is asymptotically unbiased.

Returning to the expression of the empirical expected value (1.24), let us mark along a scaled line the independent elements of a measurement series x_1, x_2, ..., x_k, ..., x_N and their arithmetic mean. Let us repeat the measurement several times, and let us calculate the arithmetic means $\overline{x}_1$, $\overline{x}_2$ .... If these means are also marked along the scaled line, it can be seen that the dispersion of the means will decrease perceptibly.

A mathematical justification may be as follows:

$$\sigma_{\overline{x}}^2 = \mathcal{E}\{(\overline{x} - \mu_x)^2\} = \frac{1}{N^2}\mathcal{E}\Big\{\sum_{k=1}^{N}(x_k - \mu_x)^2\Big\} = \frac{1}{N^2}N\sigma_x^2 = \frac{\sigma_x^2}{N}, \tag{1.32}$$

and

$$\sigma_{\overline{x}} = \frac{\sigma_x}{\sqrt{N}} \approx \frac{s_x}{\sqrt{N}}, \tag{1.33}$$

where $\sigma_{\overline{x}}$ is the empirical standard deviation of the empirical arithmetic mean $\overline{x}$. Thus, the standard deviation of the mean value of a measurement series consisting of N independent elements decreases proportionally to $1/\sqrt{N}$ when N is increased.

1.6.2 Confidence Interval

The measurement uncertainty is not fully reflected by the previously discussed estimators. There is always a random deviation between the true value and the estimated one, and its boundaries have to be expressed in some way.

A possibility is to speak not only about the estimator but also about the risk of accepting the estimation.

An estimate of a parameter given by a single number is called a *point estimate* of the parameter. If two bounds are determined to locate the true value of the parameter, the result is called an *interval estimate* of the parameter.

Interval estimates indicate also the accuracy of an estimate and are therefore preferable to point estimates.

It was already mentioned that in the case of random errors the results of the measurement and also the estimator are random variables, so they have probability distribution and probability density functions. If these functions are known, the probability can be given, that the deviation of the measured value from the true value is not greater than $\pm\epsilon_c$. (The subscript c refers to the confidence interval.) With the confidence interval a zone $\pm\epsilon_c$ is given around the measured value (the true value is not known), within which the true value can be found with a certain probability. This is illustrated in Fig. 1.12.

Let the expected value of a random variable be μ. For this, the following equation applies:

$$\mathrm{P}(\widehat{\mu}_1 \le \mu \le \widehat{\mu}_2) = 1 - p. \tag{1.34}$$

The bounds $\widehat{\mu}_1$ and $\widehat{\mu}_2$ define an interval of random position. It can be stated that the true value μ lies between the bounds $\widehat{\mu}_1$ and $\widehat{\mu}_2$, with a probability of $1-p$.

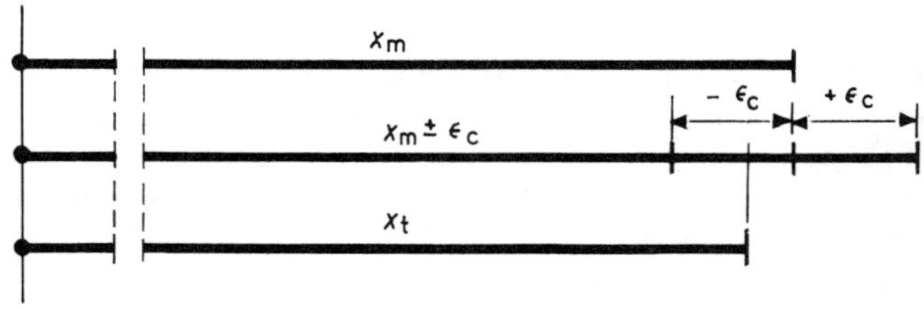

Fig. 1.12 Interpretation of the error

This statement will be true in $(1-p)\cdot 100\%$ and it will be false in $p\cdot 100\%$ of the experiments.

The interval from $\widehat{\mu}_1$ to $\widehat{\mu}_2$ is called the confidence interval, the positions of $\widehat{\mu}_1$ and $\widehat{\mu}_2$ are the confidence bounds, and the probability $(1-p)$, expressed in %, is the confidence level.

Usually confidence levels of 90, 95, and 99% are used, corresponding to the values $p = 0.1$, 0.05, and 0.01, respectively. It is not reasonable to prescribe too high a confidence level, because in doing so, the width of the interval $(\widehat{\mu}_1, \widehat{\mu}_2)$ will be enormously enlarged.

Let us transform equation (1.34) in order to make it more useful in practice. In a large number of cases the confidence limits are symmetrical. With

$$\widehat{\mu}_1 = \widehat{\mu} - \epsilon_c \tag{1.35}$$

and

$$\widehat{\mu}_2 = \widehat{\mu} + \epsilon_c\,, \tag{1.36}$$

where $\widehat{\mu}$ denotes the estimate, obtained from the measurement series,

$$\mathrm{P}(\widehat{\mu} - \epsilon_c \leq \mu \leq \widehat{\mu} + \epsilon_c) = 1 - p\,. \tag{1.37}$$

If the distribution of $\widehat{\mu}$ is known, the value of ϵ_c, belonging to a given probability $1-p$, can be determined.

Let us re-write equation (1.37) with the usual symbols used so far:

$$\mathrm{P}\big(\mu \in (\widehat{\mu} \pm \epsilon_c)\big) = 1 - p\,. \tag{1.38}$$

The symbol $\in$ in this relationship expresses that the interval $(\widehat{\mu} \pm \epsilon_c)$ contains the expected (true) value μ with a certain probability.

The confidence interval of the mean value is calculated like this: the confidence level $1-p$ is set, and on the basis of the measurement series, the estimator, $\widehat{\mu}_x = \overline{x}$ is determined (in the following the symbol $\overline{x}$ will be used for the arithmetic mean). As has been mentioned, the probability density function of $\overline{x}$ must be known (containing the unknown μ_x as a parameter), and from this the confidence limits can be determined.

In the following the confidence intervals of the most often used estimators of the expected value and the variance will be dealt with.

Example 1.2

As an example, let us consider the procedure described above for normally distributed measurement error. This is a case occurring often, since, according to the central limit theorem, the quantity affected by several small, independent additive random influences, will approximately be a normally distributed random variable, independent of the distributions of the random influences.

The normal (or Gaussian) distribution is:

$$f(x) = \frac{1}{\sigma_x \sqrt{2\pi}} e^{-\frac{1}{2}\left(\frac{x-\mu_x}{\sigma_x}\right)^2}, \tag{1.39}$$

where μ_x is the expected value, and σ_x is the standard deviation. This distribution is denoted by $N(\mu, \sigma)$.

Applying the transformation

$$z = \frac{x - \mu_x}{\sigma_x}, \tag{1.40}$$

the standard normal probability density function is obtained:

$$f(z) = \frac{1}{\sqrt{2\pi}} e^{-\frac{1}{2}z^2}, \tag{1.41}$$

which is a probability density function $N(0,1)$ (see Fig. 1.13).

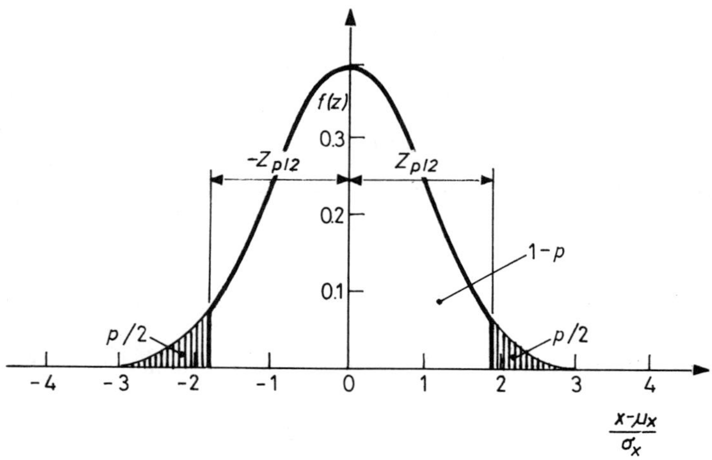

Fig. 1.13 Standardized normal probability density function

In the usual tables (Korn and Korn, 1968; Bendat and Piersol, 1971; Spiegel, 1961) the standard normal distribution function is given:

$$F(z_p) = \frac{1}{\sqrt{2\pi}} \int_{-\infty}^{z_p} e^{-\frac{1}{2}z} \, dz. \tag{1.42}$$

For our purposes:

$$2F(z_{p/2}) - 1 = 1 - p\,. \tag{1.43}$$

Some values for $1-p$, belonging to different values of $z_{p/2}$, are listed in the following table:

$z_{p/2}$	$1-p$	percentage
1	0.6827	68.27%
1.96	0.9500	95.00%
2	0.9545	95.45%
2.58	0.9900	99.00%
3	0.9973	99.73%
4	0.9999	99.99%

If $z_{p/2} = 3$, the confidence interval may be given as:

$$x_\mathrm{t} \in (\overline{x} \pm 3\sigma_x)\,,$$

and the confidence level is 99.73%.

Example 1.3

a) Assume a measuring instrument with a *known* standard deviation σ_x. For a single measured value x_i,

$$\mathrm{P}\left(x_i - z_{p/2}\sigma_x \le x_\mathrm{t} \le x_i + z_{p/2}\sigma_x\right) = 1 - p\,. \tag{1.44}$$

Thus, if x_i is an arbitrary sample, an interval $\pm z_{p/2}\sigma_x$ around this sample value will comprise the expected value μ_x with a confidence of $1-p$.

b) Let us calculate the confidence interval of the expected value μ_x in a measurement series, for a probability distribution $N(\mu, \sigma_x)$ with a *known* standard deviation σ_x.

Consider a record of N samples $x_1, x_2, \ldots x_N$, taking into account that the standard deviation of the arithmetic mean value is $1/\sqrt{N}$ times σ_x. By applying the usual substitutions

$$z = \frac{\overline{x} - \mu_x}{\sigma_x/\sqrt{N}} = \sqrt{N}\frac{\overline{x} - \mu_x}{\sigma_x}\,, \tag{1.45}$$

the empirical mean value is transformed to the normal form $N(0, 1)$. The values of $z_{p/2}$ give the confidence interval:

$$\mathrm{P}\left(\overline{x} - z_{p/2}\frac{\sigma_x}{\sqrt{N}} \le \mu_x \le \overline{x} + z_{p/2}\frac{\sigma_x}{\sqrt{N}}\right) = 1 - p\,. \tag{1.46}$$

$$x_\mathrm{t} \in \left(\overline{x} \pm z_{p/2}\frac{\sigma_x}{\sqrt{N}}\right)\,. \tag{1.47}$$

c) Let us calculate the confidence interval for the expected value μ_x, assuming a probability distribution $N(\mu_x, \sigma_x)$ and an *unknown* standard deviation. Both the mean value and the standard deviation are calculated from the samples $x_1, x_2, \ldots x_N$. Let us introduce the variable

$$t_{N-1} = \frac{\overline{x} - \mu_x}{s_x^*/\sqrt{N}} = \sqrt{N}\frac{\overline{x} - \mu_x}{s_x^*}\,, \tag{1.48}$$

which differs from the above defined variable z by applying the corrected empirical standard deviation s_x^* instead of σ_x. This is a *Student t* variable with $N-1$ degrees of freedom, differing from the normal distribution (see Fig. 1.14):

$$\mathrm{P}\left(\overline{x} - t_{N-1,p/2}\frac{s_x^*}{\sqrt{N}} \leq \mu_x \leq \overline{x} + t_{N-1,p/2}\frac{s_x^*}{\sqrt{N}}\right) = 1 - p, \tag{1.49}$$

$$x_t \in \left(\overline{x} \pm t_{N-1,\,p/2}\frac{s_x^*}{\sqrt{N}}\right). \tag{1.50}$$

The values of t_p are given in the Student distribution table (Korn and Korn, 1968; Spiegel, 1961).

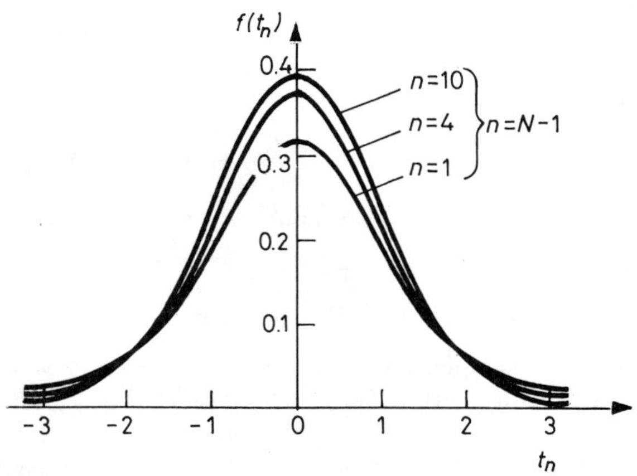

Fig. 1.14 Student t probability density function

d) Let us calculate the confidence interval for the expected value μ_x, assuming an *unknown* probability distribution and a *known* standard deviation σ_x. For an arbitrary distribution of x, the inequality

$$\mathrm{P}\left(|x - \mu_x| \leq \lambda\sigma_x\right) \geq 1 - \frac{1}{\lambda^2}, \tag{1.51}$$

holds, where λ is an arbitrary positive number. This is the well known Chebyshev inequality, meaning that the confidence level of finding the expected value μ_x in the interval $x \pm \lambda\sigma_x$ is $\left(1 - \frac{1}{\lambda^2}\right)100$ per cent. For instance, by assuming $\lambda = 3$, the probability of finding the expected value μ_x in the confidence interval $x \pm 3\sigma_x$ is at least

$$1 - \frac{1}{\lambda^2} = 1 - \frac{1}{3^2} = 88.8\%. \tag{1.52}$$

For a normal probability distribution, the confidence interval of $\pm 3\sigma_x$ provides a confidence level of 99.7%.

e) The confidence interval for the standard deviation σ_x, in the case of a probability distribution $\mathrm{N}(\mu_x, \sigma_x)$.

Taking an N-element sample from a random variable $\mathrm{N}(\mu_x, \sigma_x)$, it can be proved that the distribution of Ns_x^2/σ_x^2 has a χ^2_{N-1} (chi-square) distribution. Since this is an asymmetrical distribution, it is reasonable to choose the confidence limits

$$\mathrm{P}\left(\frac{Ns^2}{\sigma^2} > \chi^2_{1-p/2}\right) = 1 - \frac{p}{2},$$

$$\mathrm{P}\left(\frac{Ns^2}{\sigma^2} > \chi^2_{p/2}\right) = \frac{p}{2}, \tag{1.53}$$

From this it follows that

$$\mathrm{P}\left(\frac{Ns^2}{\chi^2_{N-1,\,p/2}} \le \sigma_x^2 \le \frac{Ns^2}{\chi^2_{N-1,\,1-p/2}}\right) = 1 - p, \tag{1.54}$$

and

$$\mathrm{P}\left(\sqrt{\frac{Ns^2}{\chi^2_{N-1,\,p/2}}} \le \sigma_x \le \sqrt{\frac{Ns^2}{\chi^2_{N-1,\,1-p/2}}}\right) = 1 - p.$$

The values of $\chi_{N-1,\,p/2}$, and $\chi_{N-1,\,1-p/2}$ can be taken from the table of the χ^2 distribution.

Example 1.4

Assume a constant voltage source exposed to random disturbing effects. Calculate the confidence interval of the random error at the confidence levels of 90% and 99%. Ten measurements have been made with the following results, given in volts:

151.51 148.73 150.81 148.04 153.59 147.34 149.42 150.81 151.51 150.81.

Using the relationships (1.24) and (1.30), the arithmetic mean value and the empirical standard deviation have been calculated:

$$N = 10, \qquad \overline{x} = 150.26 \qquad s^{*2} = 3.51 \qquad s^* = 1.87.$$

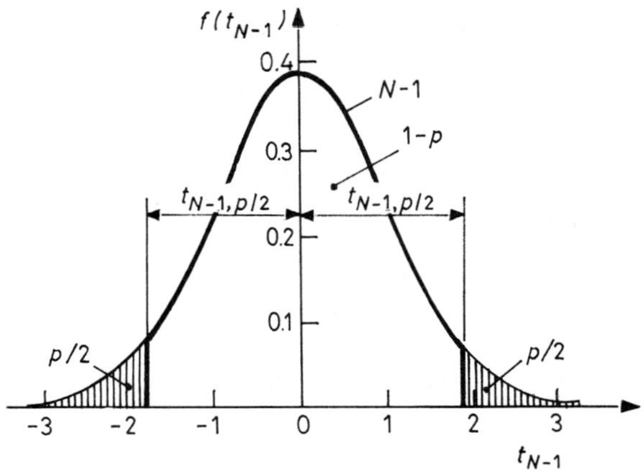

Fig. 1.15 Illustration to Example 1.4

Since both the arithmetic mean value and the empirical standard deviation have been calculated from the samples, further calculations are made using the Student distribution with a degree of freedom of $N-1$.

For a confidence level of 90%: $p=0.1$, $n=N-1$. (see Fig. 1.15).
According to equation (1.49)

$$\overline{x} \pm t_{N-1,p/2}\frac{s^*}{\sqrt{N}} = 150.26 \pm 1.833\frac{1.87}{\sqrt{10}} = 150.26 \pm 1.08$$

$$x_t \in (150.26 \pm 1.08)\,.$$

This statement will be true in the 90% of the cases, and it will be false in 10% of the cases.

For a confidence level of 99%: $p=0.01$.

$$t_{N-1,\,p/2} = t_{9,\,0.005} = 3.250$$

$$\overline{x} \pm 3.25\frac{1.87}{\sqrt{10}} = 150.26 \pm 1.92$$

$$x_t \in (150.26 \pm 1.92)\,.$$

The higher the confidence level prescribed, the broader is the confidence interval.

Example 1.5

Let us calculate also the confidence interval for the variance, from the data of the previous example. The empirical standard deviation is:

$$s^* = 1.87 \qquad s^{*2} = 3.51$$

$$\frac{(N-1)s^2}{\chi^2_{N-1,\,p/2}} \le \sigma_x^2 \le \frac{(N-1)s^{*2}}{\chi^2_{N-1,\,1-p/2}}\,.$$

Using the results of the previous example:

For a confidence level of 90%: $p=0.1$.

$$\frac{9\cdot 3.51}{16.92} \le \sigma_x^2 \le \frac{9\cdot 3.51}{3.325}$$

$$1.87 \le \sigma_x^2 \le 9.49$$

$$1.37 \le \sigma_x \le 3.08\,.$$

For a confidence level of 99%: $p=0.01$.

$$\frac{9\cdot 3.51}{23.59} \le \sigma_x^2 \le \frac{9\cdot 3.51}{1.73}$$

$$1.34 \le \sigma_x^2 \le 18.26$$

$$1.16 \le \sigma_x \le 4.27\,.$$

1.6.3 Estimated Limit of Random Error

In the preceding paragraphs the characterization of random errors has been dealt with. It has been shown that, by the use of statistical methods, they can be described correctly. However, statistical methods cannot always be applied. For example in the evaluation of measurement series, the external conditions should not change. In some cases, e. g. for the proper estimation of the confidence interval, the probability distribution of the measurement result should be known. However, this is generally not true, and the usually assumed normal distribution is often a rough approximation only. The ideal circumstances assumed for the error estimation and error accumulation (Section 1.9) are not always valid. The results have then to be evaluated by applying subjective considerations which cannot be fully justified.

Because of the above reasons, error estimation should be treated with precaution. Exact calculations are justified only when they are based on proper a priori knowledge and experience. Often rough guesses are used, resulting in some upper bounds of the error.

1.7 PRESENTATION OF MEASUREMENT RESULTS

In the preceding sections, the errors, influencing the measured value, have been discussed. After having taken into account all the errors and having determined their magnitude, their numeric value must be presented. A presentation, as simple and understandable as possible, should be striven for.

1.7.1 Absolute Error[1]

The absolute error of measurement is the measured value x_{m} minus the true value x_{t} of the measurand:

$$\text{Error} = x_{\mathrm{m}} - x_{\mathrm{t}} = \Delta x\,.$$

Instead of Δx, the symbol ϵ is often used. The values x, Δx or ϵ are to be substituted by the symbol of the measured quantity e. g. R, Z, ΔR, ΔU etc.

The absolute error can be classified in two groups.

- The known part of the error (systematic error) can often be theoretically compensated by applying appropriate correction. In such cases:

$$\widehat{x}_{\mathrm{t}} = x_{\mathrm{m}} - \Delta x\,, \tag{1.55}$$

where Δx is the correction. The value $x_{\mathrm{m}} - \Delta x$ is then the estimator of x_{t}. Since Δx is a concrete numerical value, this is called the *point estimator.*

[1] The absolute error has a sign, it should not be confused with the absolute value of the error, characterizing a measured parameter.

If the value Δx is not given as a number, but it has an upper limit, a zone of width $\pm\Delta x$ ($\pm\epsilon$) is obtained, containing the true value x_t (generally with high probability):

$$x_t \in (x_m \pm \Delta x). \tag{1.56}$$

$x_m \pm \Delta x$ is called the *interval estimator,* the error modelled by it is called *uncertainty* (See Fig. 1.12).

1.7.2 Relative Error in Terms of the Measured Value

It is very often practical to give the relative error:

$$\widehat{x}_t = x_m\left(1 - \frac{\Delta x}{x_m}\right) = x_m(1 - \epsilon_r)$$

or

$$x_t \in x_m\left(1 \pm \frac{\Delta x}{x_m}\right) = x_m(1 \pm \epsilon_r), \tag{1.57}$$

where the relative error is

$$\epsilon_r = \frac{\Delta x}{x_m}. \tag{1.58}$$

The relative error is evidently a dimensionless quantity.

In most cases, the relative error is expressed in percent, per mill or parts per million (ppm):

$$\epsilon_r = \epsilon_r \cdot 10^2 \,\text{per cent} = \epsilon_r \cdot 10^3 \,\text{per mil} = \epsilon_r \cdot 10^6 \,\text{parts per million}.$$

1.7.3 Relative Error in Terms of the Full Scale

In some instruments, where the effect causing the error is independent of the deflection of the instrument (e. g. friction), the relative error is expressed as a percentage of the full scale. For example, in case of an electromechanical voltmeter with a full scale of 150 V and with an accuracy class of 1, the maximum error (uncertainty) is $\pm1\%$ of the full scale that is ±1.5 V along the scale. Thus, the smaller the value measured, the greater is the relative error in terms of the measured value (the actual deflection). Therefore, when measuring with such an instrument, it is desirable to utilize the measurement range as much as possible.

1.7.4 Interpretation of Errors

The overall errors, introduced in Sections 1.5 and 1.6, should be interpreted somehow. One can speak, e. g., about the limit of error, the uncertainty, the total uncertainty, and the accuracy.

- The interpretation of the *limit of error* is in general that the error is thought not to exceed the limits of the given interval. This implies some guarantee-like character in the sense that the manufacturer guarantees that, if the reference conditions are fulfilled, the error will not exceed the limit of error.
- The *uncertainty* points to the fact already by its name that the given error is uncertain: the true value (x_t) is expected to be within the given interval, but it is not quite certain, the 'confidence' attached to this expectation has some risk. This risk can be expressed by a number, based on the confidence interval, if the circumstances make such an estimation possible.
- The *total uncertainty* means the resulting uncertainty. Its presentation may look like this:

$$\begin{aligned} x_\mathrm{t} \in (x_\mathrm{m} \pm \Delta x) = \big(x_\mathrm{m} \pm (&\text{estimated limit of systematic error} \\ &+ 2 \times \text{the standard deviation of random error})\big) \,. \end{aligned} \tag{1.59}$$

  The term '2 × the standard deviation of random error' indicates a confidence interval of about 95% (at least for the Gaussian case). When giving such a quantity, the definition actually used is to be given.
- Another concept also concerning the amount of measurement error is *the accuracy of measurement,* which specifies the closeness of the agreement between the result of a measurement and the true value of the measurand. This is complementary related to the 'error' and 'uncertainty' introduced so far, because a small accuracy belongs to a large error and vice-versa. The accuracy as a qualifier of the measurement is often used, however, without giving a numerical value. Of course, no objection can be made against saying 99% accuracy instead of 1% uncertainty, but it is seldom said.
  Sometimes, accuracy is also used in the sense of uncertainty or error.

We still have the problem of how to sum the two types of error component: whether it should be done linearly or quadratically (see Section 1.9). A stiff rule for this can not be applied. For the interpretation of the different kinds of error and their summation rules, different opinions are published in the literature (Luppold, 1969; Plourde, 1962; Youden, 1962). One thing is sure: linear summation gives a higher numerical value, so it gives better safety than the quadratic method.

The choice of the proper method can be made either on the basis of some technically supported argument, or on the basis of the difference between the above mentioned error limits.

The measurement error is given by digits. It is a basic rule that the measurement result should be given using such a number of digits that the least significant digit — even if it is a zero — gives information about the uncertainty of the measurement. In other words: the last digit given should represent the magnitude of uncertainty.

1.8 QUANTITIES INFLUENCING THE MEASUREMENT ERROR

A quantity which is not the subject of measurement, but influences the value of the measurand or the indication of the measuring instrument, is called the *influence quantity.* The most frequent influencing quantities affecting the indicated measurement result are temperature, humidity, external magnetic and electrical fields, waveform of signals, etc.

The boundary values of the influencing quantities between which the technical data of a measurement system are valid, are expressed by the *reference conditions.* For instance, the reference temperature interval of an electromechanical instrument may be $20 \pm 1°C$. Reference conditions of influencing quantities are defined by standards for commonly used electronic instruments.

The measurement error can be stated even outside the reference interval by assuming that the dependence of the indicated measurement result on the influencing quantities is known. Evidently, the reference interval should not be exceeded arbitrarily because above a given value of the influencing quantity, the error changes become nonlinear, and cannot be calculated. Within the *nominal application range,* the dependence of the measurement error on the influencing quantities can be stated. For electromechanical instruments, these boundaries of the influencing factors are given by standards. In addition to the nominal application range, the following instrument parameters are also defined by some standards.

Nominal operating range. Within this range, the instrument is capable of operating. For instance, the semiconductors may become nonconductive below a given temperature.

Transportation and storage range. Outside this range, the instrument may break down even in the switched-off state, e. g. due to melting of components at high temperatures.

Repeatability. This term means the closeness of the agreement between the results of successive measurements of the same measurand carried out subject to all of the following conditions:

- the same method of measurement;
- the same observer;
- the same measuring instrument;
- the same location;

- the same conditions of use;
- repetition over a short period of time.

Repeatability depends on the random error of a given instrument within a given time interval, and can be expressed, e. g., by the confidence interval calculated from the indicated measurement results. It can be expected that for an instrument in good operating condition, this confidence interval will be smaller than the specified error interval of the instrument.

Reproducibility. This is the closeness of the agreement between the results of measurements of the same measurand, where the individual measurements are carried out changing the conditions such as

- the method of measurement,
- the observer,
- the measuring instrument,
- the location,
- the conditions of use,
- the time.

Reproducibility can also be characterized by the confidence interval calculated from the indicated measurement results. It is seen that reproducibility includes not only the instrument error but also the variation of measurement results due to the influencing quantities (DIN, 1984).

1.9 ACCUMULATION OF ERRORS

The measured values, bearing individual errors, are frequently substituted into mathematical formulae. A similar problem arises, when a measured value is associated with more kinds of error. These can be given one by one, or they are summed in some way. The following may support such summations as well.

In different measurement problems the measured value is determined as a function of one or more directly measurable quantities. If the parameter to be measured is y and the directly measurable quantities are $x_1, x_2, \ldots, x_N$,

$$y = y(x_1, x_2, \ldots, x_\mathrm{N}) . \tag{1.60}$$

The key question of the error propagation problem or — as it will be called here — of practical error calculation is how to determine the error of y, knowing the measurement errors of the quantities $x_1, x_2, \ldots, x_N$. The answer can be given separately for the case of systematic and random errors.

Error propagation in the case of systematic errors

Assume that the relative systematic errors of the quantities $x_1, x_2, \ldots, x_N$ are small and known. The change of y can be expressed as a total differential:

$$\Delta y = \sum_{i=1}^{N} \frac{\partial y}{\partial x_i}\Big|_{x_i=x_{i0}} \Delta x_i \,, \tag{1.61}$$

$$y_0 = y_0(x_{10}, x_{20}, \ldots, x_{N0}) \,, \tag{1.62}$$

where y_0 and $x_{10}, x_{20}, \ldots, x_{N0}$ are the error-free values of y and $x_1, x_2, \ldots, x_N$, respectively. The relative error is:

$$\frac{\Delta y}{y_0} = \sum_{i=1}^{N} \frac{\partial y}{\partial x_i}\Big|_{x_i=x_{i0}} \frac{\Delta x_i}{y_0} \,. \tag{1.63}$$

The condition for the application of the relationship (1.63) is that the function $y = y(x_1, x_2, \ldots, x_N)$ should be smooth in the environment of the investigated place, further the changes Δx_i should not be large, to keep the approximation error of the total differential small. If these conditions are not fulfilled, an exact evaluation of the function by the substitution of the values $x_1, x_2, \ldots, x_N$, containing also the errors, can lead to the result.

If an error limit of Δx_i is known only, the absolute values of the terms in (1.63) are to be summed.

Error propagation in the case of random errors

If the quantities $x_1, x_2, \ldots, x_N$ have random error, they can be regarded as random variables. In accordance with this, the quantity y to be measured is a random variable too, since it is generated as a function of random variables. It is a matter of common knowledge that the characteristics of y are described by its probability distribution or density function, which can theoretically be derived if the distributions of the quantities $x_1, x_2, \ldots, x_N$ are known.

In some cases, it can be supposed that $x_1, x_2, \ldots, x_N$ are independent of each other. Then, if the function y is approximated around y_0 by its total differential, the variance can be determined by the use of a known theorem of probability theory: the variance of the linear combination of independent random variables can be determined by taking the sum of the variances, weighted by the square of the coefficients of the linear combination. Thus:

$$\sigma_y^2 = \sum_{i=1}^{N} \left(\frac{\partial y}{\partial x_i}\Big|_{x_i=x_{i0}} \right)^2 \sigma_{x_i}^2 \,. \tag{1.64}$$

If the random error is described by the confidence interval, derived by the multiplication of the standard deviations by a constant

$$\Delta y = \left(\sum_{i=1}^{N} \left(\frac{\partial y}{\partial x_i}\Big|_{x_i = x_{i0}} \right)^2 \Delta x_i^2 \right)^{\frac{1}{2}} . \tag{1.65}$$

or, expressing this with the relative values:

$$\frac{\Delta y}{y_0} = \left(\sum_{i=1}^{N} \left(\frac{\partial y}{\partial x_i}\Big|_{x_i = x_{i0}} \frac{\Delta x_i}{y_0} \right)^2 \right)^{\frac{1}{2}} . \tag{1.66}$$

It can be seen that these relationships may only be applied if the variables x_1, x_2, ..., x_N are independent. For correlated variables more complex formulas can be derived.

Note 1.6

a) Assume that y can be expressed as the product of the functions of $x_1, x_2, \ldots, x_N$:

$$y = f_1(x_1) f_2(x_2) \cdots f_N(x_N) . \tag{1.67}$$

By taking the logarithm of both sides:

$$\ln y = \sum_{i=1}^{N} \ln f_i(x_i) , \tag{1.68}$$

and by differentiating both sides, the expression

$$\frac{\Delta y}{y_0} = \sum_{i=1}^{N} \frac{f_i'(x_i)}{f_i(x_i)}\Big|_{x_i = x_{i0}} \Delta x_i \tag{1.69}$$

is obtained, which, in some cases, allows more straightforward calculations than (1.63).

b) For the validity of the relationships (1.65) and (1.66), given on the basis of a stochastic model, the independence of the measurement results x_1, x_2, ..., x_N are required, and for the estimation of the confidence interval the knowledge of the probability distribution is required. Since the fulfilment of these preconditions is not always provided, for the estimation of the relative error, the formula expressing the summation of the absolute values of individual error components is often used:

$$\frac{\Delta y}{y_0} = \sum_{i=1}^{N} \left| \frac{\partial y}{\partial x_i}\Big|_{x_i = x_{i0}} \frac{\Delta x_i}{y_0} \right| . \tag{1.70}$$

This is the worst case error. Since it gives a greater error than that calculated from equation (1.66), its use results in a 'safer' estimation. This expression is to be used when it is not known how the error limits were determined.

c) The weighting factor

$$\frac{\partial y}{\partial x_i}\Big|_{x_i = x_{i0}}$$

arises in the weights of the individual parameters in the summation of errors. It is a sensitivity term, showing how much the measurement error of the individual parameters influence the calculated measurement result.

Example 1.6

a) Let us calculate the relative error of y in the expression $y=x_1 \pm x_2$, assuming that the errors of x_1 and x_2 are systematic errors with known magnitudes and signs. According to Eq. (1.63),

$$\frac{\Delta y}{y_0} = \frac{\Delta x_1 \pm \Delta x_2}{x_{10} \pm x_{20}} = \frac{\dfrac{\Delta x_1}{x_{10}} \pm \dfrac{\Delta x_2}{x_{20}} \dfrac{x_{20}}{x_{10}}}{1 \pm \dfrac{x_{20}}{x_{10}}} . \tag{1.71}$$

If the errors are random, and they are expressed by standard deviations,

$$\frac{\Delta y}{y_0} = \frac{\sqrt{\Delta x_1^2 + \Delta x_2^2}}{x_{10} \pm x_{20}} = \frac{1}{1 \pm \dfrac{x_{20}}{x_{10}}} \sqrt{\left(\frac{\Delta x_1}{x_{10}}\right)^2 + \left(\frac{\Delta x_2}{x_{20}} \cdot \frac{x_{20}}{x_{10}}\right)^2} . \tag{1.72}$$

In the case of Note 1.6b,

$$\frac{\Delta y}{y_0} = \frac{|\Delta x_1| + |\Delta x_2|}{x_{10} \pm x_{20}} = \frac{\dfrac{|\Delta x_1|}{x_{10}} + \dfrac{|\Delta x_2|}{x_{20}} \dfrac{x_{20}}{x_{10}}}{1 \pm \dfrac{x_{20}}{x_{10}}} . \tag{1.73}$$

Note that the difference between x_{10} and x_{20} may appear in the denominator. For small differences the denominator tends to zero, and the relative error may become extremely high. This implies that measurement methods, which require the difference building of two nearly equal quantities, should be avoided.

b) Let us calculate the relative error of y in the expression $y=x_1^n x_2^m$ by assuming that the errors of x_1 and x_2 are systematic errors of known magnitudes and signs.

According to Eq. (1.58).

$$\frac{\Delta y}{y_0} = \left.\frac{n x_1^{n-1} x_2^m}{x_1^n x_2^m}\right|_{x_1=x_{10}} \Delta x_1 + \left.\frac{x_1^n m x_2^{m-1}}{x_1^n x_2^m}\right|_{x_2=x_{20}} \Delta x_2 = n\frac{\Delta x_1}{x_{10}} + m\frac{\Delta x_2}{x_{20}} . \tag{1.74}$$

It is easier to derive this result by utilizing expressions (1.68) and (1.69)

$$\ln y = n \ln x_1 + m \ln x_2 \tag{1.75}$$

and the relative error follows directly from the following relation:

$$\frac{\Delta y}{y_0} = n\frac{\Delta x_1}{x_{10}} + m\frac{\Delta x_2}{x_{20}} . \tag{1.76}$$

If the errors are random, (1.64) can be applied:

$$\frac{\Delta y}{y_0} = \left(\left(n\frac{\Delta x_1}{x_{10}}\right)^2 + \left(m\frac{\Delta x_2}{x_{20}}\right)^2\right)^{\frac{1}{2}} . \tag{1.77}$$

Finally, in the case of Note 1.6b,

$$\frac{\Delta y}{y_0} = \left| n \frac{\Delta x_1}{x_{10}} \right| + \left| m \frac{\Delta x_2}{x_{20}} \right| , \tag{1.78}$$

which gives the worst case estimation.

c) Let us consider the measurement of the power across impedance Z by an electrodynamic wattmeter (see Fig. 3.9), with insertion of a current transformer and a voltage transformer, and assume that the systematic ratio and phase errors of the measuring transformers are known. Calculate the relative error $\Delta P_2/P_2$ of the power measurement, introduced by these systematic errors.

The error-free power, indicated by the wattmeter, is given by

$$P_2 = U_2 I_2 \cos \phi_2 . \tag{1.79}$$

Applying the presentation method according to (1.68), we have

$$\ln P_2 = \ln U_2 + \ln I_2 + \ln \cos \phi_2 \tag{1.80}$$

and according to (1.69),

$$\frac{\Delta P_2}{P_2} = \frac{\Delta U_2}{U_2} + \frac{\Delta I_2}{I_2} - \Delta\phi \tan \phi_2 , \tag{1.81}$$

where $\Delta U_2/U_2$ and $\Delta I_2/I_2$ are the ratio errors of the voltage and current transformers, respectively, ϕ_2 is the angle between U_2 and I_2 and $\Delta\phi$ is the difference between the angle errors of the measuring transformers.

d) A resonance circuit has L and C elements bearing known errors. What is the error of the resonant frequency?

$$f = \frac{1}{2\pi} \cdot \frac{1}{\sqrt{LC}} , \tag{1.82}$$

$$\ln f = -\left(\ln 2\pi + \frac{1}{2}(\ln L + \ln C) \right) . \tag{1.83}$$

According to (1.65),

$$\frac{\Delta f}{f} = \frac{1}{2}\sqrt{\left(\frac{\Delta L}{L}\right)^2 + \left(\frac{\Delta C}{C}\right)^2} , \tag{1.84}$$

and according to (1.70),

$$\frac{\Delta f}{f} = \frac{1}{2}\left(\left|\frac{\Delta L}{L}\right| + \left|\frac{\Delta C}{C}\right| \right) . \tag{1.85}$$

References

Bendat, J. S. and Piersol, A. G. (1971): Random Data: Analysis and Measurement Procedures. New York, John Wiley & Sons.

Courtois, P. I. (1985): On Time and Space Decomposition of Complex Structures. *Communications of the ACM,* Vol. 28, No. 6, pp. 590–603, June 1985.

DIN (1984): International Vocabulary of Basic and General Terms in Metrology. Edited by DIN. Deutsches Institut für Normung E. V. Berlin, 1984. Published in the name of the organizations: BIPM, IEC, ISO and OIML.

Eykhoff, P. (1974): System Identification. London, John Wiley & Sons.

Finkelstein, L. (1975): Fundamental Concepts of Measurement. *Measurement and Control,* Vol. 8, pp. 105–111.

Gihman, I. I. and Skorohod, A. V. (1979): The Theory of Stochastic Processes. New York, Springer-Verlag.

Goodwin, G. C. and Payne, R. L. (1977): Dynamic System Identification: Experiment Design and Data Analysis. New York, Academic Press.

Korn, G. A. and Korn, T. M. (1968): Mathematical Handbook for Scientists and Engineers: Definitions, Theorems and Formulas for Reference and Review. 2nd ed. London, McGraw-Hill.

Loève, M. (1977): Probability Theory I-II. New York, Springer-Verlag.

Luppold, D. S. (1969): Precision dc Measurements and Standards. Reading, MA, Addison-Wesley.

Norton, J. P. (1986): An Introduction to Identification. London, Academic Press.

Oppenheim, A. V., Willsky, A. S. and Young, I. T. (1983): Signals and Systems. Englewood Cliffs, NJ, Prentice-Hall.

Papoulis, A. (1965): Probability, Random Variables and Stochastic Processes. New York, McGraw-Hill.

Papoulis, A. (1977): Signal analysis. New York, McGraw-Hill.

Pfanzagl, J. (1968): Theory of Measurement. New York, John Wiley & Sons.

Plourde, A. J. (1962): Measurement Errors — Identification, Detection, Evaluation and Expression. *Proceedings of the 1962 Standards Laboratory Conference.* NBS Miscellaneous Publications No. 248, pp. 104-106.

Sage, A. P. and Melsa, J. L. (1971): Estimation Theory with Applications to Communications and Control. New York, McGraw-Hill.

Spiegel, M. R. (1961): Theory and Problems of Statistics. New York, Schaum Publishing Co.

Urkowitz, H. (1983): Signal Theory and Random Processes. Norwood, MA, Artech House.

Van Trees, H. L. (1968): Detection, Estimation and Modulation Theory. Part I. New York, John Wiley & Sons.

Youden, W. J. (1962): Uncertainties in Calibration. *IRE Trans. on Instrumentation,* Vol. I–II, No. 3–4, pp. 133–138.

Comprehensive works on the topics of this chapter:

Bendat, J. S. and Piersol, A. G. (1971): Random Data: Analysis and Measurement Procedures. New York, John Wiley & Sons.

Coombs, C. F. (1972): Basic Electronic Instrument Handbook. New York, McGraw-Hill.

Cooper, W. D. (1970): Electronic Instrumentation and Measurement Techniques. Englewood Cliffs, NJ, Prentice-Hall.

Eykhoff, P. (1974): System Identification. London, John Wiley & Sons.

Frühauf, U. (1977): Grundlagen der elektronischen Meßtechnik (Basic Concepts of Electronic Measurement). Leipzig, Geest & Portig KG. (In German)

Gihman, I. I. and Skorohod, A. V. (1979): The Theory of Stochastic Processes. New York, Springer-Verlag.

Goodwin, G. C. and Payne, R. L. (1977): Dynamic System Identification: Experiment Design and Data Analysis. New York, Academic Press.

Jazwinski, A. H. (1970): Stochastic Processes and Filtering Theory. New York, Academic Press.

Jones, L. and Chin, A. F. (1983): Electronic Instruments and Measurements. Chichester, John Wiley & Sons.

Karlin, S. and Taylor, H. M. (1975): A First Course in Stochastic Processes. New York, Academic Press.

Loève, M. (1977): Probability Theory I-II. New York, Springer-Verlag.

Norton, J. P. (1986): An Introduction to Identification. London, Academic Press.

Oliver, B. M. and Cage, J. M. (1971): Electronic Measurements and Instrumentation. New York, McGraw-Hill.

Oppenheim, A. V., Willsky, A. S. and Young, I. T. (1983): Signals and Systems. Englewood Cliffs, NJ, Prentice-Hall.

Papoulis, A. (1965): Probability, Random Variables and Stochastic Processes. New York, McGraw-Hill.

Papoulis, A. (1977): Signal analysis. New York, McGraw-Hill.

Parzen, E. (1962): Stochastic Processes. San Francisco, Holden-Day.

Pfanzagl, J. (1968): Theory of Measurement. New York, John Wiley & Sons.

Sage, A. P. and Melsa, J. L. (1971): Estimation Theory with Applications to Communications and Control. New York, McGraw-Hill.

Sydenham, P. H. ed. (1982): Handbook of Measurement Science I-II. New York, John Wiley & Sons.

Tran Tien Lang (1987): Electronics of Measuring Systems. Chichester, John Wiley & Sons.

Urkowitz, H. (1983): Signal Theory and Random Processes. Norwood, MA, Artech House.

Van Trees, H. L. (1968): Detection, Estimation and Modulation Theory. Part I. New York, John Wiley & Sons.

2

Measurement of Current and Voltage

I. Zoltán

Editorial introduction

This chapter deals with the measurement of ac-dc currents and voltages. As will be seen in the following chapters, measurement methods of these quantities are very important in power and impedance measurements as well.

Basic measuring networks and sources of interference will be introduced, since they are in the focus of our interest. Construction, accuracy and interference-sensitivity of signal converters will also be analyzed in detail. Finally, a survey of instruments for current and voltage measurements will be presented.

2.1 BASIC CONSIDERATIONS

Electrical measurements are usually based on current and voltage measurements. There are several methods and instruments for the measurement of these two basic quantities.

Current and voltage measurements are carried out in an extremely wide range (Fig. 2.1).

Very sensitive electrometers can measure even as low a current as 10 aA, produced by 62 electrons per second.

Using a special current transformer for range extension, even 100 kA can be measured. This extremely high current produces 10 kW dissipation in a copper bus-bar with the following parameters:

Length:	1 m
Cross section:	1.78 dm^2
Resistance:	1 $\mu\Omega$.

The ranges of voltage measurements are shown in Fig. 2.2.

Technology of Electrical Measurements, Edited by L. Schnell
© 1993 John Wiley & Sons Ltd.

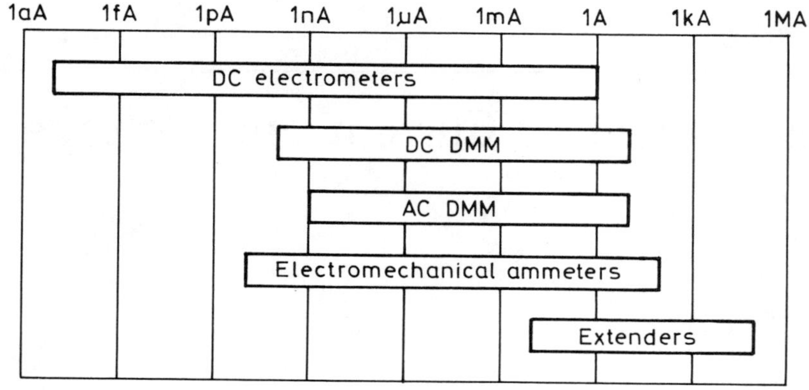

Fig. 2.1 Ranges of current measurements

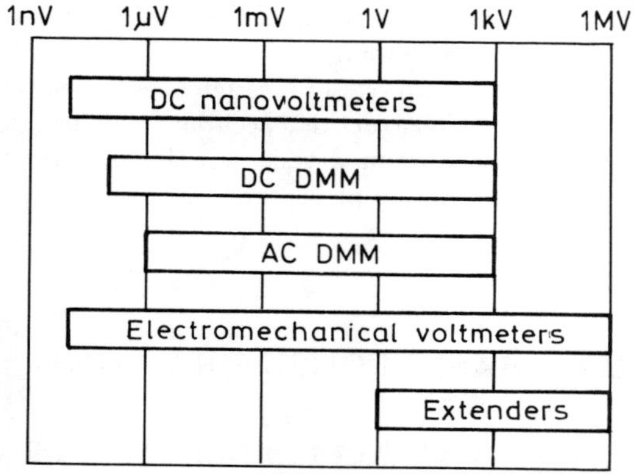

Fig. 2.2 Ranges of voltage measurements

Nanovoltmeters can measure 10 nV, a very low voltage compared with the thermal noises. The rms thermal noise voltage between the terminals of a resistor, having 10 kΩ resistance, is 1.2 µV at room temperature within the bandwidth of 10 kHz.

Using special high-voltage capacitance dividers for range extension, very high voltages up to 1 MV can be measured. Such a high-voltage capacitor can be very spectacular. For example a compressed gas standard capacitor for 1 MV is higher than 7 m (Tettex, 1987).

This chapter deals with the measurement of the following parameters:

- Direct voltage
- Direct current
- Mean value
- Mean absolute value
- Root-mean-square (rms) value

- Peak value
- Instantaneous value
- Vector components
- Frequency spectrum.

Frequency spectrum analysis methods are discussed in more detail in Chapter 6.

In practice different kinds of interference may disturb the measurement of the above parameters, such as high-voltage, high-current sources, electric and magnetic fields, common-mode signals, offsets, thermal noises, thermoelectric voltages etc; consequently, extreme care must be taken when current and voltage measurements are carried out.

In order to obtain a general impression of the problems that arise, let us consider the model of Fig. 2.3.

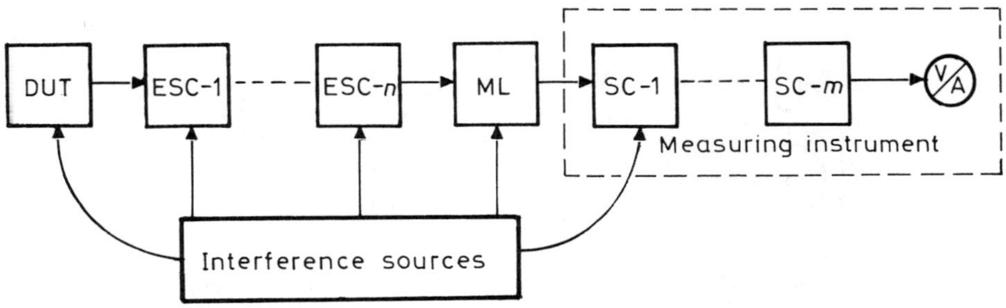

Fig. 2.3 Model of a current or voltage measuring system

The current or voltage to be measured is produced by the device under test (DUT). The measuring instrument is connected to the DUT via external signal converters (ESC) and the measuring leads (ML). The measuring instrument consists of internal signal converters (SC) and meter (V or A). Each part of the system can be disturbed by the interference sources.

The external and internal signal converters change the form or the value of a signal, e. g. to achieve range extension. Our investigations are focused on these signal converters because they play a pivotal part in the measurements. Construction, accuracy and interference-sensitivity of signal converters will be analyzed in detail in the relevant subsections.

Finally, current and voltage measuring methods and instruments will also be discussed.

2.2 MEASURING NETWORKS

In current and voltage measurement technology, the proper design of measuring networks is vital. Their behaviour in the presence of interference will be analyzed in the following subsections.

2.2.1 Basic Measuring Networks

The simplest model of a measuring network is shown in Fig. 2.4.

Fig. 2.4 Model of the measuring network

Let us analyze the main parts of the measuring network. Fig. 2.5 shows the simplest models of the source.

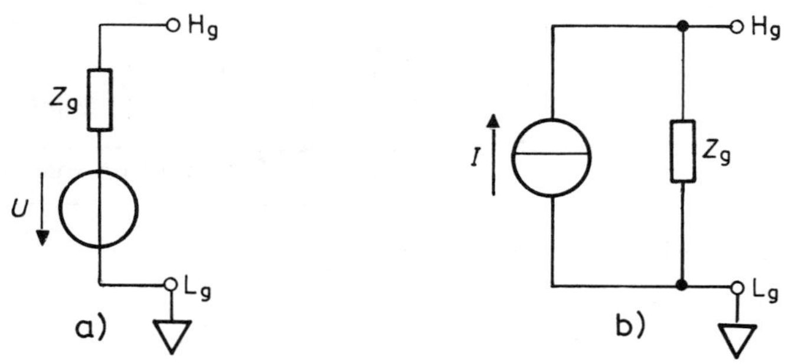

Fig. 2.5 Models of signal sources *a)* voltage source *b)* current source

As can be seen, Z_g represents the internal impedance of the source, grounded at terminal L_g. The current or voltage source to be measured is connected to the measuring instrument via measuring leads. The physical representation of the measuring lead is shown in Fig. 2.6.

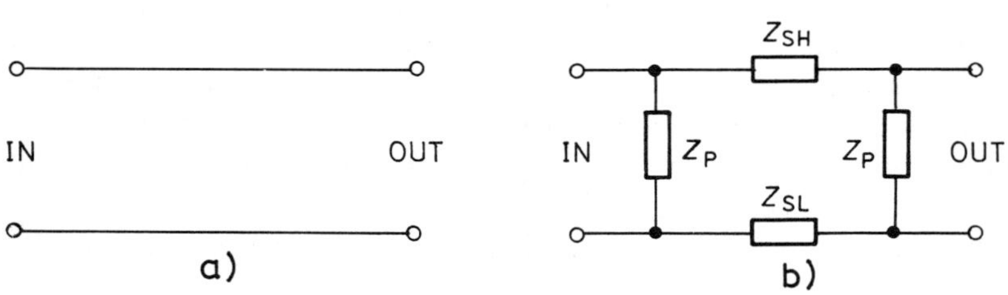

Fig. 2.6 Model of the measuring lead *a)* ideal *b)* nonideal

Z_{SH} and Z_{SL} are series parasitic impedances of the lead. The parallel impedances (Z_P) are concentrated at the terminals.

The simplest models of the measuring instrument having internal impedance Z_{in} are shown in Fig. 2.7.

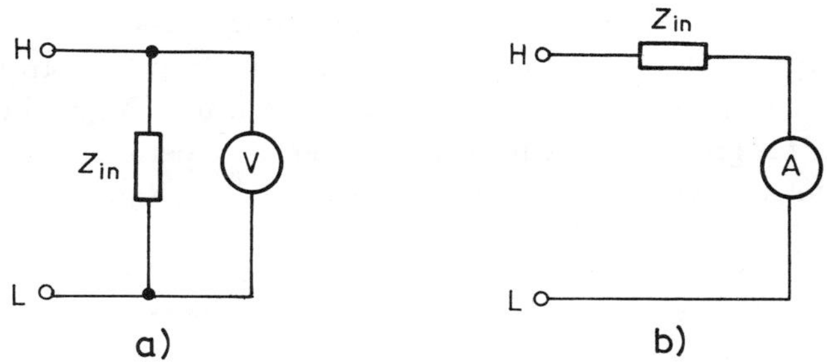

Fig. 2.7 Models of V/A meters *a)* voltmeter *b)* ammeter

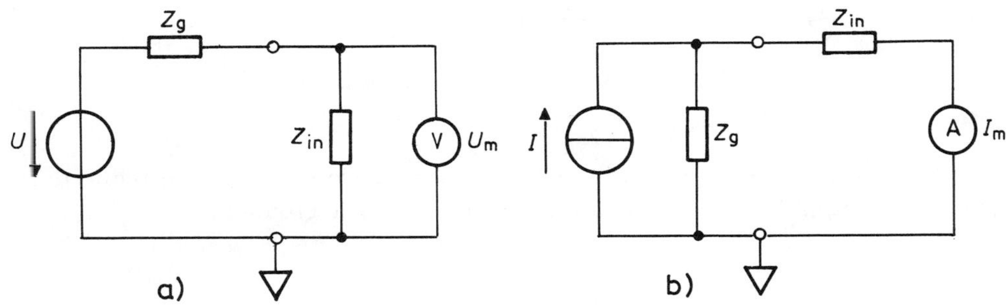

Fig. 2.8 Loading error *a)* in voltage measurement *b)* in current measurement

Internal impedances of the source and the meter cause loading errors as shown in Fig. 2.8.

The measured voltage and current are expressed by

$$U_{\mathrm{m}} = U \frac{Z_{\mathrm{in}}}{Z_{\mathrm{g}} + Z_{\mathrm{in}}} \tag{2.1}$$

and

$$I_{\mathrm{m}} = I \frac{Z_{\mathrm{g}}}{Z_{\mathrm{g}} + Z_{\mathrm{in}}} . \tag{2.2}$$

The relative errors due to the loading effect are

$$\epsilon_{\mathrm{ur}} = \frac{U_{\mathrm{m}} - U}{U} = -\frac{Z_{\mathrm{g}}}{Z_{\mathrm{g}} + Z_{\mathrm{in}}} \tag{2.3}$$

and

$$\epsilon_{\mathrm{ir}} = \frac{I_{\mathrm{m}} - I}{I} = -\frac{Z_{\mathrm{in}}}{Z_{\mathrm{g}} + Z_{\mathrm{in}}} , \tag{2.4}$$

respectively.

It can be seen from these expressions that an ideal voltmeter has infinite internal impedance, and the internal impedance of an ideal ammeter is zero.

Next, let us consider the asymmetric measuring network, shown in Fig. 2.9. The terminal H_g of the grounded[1] voltage source is connected to the grounded, asymmetric voltmeter by an asymmetric measuring lead. The main source of the error is the multiple grounding. Any potential difference between the source ground and measurement ground causes an additive error voltage U_d.

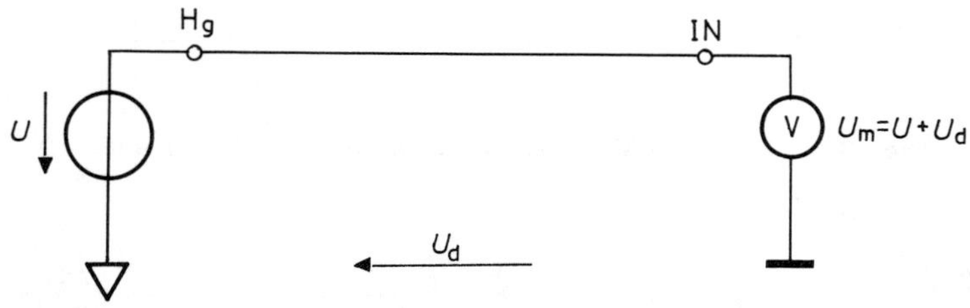

Fig. 2.9 Asymmetric measuring network

Using a symmetric measuring network, the above error can be reduced or eliminated. Fig. 2.10b shows the best solution, the single grounding.

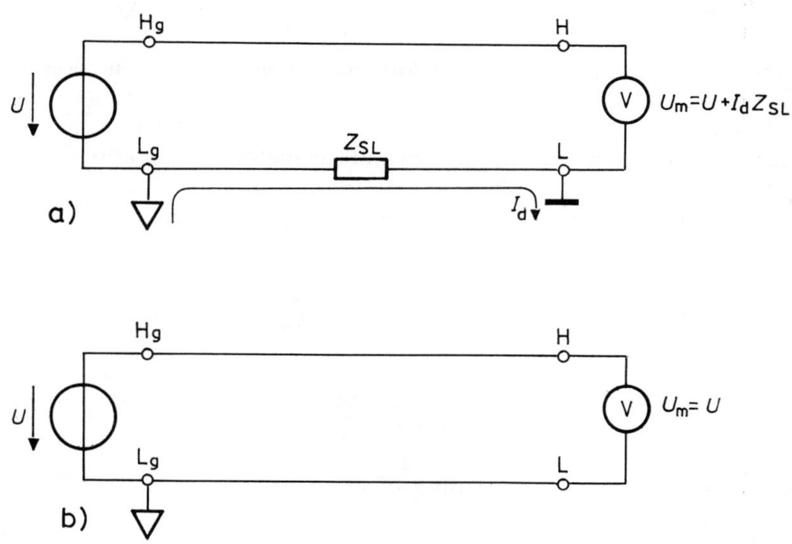

Fig. 2.10 Symmetric measuring networks *a)* multiple grounded *b)* single grounded

Using multiple grounding, any potential difference between the source ground and measuring ground produces a disturbing current I_d. The disturbing voltage is

[1]Ground (earth) is a conducting connection, whether intentional or accidental, by which an electrical circuit or electrical equipment is connected to the earth, or to some conducting body of relatively large extent that serves in place of earth (IEEE, 1977).

caused by I_d flowing through Z_{SL}. The longer the measuring lead, the higher the series parasitic impedance Z_{SL} and the disturbing voltage will be.

In practice the voltage or current source is often floating on top of a common-mode voltage, as shown in Fig. 2.11.

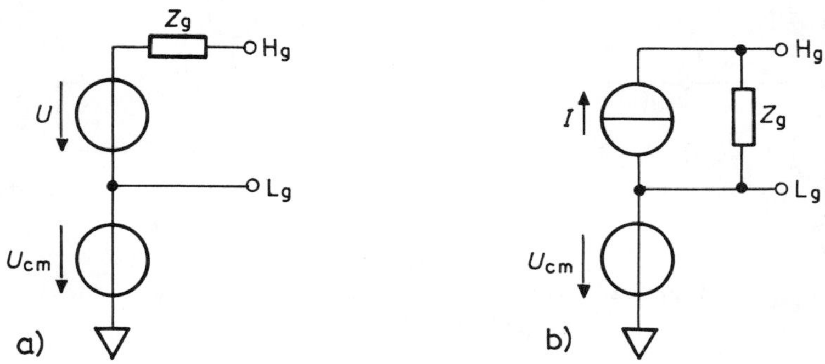

Fig. 2.11 Floating sources *a)* voltage source *b)* current source

The main source of the measurement error is the common-mode voltage, because the instrument is also floating on U_{cm} as illustrated in Fig. 2.12.

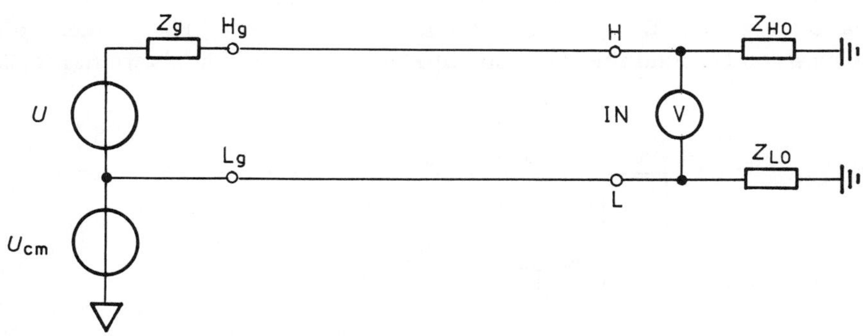

Fig. 2.12 Symmetric measuring network for floating voltage source

Because of the finite terminal-to-ground impedance Z_{H0} of the voltmeter, the input voltage of the measuring instrument is:

$$U_{in} = U - U_{cm} \frac{Z_g}{Z_g + Z_{H0}} . \tag{2.5}$$

The aim is that the common-mode voltage U_{cm} should not disturb the measurement. This can be reached only if $Z_g = 0$.

In the case of $Z_g > 0$ and if better common-mode voltage rejection (cmr) is necessary, the instrument must be guarded, like in Fig. 2.13.

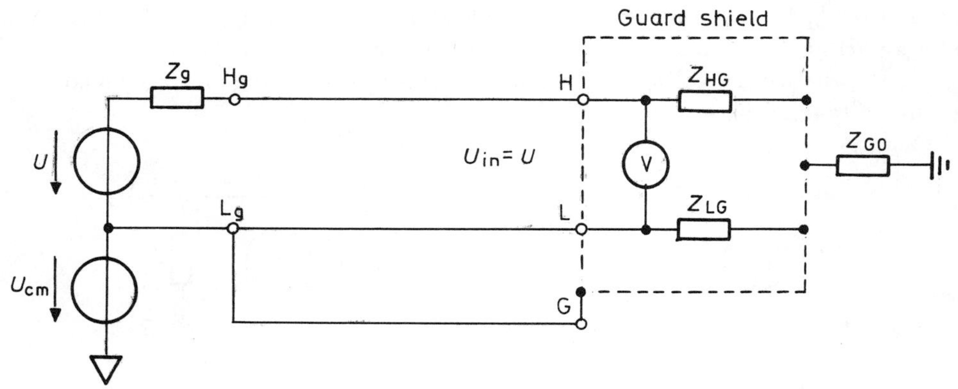

Fig. 2.13 Guarding of the measuring instrument

The guard shield of the instrument is connected to the terminal L_g, that is, the shield is also floating on the common-mode voltage U_{cm}. It is easy to see from the figure that the common-mode voltage cannot penetrate into the instrument. Using this guarding technique, perfect common-mode rejection can be reached in an ideal case. However, the measuring instrument is floating on potential U_{cm}, which cannot always be allowed for safety reasons. On the other hand, a metal box, floating on potential U_{cm}, can be a source of interference. The problem can be solved by a second ground shield surrounding the guard shield as shown in Fig. 2.14.

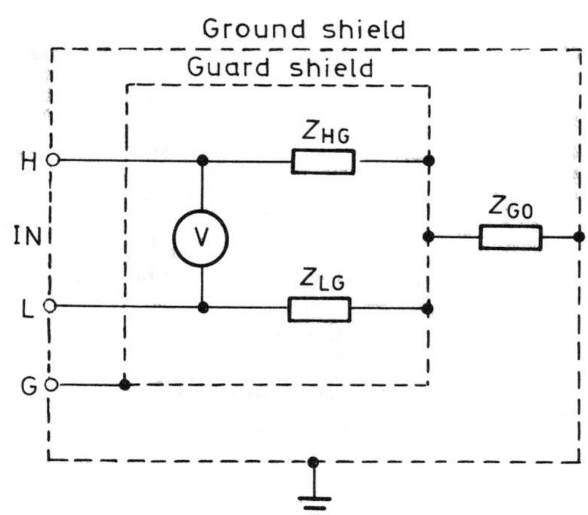

Fig. 2.14 Double shielding of the measuring instrument

For safety reasons the ground shield is connected to the power ground. At the same time, this grounding eliminates the source of interference.

2.2.2 Sources of Interference

The interference source and the measurement system can be coupled in a capacitive, conductive or inductive way. The interference penetrates into the measurement system by an electric field and/or by a magnetic field, produced by the interference source. The three types of coupling are shown in Fig. 2.15.

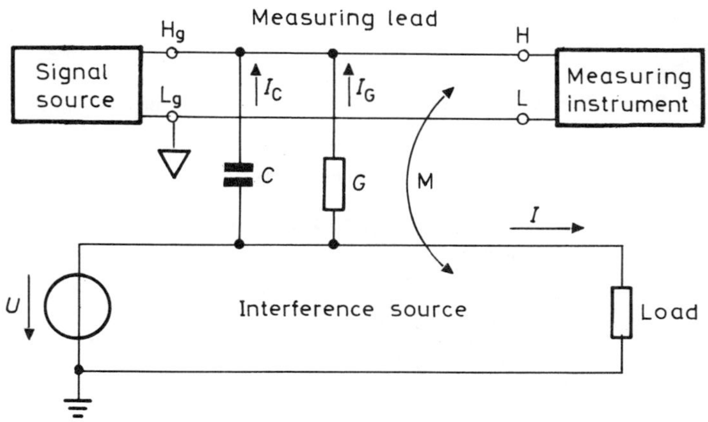

Fig. 2.15 Capacitance, conductance and inductance coupling

The interference source consists of a voltage source and the load. The current of the load can be alternating current or varying direct current. The interference currents are I_C and I_G, flowing through the stray capacitance C and stray conductance G, respectively. Due to the mutual inductance M, an interference voltage is also induced in the measuring system.

There are two basic methods for the reduction or elimination of the capacitance, conductance and inductance coupling:

- static and magnetic shielding of the measuring system,
- static and magnetic shielding of the interference source.

Using a cable composed of two small insulated conductors, twisted together with a common covering, the coupling between the interference source and the system can be reduced, as shown in Fig. 2.16.

The interference currents I_C and I_G flow into the ground. However, at higher frequencies the serial parasitic impedance Z_{SL} of the common covering reduces the effectiveness of the shielding. The twisting of the two inner conductors reduces the area produced by the conductors thus the sensitivity to the alternating or varying magnetic field (B) can be reduced. Fig. 2.17 shows the cross section of a high-quality cable.

The double shielding and the MUMETALL magnetic shield of high permeability guarantee very effective interference protection. Such a cable is most suitable for a measuring instrument, with double shielding. To achieve a very low interference, the signal source must also be shielded, similarly to the measuring instrument.

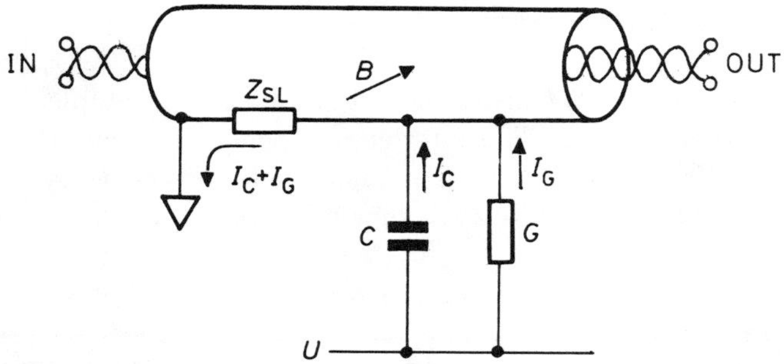

Fig. 2.16 Reduction of capacitive, conductive and inductive coupling

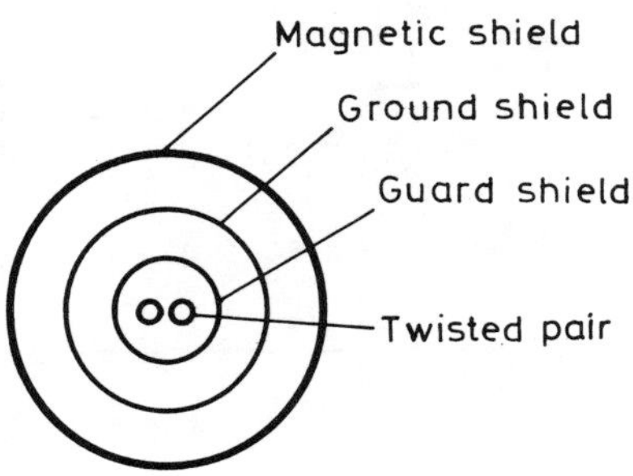

Fig. 2.17 High-quality cable

Very special measurements can be carried out in a shielded room covered by copper and MUMETALL plates (Boll, 1990).

The second method for reduction of the coupling is the static and magnetic shielding of the interference source. Fig. 2.18 shows how the magnetic field of a long, current carrying lead can be reduced by using coaxial cable.

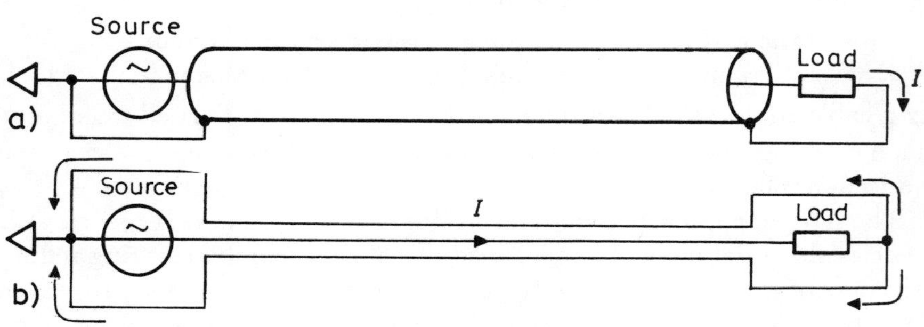

Fig. 2.18 Reduction of the stray magnetic field *a)* using coaxial cable *b)* using coaxial cable combined with full shielding

With coaxial connection of the source and the load, even the leakage magnetic and electric fields can be eliminated.

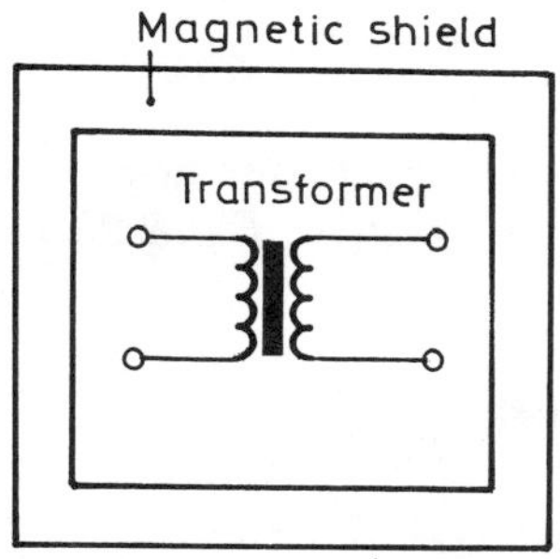

Fig. 2.19 Shielded transformer

Transformers are typical sources of magnetic interference. A magnetic shield surrounding the transformer can be very effective. Fig. 2.19 shows a transformer with magnetic shield.

The leakage magnetic field can be further reduced by using a toroidal core.

Thermoelectric voltages are the most common sources of error in low-voltage measurements. These voltages are produced by junctions of different materials having different temperatures (Keithley, 1984). For example, the thermoelectric potential of Pb/Sn solder with respect to copper is $1.3\,\mu V/^\circ C$. Using the same material for conductors in the whole circuit, the thermoelectric voltages can be minimized. Good thermal coupling of all junctions to a common heat sink minimizes temperature gradients within the circuit, thus the thermal voltages can be reduced. In a constant ambient temperature any remaining thermoelectric voltage can be compensated by zero controls provided on the meter.

The fundamental limit to current and voltage measurement is the Johnson noise in the resistances. It is caused by the movement of charged particles due to thermal energy. The Johnson or thermal voltage noise developed in a resistor R is

$$U_{\mathrm{rms}} = \sqrt{4kTR\Delta f}\,, \tag{2.6.}$$

where k is Boltzmann's constant ($1.38 \cdot 10^{-23}$ J/°K), T is the temperature in Kelvin, and Δf is the noise bandwidth in Hz (Keithley, 1984).

For more information about EMC problems, see Stoll (1976).

2.3 SIGNAL CONVERTERS

Signal converters are used for changing a signal from one form or value to another form or value. The aim of signal conversion can be range extension or obtaining another signal form which is more suitable for signal processing or signal transmission than the original signal.

2.3.1 Errors of Signal Converters

The accuracy of current and voltage measurement depends on the accuracy of signal converters used in the measuring system. Fig. 2.20 shows a signal converter having a transfer ratio H.

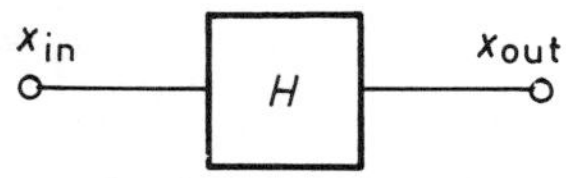

Fig. 2.20 Signal converter

In the ideal linear case the ratio between the output quantity and input quantity is given by

$$H_n = \frac{X_{out}}{X_{in}} . \tag{2.7}$$

In practice, however, the output quantity deviates from the nominal value. For sinusoidal signals the following equation can be written:

$$\frac{X_{out}}{X_{in}} = H_n + \alpha + j\beta , \tag{2.8}$$

where H_n is the nominal transfer ratio, α is the in-phase transfer ratio error and β is the quadrature transfer ratio error.

Typical error components of the transfer ratio of signal converters are shown in Fig. 2.21. These error components are as follows:

Offset

The constant component of the error, that is, the one independent of the input, is called offset.

Calibration error

Difference between the actual transfer ratio of a converter and the one predicted by the ideal transfer ratio expression.

Linearity error

The deviation of the output quantity from a specified linear reference curve.

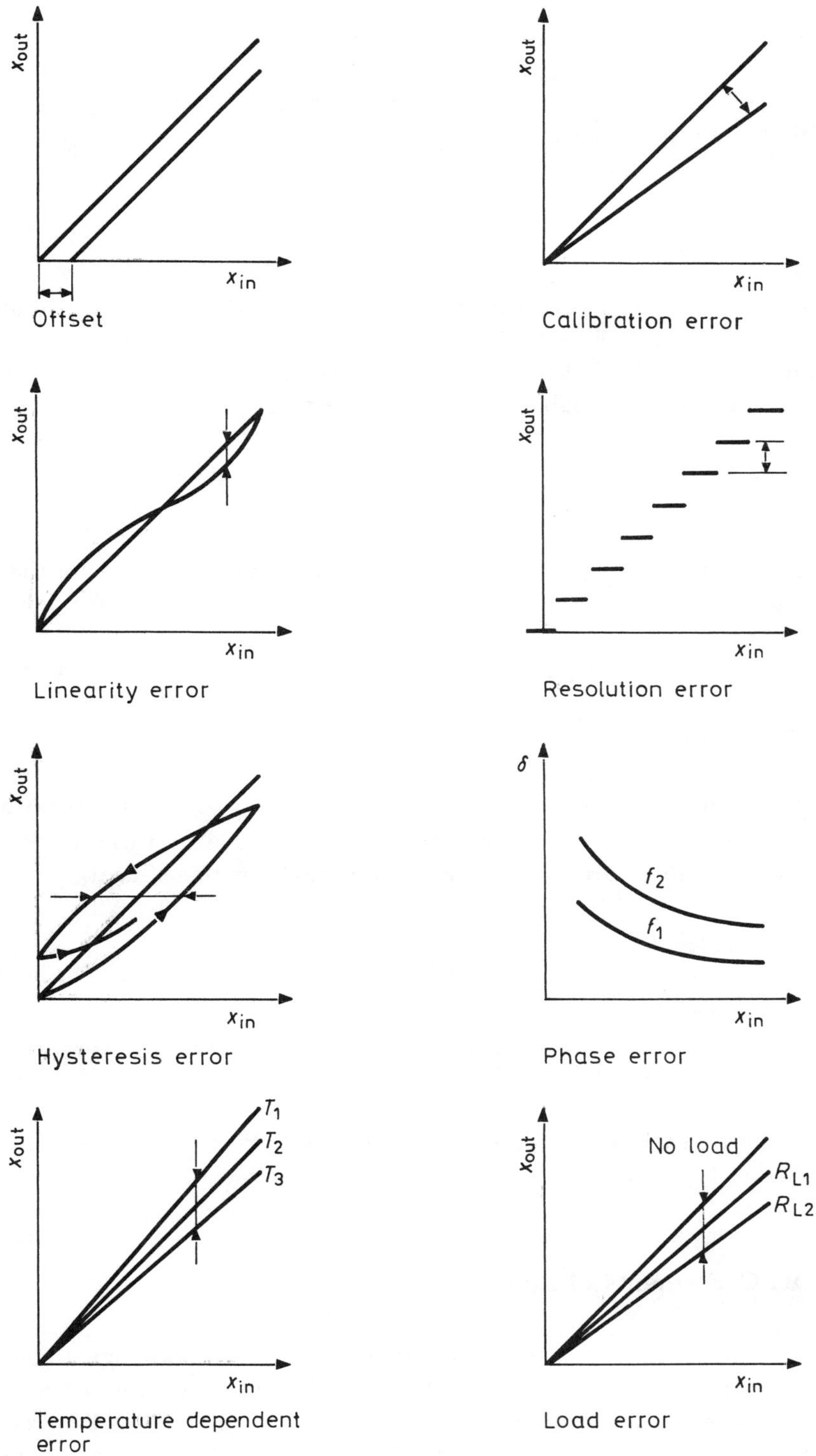

Fig. 2.21 Typical error components

Resolution error

The error due to the inability of a converter to respond to changes of a variable smaller than a given increment.

Hysteresis error

The difference between the increasing input value and the decreasing input value that produce the same output value is called hysteresis. Hysteresis error is the maximum separation due to hysteresis between up-scale-going and down-scale-going indications of the input variable after the transient have decayed.

Phase error

Phase error δ is the phase angle of the output minus the phase angle of the input, for sinusoidal input and output of the same frequency. The phase error depends on the frequency.

Temperature dependent error

Temperature dependent errors are characterized by the so-called temperature stability, which means low sensitivity of the output to temperature variation: a change in the ambient temperature or temperature variation due to self-heating.

Loading error

Loading error is due to the effect of a load impedance upon the converter or signal source driving it.

The above error components are well known in measurement technology, but there are many other types as well.

2.3.2 RLC Elements of Signal Converters

Resistors, inductors and capacitors are fundamental elements of signal converters used in measuring instruments. Zinke and Seither (1982) describe the characteristics of these components, and present different models of them. In this subsection we are going to briefly discuss these models. The behaviour of the above components is treated also in Chapter 4, from the point of view of their measurement.

Resistors

The model of a resistor is shown in Fig. 2.22.

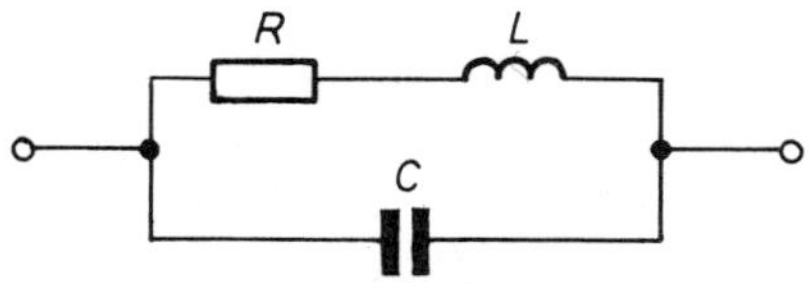

Fig. 2.22 Model of a resistor

L is the series inductance of the resistor, and C is the parallel capacitance between the terminals of the resistor. The effective series impedance is:

$$Z_S = \frac{R + j\omega\Big(L(1-\omega^2 LC) - R^2 C\Big)}{(1-\omega^2 LC)^2 + (\omega RC)^2}. \tag{2.9}$$

At low frequencies, where the terms in ω^2 are negligible, the following simplified equation can be written:

$$Z_s \approx R + j\omega(L - R^2 C). \tag{2.10}$$

For low resistance values the resistor may be represented by R in series with an inductance equal to $L-R^2C$.

The effective parallel admittance of the resistor is given by:

$$Y_P = \frac{G + j\omega\Big(C - LG^2(1-\omega^2 LC)\Big)}{1 + (\omega LG)^2} \approx G + j\omega(C - LG^2), \tag{2.11}$$

where $G=1/R$. At low frequencies the terms in ω^2 are negligible.

For high resistance values the resistor may be represented by R in parallel with a capacitance equal to $C-LG^2$.

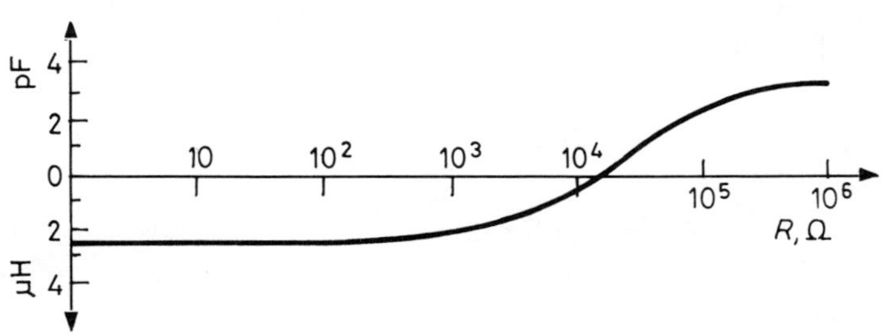

Fig. 2.23 Typical inductance and capacitance of wire-wound resistors

At higher frequencies, if the inductance and capacitance of the resistor are both small, the skin effect becomes predominant, causing an increase of resistance value.

Typical specifications of wire-wound resistors are as follows:

Range of resistance values:	from $0.1\,\Omega$ to $20\,\mathrm{M}\Omega$
Resistance tolerances:	from $\pm 0.05\%$ to $\pm 0.005\%$
Temperature coefficient:	from $\pm 25\,\mathrm{ppm}/^\circ\mathrm{C}$ to $\pm 5\,\mathrm{ppm}/^\circ\mathrm{C}$
Long-term stability:	$\pm 0.005\%/\mathrm{year}$
Inductance tolerance:	$\pm 1.5\,\mu\mathrm{H}$ typically
Capacitance tolerance:	$\pm 1.5\,\mathrm{pF}$ typically.

Using ultra-precision film resistors some parameters can be improved:

Temperature coefficient:	from $10\,\mathrm{ppm}/^\circ\mathrm{C}$ to $1\,\mathrm{ppm}/^\circ\mathrm{C}$
Inductance:	$80\,\mathrm{nH}$ typically
Capacitance:	$0.5\,\mathrm{pF}$ typically.

Inductors

The model of an air-core inductor is given in Fig. 2.24.

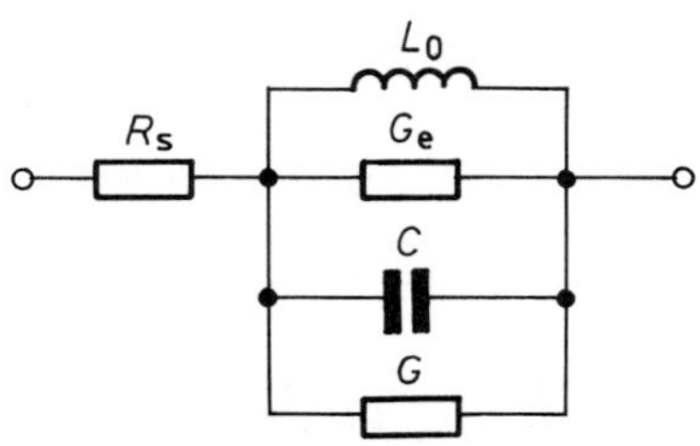

Fig. 2.24 Model of an air-core inductor

R_S is the series copper resistance, G_e is the conductance due to eddy-current loss in the winding, C is the winding capacitance and G represents the dielectric loss of capacitance C.

The frequency dependent change of the inductance is caused by the winding capacitance C. The effective inductance is as follows:

$$L = \frac{L_0}{1 - \omega^2 L_0 C}, \tag{2.12}$$

where L_0 is the inductance defined for dc.

The dissipation factor or loss tangent of an air-core inductor, represented by the above equivalent circuit, is as follows:

$$D = \tan\delta = \frac{R_S}{\omega L} + \omega L(G_e + G)\Big(1 + R_S(G_e + G)\Big). \tag{2.13}$$

In practice $R_e(G_e+G) \ll 1$, thus it is negligible. The dissipation factor consists of three components, which are the ohmic loss, the eddy-current loss and the dielectric loss. Using special alloys of high permeability, larger inductance can be constructed. However, the eddy currents and hysteresis loss in the core cause an increase in the dissipation factor.

To achieve a very small stray magnetic field, the toroidal inductor is preferable to the solenoid. An air-core inductor has the highest stability and a negligible variation of inductance with current. The iron-core inductor is less stable than the air-core inductor, because the permeability of the core can change with age and with current. The performance of the iron-core inductor can be improved by the formation of an air gap in the core.

Capacitors

The commonly used model of a capacitor is given in Fig. 2.25. (Another model is briefly discussed in Section 4.5.)

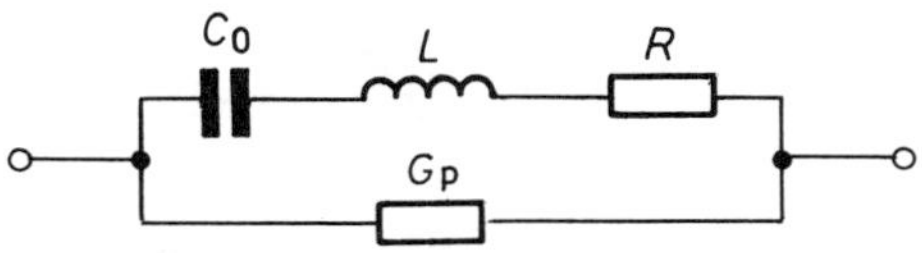

Fig. 2.25 Model of a capacitor

R represents the metallic resistance in the leads and plates, L represents the series inductance of the leads and plates. The losses in the air or solid dielectric between capacitor plates and the dc leakage conductance are represented by the conductance G_p.

The change of the capacitance with frequency is caused by the series inductance L. The effective capacitance is as follows:

$$C = \frac{C_0}{1-\omega^2 LC_0}, \tag{2.14}$$

where C_0 is the capacitance defined for dc. Another source of frequency dependent capacitance change is dielectric absorption.

The dissipation factor or loss tangent of a capacitor is as follows:

$$D = \tan\delta = \frac{G_p}{\omega C} + \omega CR(1+G_pR). \tag{2.15}$$

In practice $G_pR \ll 1$, thus it is negligible. At low frequencies only the dielectric losses, represented by G_p, are important. At higher frequencies the series resistance R becomes predominant.

Due to the skin effect, the series resistance increases at higher frequencies with the square root of frequency.

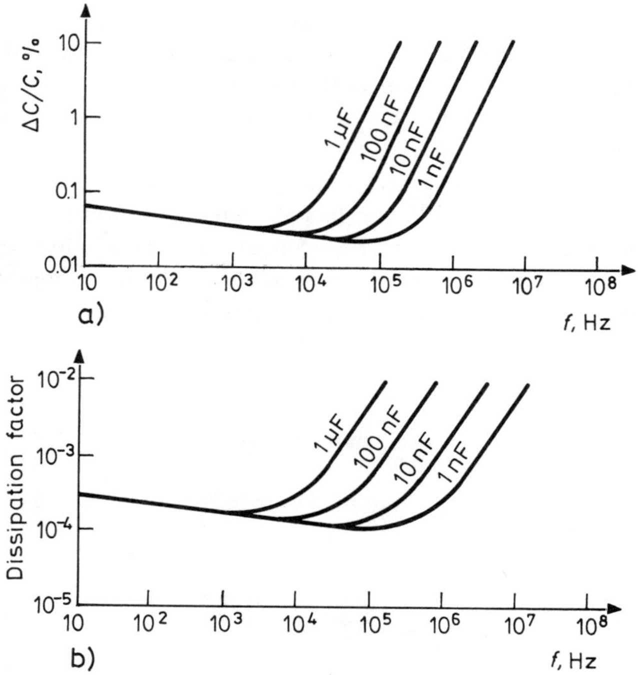

Fig. 2.26 Frequency dependence of mica capacitors *a)* change of the capacitance *b)* change of the dissipation factor

In Fig. 2.26 the variation of the capacitance and the dissipation factor for mica capacitors are shown as a function of frequency.

The increase of the capacitance at low frequencies is due to the dielectric absorption. At high frequencies the series inductance becomes predominant, causing an increase in the effective capacitance.

Using mica capacitors, a long-term stability of 0.01%/year can be achieved. The temperature coefficient of mica capacitors is typically +35 ppm/°C.

The low frequency change in the capacitance for a polystyrene capacitor is much smaller than for mica ones. However, the temperature coefficient of a polystyrene capacitor is approximately –140 ppm/°C: much larger than that of mica capacitors.

The air capacitor approaches the ideal reactance as it has very low loss and very small changes with time, frequency and voltage. Using low temperature-coefficient materials in the capacitor, the changes with temperature can be smaller than a few ppm/°C. The maximum capacitance, however, is limited by the size and working voltage of the capacitor.

2.3.3 Resistance Voltage Dividers

Voltage dividers are used for range extension. The simplest resistance voltage divider is shown in Fig. 2.27.

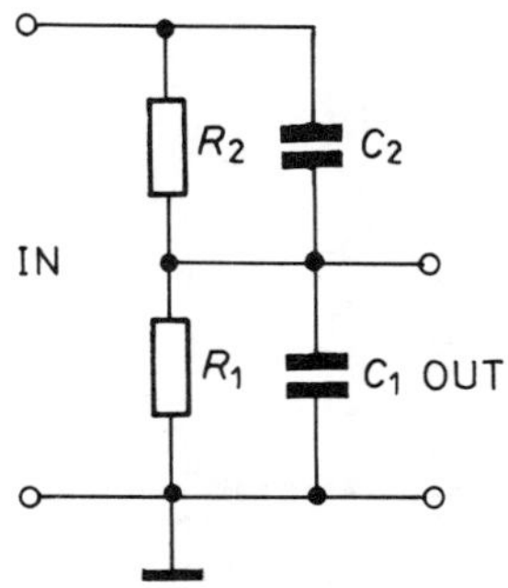

Fig. 2.27 Resistance voltage divider

The transfer factor or attenuation of the divider is

$$H = \frac{U_{\text{out}}}{U_{\text{in}}} = \frac{R_1}{R_1 + R_2}, \tag{2.16}$$

if

$$R_1 C_1 = R_2 C_2, \tag{2.17}$$

where C_1 and C_2 are the parallel stray capacitances.

At high input voltages, however, the stray capacitance between the resistor R_2 and the ground becomes predominant. A special shielding arrangement, shown in Fig. 2.28, can be used in high-voltage dividers to overcome this problem.

The resistor R_2 is broken into three sections, each enclosed in its own shield, and each shield is held at the mid-potential of the section it encloses, by means of an auxiliary voltage divider. Using this shielding, the capacitive current from the resistor to the shield can be reduced. The higher the number of sections, the lower is the capacitance error of the divider.

Fig. 2.29. shows a variable resistance divider which attenuates by a factor of

$$H = \frac{k}{10}, \tag{2.18}$$

where $k = 0, 1, 2, \ldots, 10$.

For fine setting of the attenuation, the cascade resistance divider can be used. Fig. 2.30 shows a two-decade cascade resistance divider. The input resistance of the second stage is $9R$, thus its current is $0.1\,I$.

The attenuation of the divider is:

$$H = \frac{100}{99} \sum_{i=1}^{2} \frac{k_i}{10^i}, \tag{2.19}$$

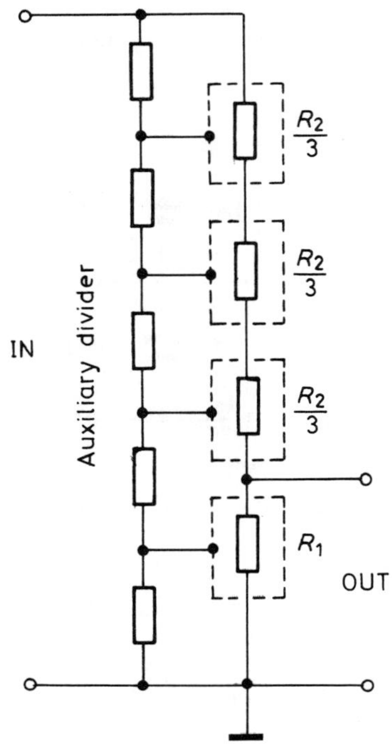

Fig. 2.28 Shielded high-voltage divider

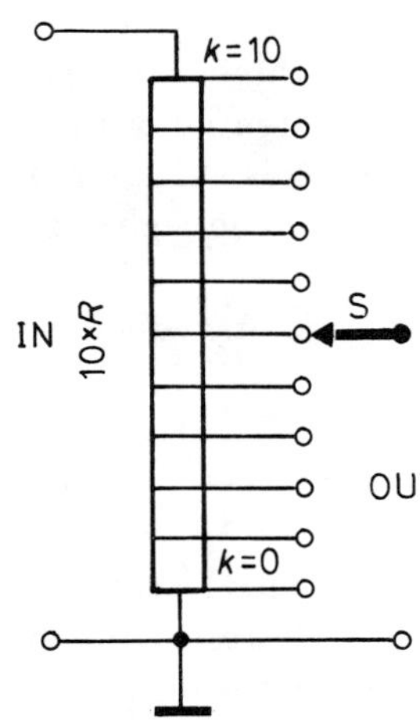

Fig. 2.29 Variable resistance divider

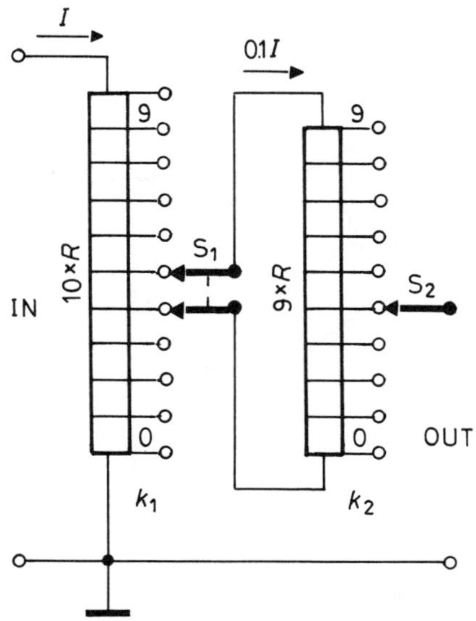

Fig. 2.30 Cascade resistance divider

where k_1, $k_2=0, 1, 2, \ldots, 9$.

The transfer factor can be adjusted by means of the switches S_1 and S_2. In practice even a 5-decade cascade divider can be realized.

The attenuation of the three-decade series resistance divider, shown in Fig. 2.31 is similar to that of the cascade divider.

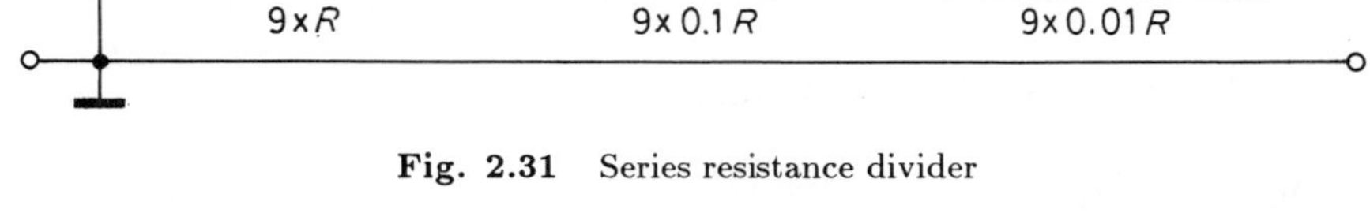

Fig. 2.31 Series resistance divider

A ladder network can also be used as a variable voltage divider, see Fig. 2.32. The resulting transfer factor is

$$H=\frac{2}{3}\sum_{i=1}^{n}\frac{k_i}{2^i}, \tag{2.20}$$

where $k_i=0$ or $k_i=1$.

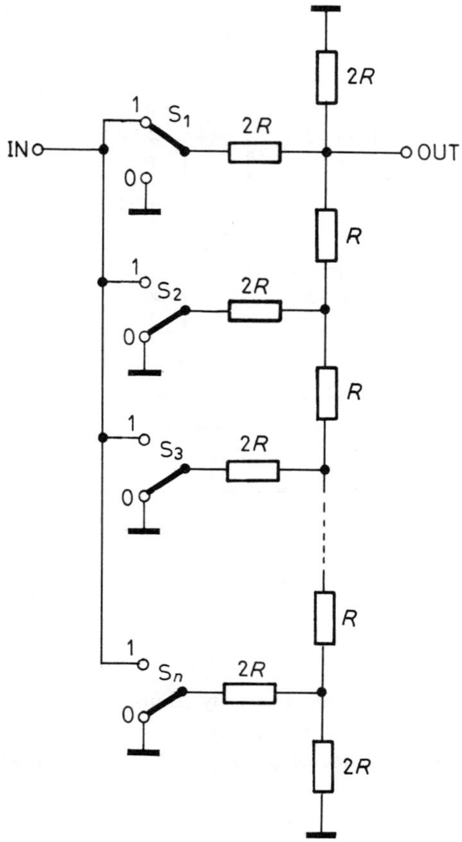

Fig. 2.32 Ladder network as variable voltage divider

This network may also act as a multiplying digital-to-analog converter, which generates the product of the digital input and the (maybe changing) reference voltage.

2.3.4 Inductance Voltage Dividers

The simplest inductance voltage divider consists of a core with a tapped winding (Fig. 2.33), with an attenuation in the ideal case:

$$H = \frac{U_{\text{out}}}{U_{\text{in}}} = \frac{N_1}{N_1 + N_2}. \tag{2.21}$$

Using a high-quality toroidal core, a very accurate divider can be constructed: 1 ppm accuracy can be achieved at 1 kHz.

The error of the ratio is caused by the leakage inductances and copper resistances. At higher frequencies, however, the winding capacitances become predomi-

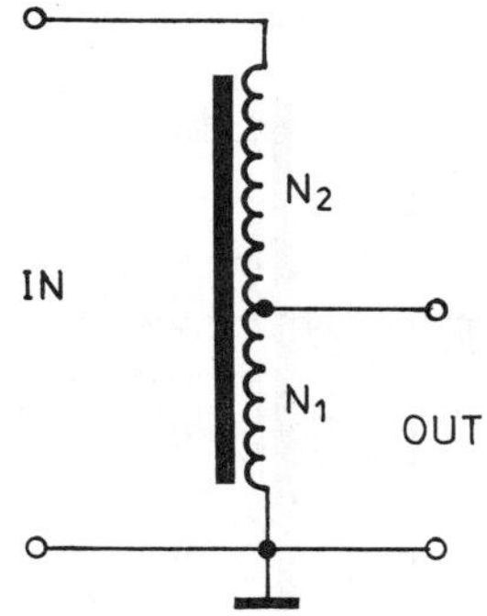

Fig. 2.33 Inductance voltage divider

nant. Due to the low output impedance, the inductance voltage divider has a much lower loading error than the resistance divider.

Fig. 2.34 shows a variable inductance divider. The transfer factor of the divider is

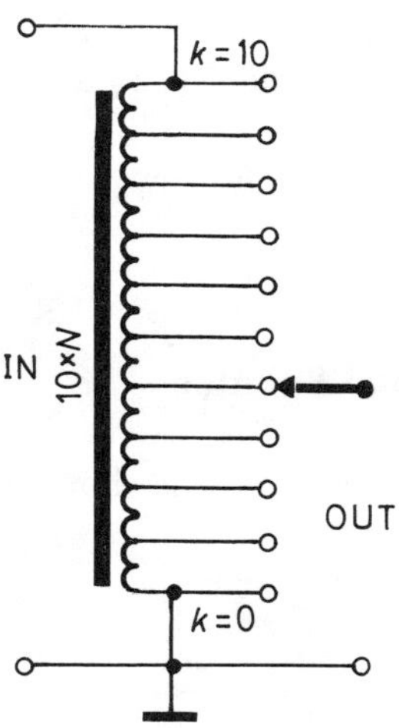

Fig. 2.34 Variable inductance divider

$$H = \frac{k}{10}, \tag{2.22}$$

where $k = 0, 1, 2, \ldots, 10$.

For fine setting of the transfer factor, a cascade inductance divider can be used. A three-decade cascade divider is shown in Fig. 2.35. Its attenuation is

$$H = \sum_{i=1}^{3} \frac{k_i}{10^i}, \tag{2.23}$$

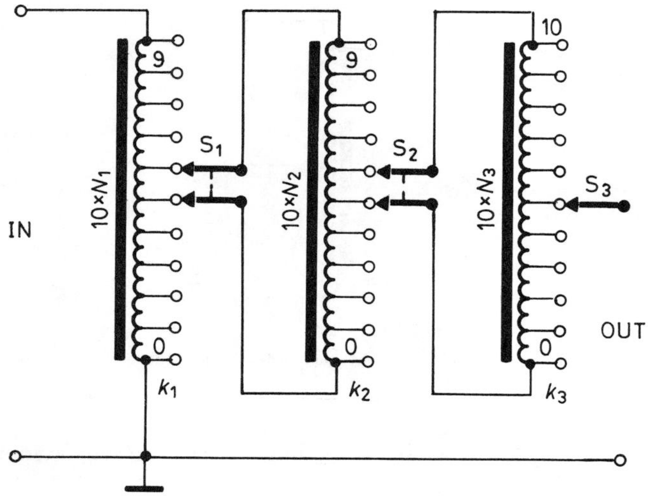

Fig. 2.35 Cascade inductance divider

where k_1, $k_2 = 0, 1, 2, \ldots, 9$, $k_3 = 0, 1, 2, \ldots, 10$.

H depends on the position of the S_1, S_2 and S_3. In practice, even a 7–decade divider can be realised.

The above inductance dividers are sensitive to external magnetic fields, thus they have to be shielded.

2.3.5 Capacitance Voltage Dividers

The simplest capacitance voltage divider is shown in Fig. 2.36.

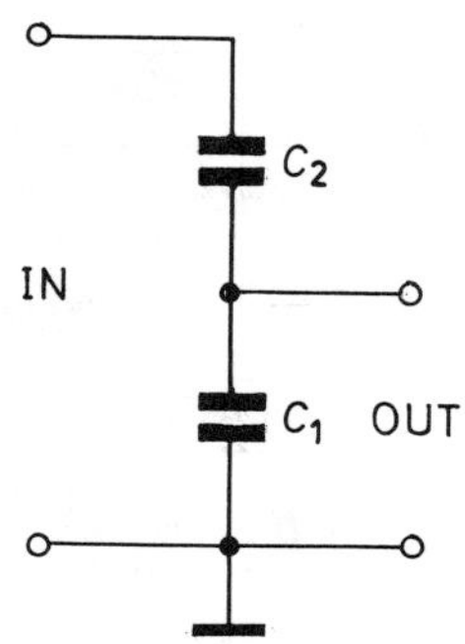

Fig. 2.36 Capacitance voltage divider

The attenuation is in the ideal case:

$$H = \frac{U_{\text{out}}}{U_{\text{in}}} = \frac{C_2}{C_1 + C_2} \,. \tag{2.24}$$

The main sources of error are the difference of the dissipation factors of the capacitors, and the voltage dependence of the capacitances.

Capacitance voltage dividers are very suitable for high voltage measurements. Their favourable properties are a high input impedance and the negligible dissipation. Fig. 2.37. shows a variable capacitance divider.

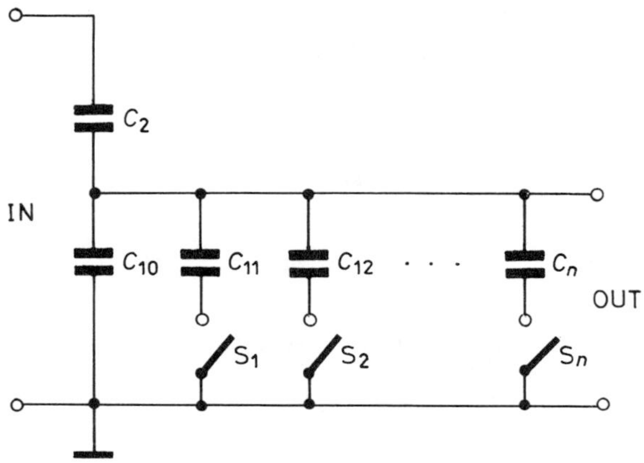

Fig. 2.37 Variable capacitance divider

The attenuation of the divider can be varied by means of switches as follows:

$$H = \frac{C_2}{C_{10} + C_2 + \sum_{i=1}^{n} k_i C_{1i}}, \tag{2.25}$$

where $k_i = 0$ or $k_i = 1$.

For high-voltage measurements three-terminal capacitors are used. If the guard ring of the three-terminal capacitor is at the same potential as the low-voltage electrode, the electric field within the capacitor will be homogeneous, and the capacitance can be calculated with high accuracy. Fig. 2.38 shows a high-voltage capacitance divider.

The guard ring is driven by the unity gain voltage follower, thus it is at the same potential as the low-voltage electrode. The high input and low output impedances of the amplifier guarantee a negligible loading error.

Driving the shield of the connecting cable by the unity gain voltage follower, the voltage difference between the inner conductor and the shield can also be minimized. This shield driving eliminates the loading effect of the cable capacitance.

Using voltage dividers, special attention should be paid to the possibility of eventual surface discharges and partial discharges. At high voltage, surface discharges occur around the edges of the electrodes, unless special precautions are taken. Even at 1 kV, with modern thin film capacitors the stress may be sufficient to cause internal partial discharges. Both types of discharge may cause severe errors and interference.

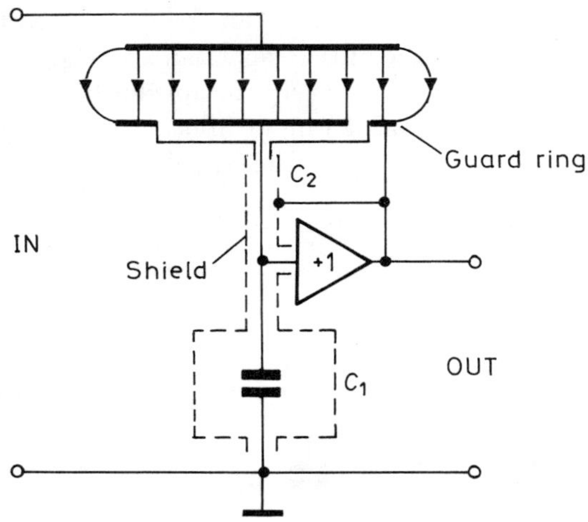

Fig. 2.38 High-voltage capacitance divider

2.3.6 Pwm dc Potentiometer

The pulse width modulation (Pwm) dc potentiometer consists of a switch S and a lowpass filter, as shown in Fig. 2.39.

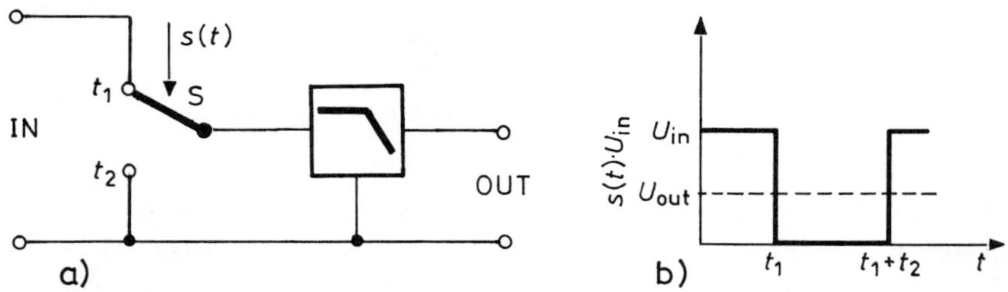

Fig. 2.39 Pwm dc potentiometer *a)* block diagram *b)* waveform

The lowpass filter is periodically connected to the input by the switch for a time t_1. Supposing that the transfer ratio of the lowpass filter is unity at dc, the transfer ratio of the dc potentiometer is:

$$\frac{U_{\text{out}}}{U_{\text{in}}} = \frac{t_1}{t_1 + t_2}, \tag{2.26}$$

where $t_1 + t_2$ is the period length of the switching signal $s(t)$.

Using a digital control unit for the generation of the switching signal, the transfer ratio of the dc potentiometer can be controlled very finely and accurately.

2.3.7 Voltage Transformers

Principle of the voltage transformer

Voltage transformers are intended for range extension and isolation. The voltage transformer (VT) has primary and secondary windings located on the core as shown in Fig. 2.40.

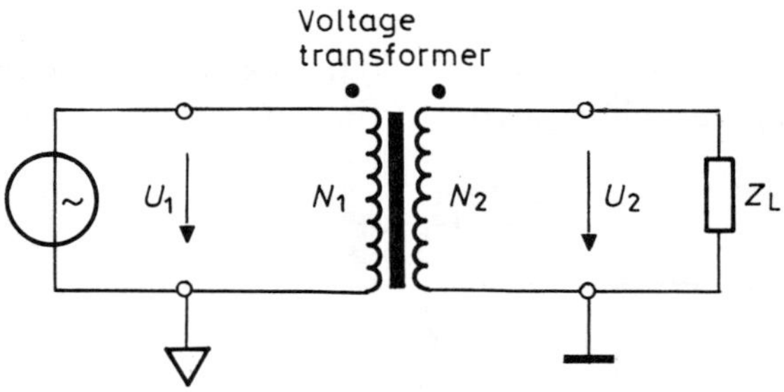

Fig. 2.40 Voltage transformer

U_1 is the primary voltage to be measured, Z_L is the load impedance or burden. The transfer factor of the ideal voltage transformer is as follows:

$$H = \frac{U_2}{U_1} = \frac{N_2}{N_1}, \tag{2.27}$$

where N_1 and N_2 are the numbers of turns of the primary and secondary windings, respectively. (The turn ratio of a VT is defined as N_1/N_2).

For the estimation of the complex error of H, the model of the VT is given in Fig. 2.41. It is supposed for simplicity that $N_1 = N_2$.

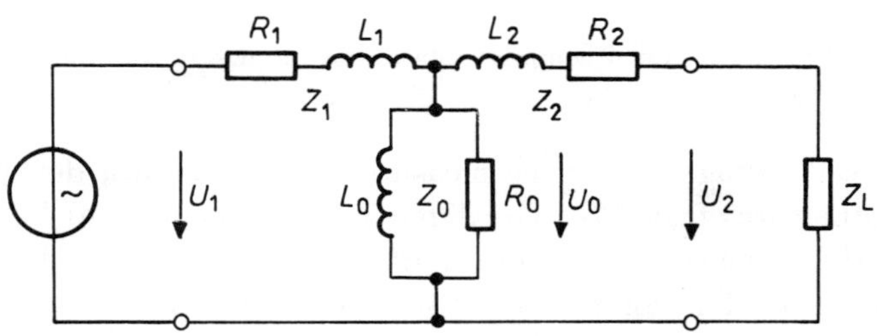

Fig. 2.41 Model of the voltage transformer

Z_1 and Z_2 are the primary and secondary leakage impedances, respectively. The leakage impedance consists of the copper resistances R_1, R_2, and leakage inductances L_1, L_2. The components of impedance Z_0 are the magnetizing inductance

L_0 and the core-loss resistance R_0. It is easy to see from the figure that the relative error of H is given by

$$\epsilon_r = \frac{U_2 - U_1}{U_1} = -\frac{Z_1}{Z_1 + Z_0}, \tag{2.28}$$

if the transformer is unloaded. The error is caused by the primary leakage impedance.

Error compensation of voltage transformers

The above error can be compensated for by means of a compensation transformer (T) which measures the voltage across the impedance Z_1, as shown in Fig. 2.42.

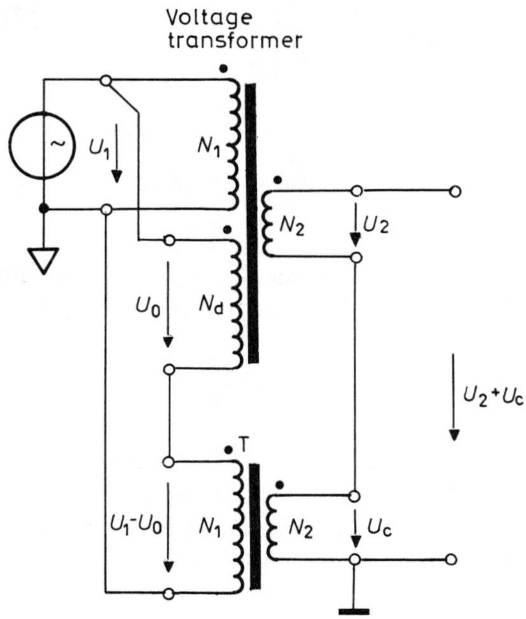

Fig. 2.42 Error compensation of the voltage transformer

The induced voltage U_0 can be measured by means of a detection winding N_d located on the core of the VT. The difference between U_1 and U_0, which is the voltage across the primary leakage impedance Z_1, is transformed by the compensation transformer T. The compensating voltage U_c is then added to U_2 resulting in the error compensation.

Practical solutions and specifications

Fig. 2.43 shows a multi-ratio voltage transformer with tapped primary and tapped secondary windings.

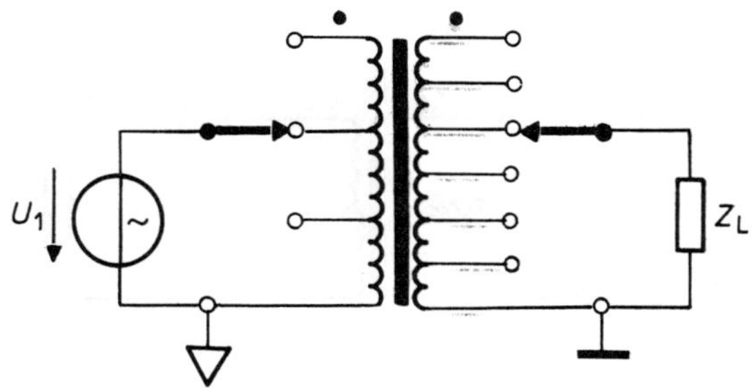

Fig. 2.43 Multi-ratio voltage transformer

The accuracy class of voltage transformers for industrial purposes is 0.1-5% at line frequency. Their rated secondary voltages are typically 100, 110, 115, 120, 200 and 230 V. These ratings, divided by $\sqrt{3}$, are also standard values.

Standard voltage transformers are applicable up to 400 kV between 40 Hz and 300 Hz. Their ratio uncertainty is about 5 ppm.

Ratio transformers are designed to produce accurate voltage ratios, as shown in Fig. 2.44.

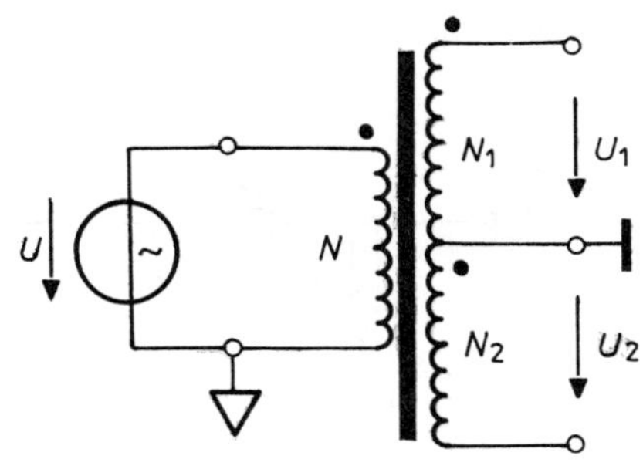

Fig. 2.44 Ratio transformer

In the ideal case

$$\frac{U_1}{U_2} = \frac{N_1}{N_2}\,. \tag{2.29}$$

As can be seen, the voltage ratio is determined by the turn ratio. The ratio transformer is a very accurate device, used in impedance measurements (see Chapter 4).

2.3.8 Current Transformers

Principle of the current transformer

Current transformers (CT-s) are used for range extension and isolation. The primary and secondary windings are located on the same core as shown in Fig. 2.45.

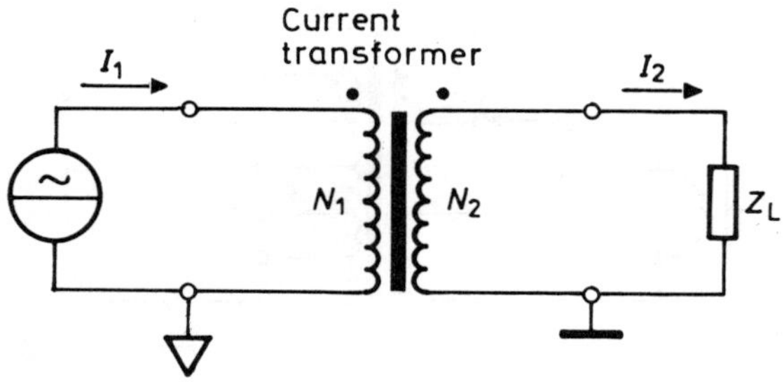

Fig. 2.45 Current transformer

I_1 is the primary current to be measured, Z_L is the load impedance or burden. The transfer factor of the ideal current transformer is as follows:

$$H = \frac{I_2}{I_1} = \frac{N_1}{N_2}, \tag{2.30}$$

where N_1 and N_2 are the numbers of turns of the primary and secondary windings, respectively. (The turn ratio of a CT is defined as N_2/N_1).

The model of the current transformer is given in Fig. 2.46. It is assumed for simplicity that $N_1 = N_2$.

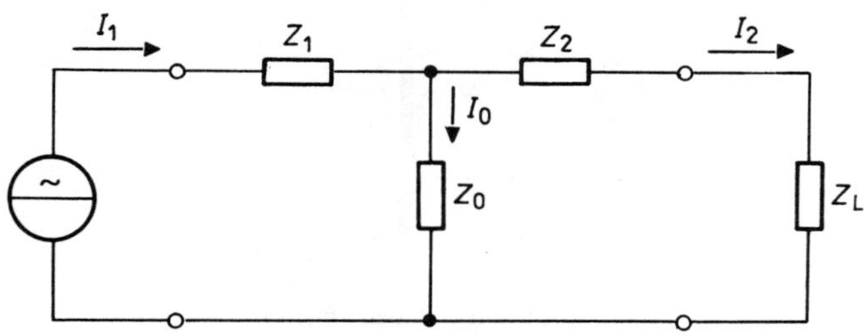

Fig. 2.46 Model of the current transformer

Z_1 and Z_2 are primary and secondary leakage impedances, respectively. The main source of error is the magnetizing impedance Z_0 or rather the magnetizing current I_0. The relative error of H is given by

$$\epsilon_r = \frac{I_2 - I_1}{I_1} = -\frac{Z_2 + Z_L}{Z_2 + Z_L + Z_0}. \tag{2.31}$$

The error of the unloaded CT is smaller, but not zero, because the secondary leakage impedance Z_2 is not negligible. (A CT is unloaded if $Z_L = 0$).

Winding capacitances are further sources of error in current transformers (Zoltán, 1989). At higher frequencies or at higher working voltages the capacitive error becomes predominant.

The two-stage current transformer

Fig. 2.47 shows a two-stage current transformer of high accuracy.

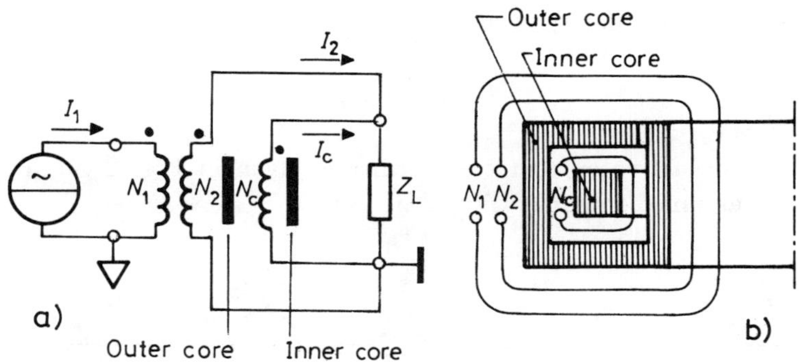

Fig. 2.47 Two-stage current transformer *a)* circuit diagram *b)* core construction

The second stage composed of inner core and compensation winding N_c is excited by the magnetizing current of the first stage. The compensating current I_c, generated by the above excitation, is approximately equal to the magnetizing current, if $N_c = N_2$ and the error of the second stage is negligible. The compensating current is added to the secondary current of the first stage, thus the error is reduced.

Active current transformer

The passive current transformers above make an energy transfer between the source and the load. Their cores have to be designed for this purpose. However, a much smaller core can be used in an active current transformer, composed of a current comparator and power amplifier, shown in Fig. 2.48.

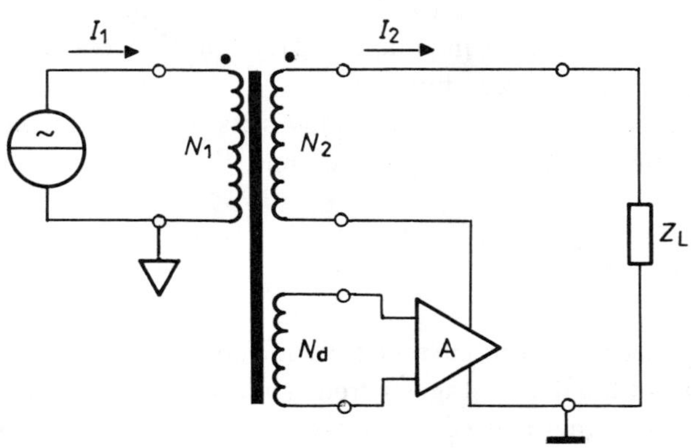

Fig. 2.48 Active current transformer

N_1, N_2 and N_d are primary, secondary and detector windings, respectively of the current comparator. The residual flux in the core produces a voltage in the detector winding which is amplified by the power amplifier A. The power amplifier outputs the secondary current I_2. In the balanced state the primary excitation is compensated by the secondary excitation, that is

$$I_1 N_1 \approx I_2 N_2 \,. \tag{2.32}$$

The model of the active current transformer is given in Fig. 2.49. For sake of simplicity it is assumed that $N_1 = N_2 = N_d$.

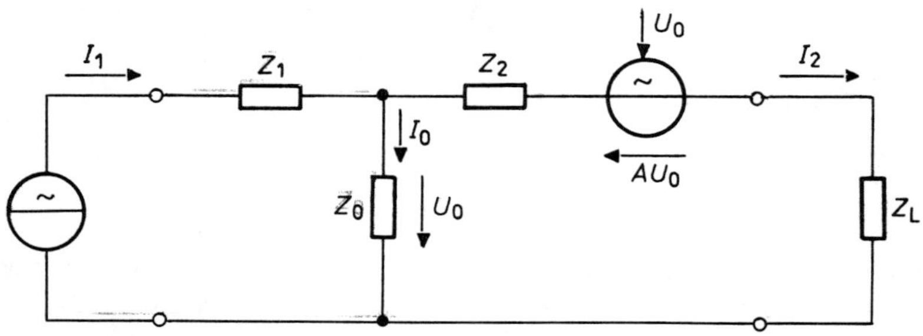

Fig. 2.49 Model of the active current transformer

The power amplifier is represented by a controlled voltage source. The relative error of the active current transformer can be determined by the above model and is as follows:

$$\epsilon_r = \frac{I_2 - I_1}{I_1} = -\frac{Z_2 + Z_L}{Z_2 + Z_L + Z_0(1 + A)} \,, \tag{2.33}$$

where A is the gain factor of the power amplifier.

Active current transformers have a very low input impedance because the residual flux in the core is near to zero. The input impedance is approximately equal to the primary leakage impedance Z_1. For this reason an active current transformer is also called a zero-flux CT.

Active current transformers are highly sensitive to external magnetic fields, thus their magnetic shielding is very important.

Hybrid current transformer

For very high accuracy, hybrid current transformers can be used. A hybrid current transformer is a modified two-stage current transformer as shown in Fig. 2.50. An active current transformer is used in the second stage for error compensation of the passive first stage. The energy-carrying outer core acts as a magnetic shield for the inner core, reducing the sensitivity to external magnetic fields.

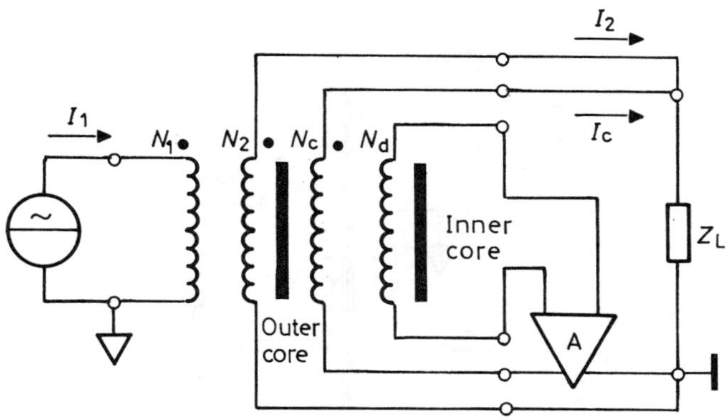

Fig. 2.50 Hybrid current transformer

Practical solutions and specifications

More than one ratio can be obtained by the use of taps on the secondary and primary windings, as shown in Fig. 2.51.

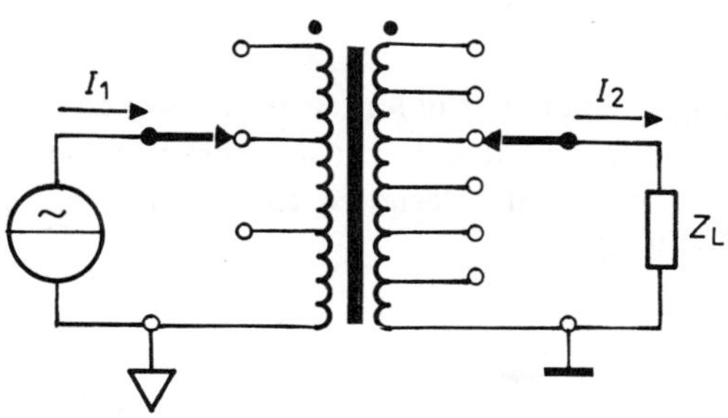

Fig. 2.51 Multi-ratio current transformer

The multi-input current transformer can be used for weighted summation and subtraction of currents as shown in Fig. 2.52.

The output current I_2 is given by the following equation:

$$I_2 = \sum_{i=1}^{n} I_{1i} \frac{\pm N_{1i}}{N_2} . \tag{2.34}$$

As it can be seen from the figure, even the inputs can be isolated from each other.

The accuracy class of current transformers for industrial purposes is 0.1–5% at line frequencies. Their rated secondary currents are typically 1, 2 and 5 A. The rated secondary current of current transformers used in electronic measuring instruments is between 1–100 mA.

Standard current transformers are applicable up to 100 kA between 40 Hz and 300 Hz. Their ratio uncertainty is about 1 ppm.

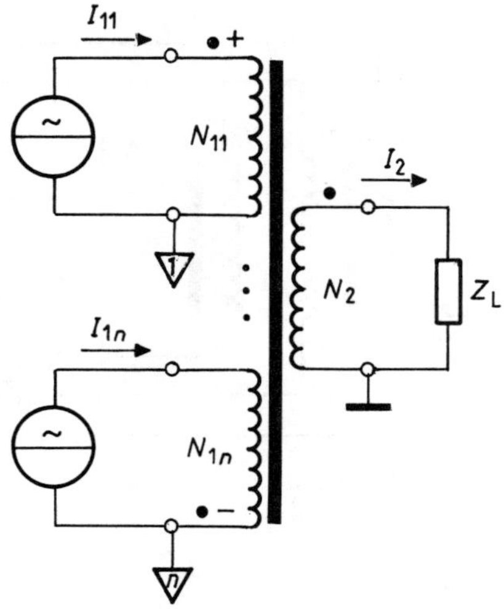

Fig. 2.52 Multi-input current transformer

2.3.9 Dc Current Transformer

The lower the operating frequency, the larger is the necessary core of an ac current transformer. For this reason ac current transformers are used down to 16 Hz only.

Dc current transformers are designed for operation from 0 Hz. A typical solution is given in Fig. 2.53.

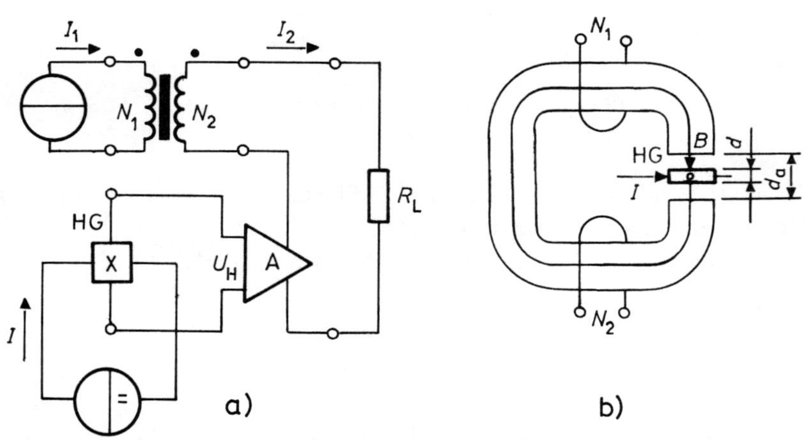

Fig. 2.53 Dc current transformer

The dc current transformer is composed of a two-winding transformer with a Hall generator (HG), located in the air gap of the core, and a power amplifier A. The magnetic induction B in the air gap is detected by the Hall generator. The voltage U_H, generated in the Hall plate due to the Hall effect, is amplified by the power amplifier which generates the secondary current I_2. In the balanced state the primary excitation is compensated by the secondary excitation, that is,

$$I_1 N_1 \approx I_2 N_2 . \tag{2.35}$$

If the magnetic resistance of the core can be neglected, the magnetic induction in the air gap can be expressed by the following equation:

$$B = (I_1 N_1 - I_2 N_2) \frac{\mu_0}{d_a} , \tag{2.36}$$

where μ_0 is the permeability of the vacuum, and d_a is the width of the air gap.

The Hall voltage is

$$U_H = \frac{R_H}{d} I B , \tag{2.37}$$

where R_H is the Hall constant, I is the controlling current of the Hall generator and d is the thickness of the Hall plate.

The secondary current is given by

$$I_2 = \frac{A U_H}{R_2 + R_L} , \tag{2.38}$$

where A is the gain of the power amplifier, R_2 is the copper resistance of the secondary winding, and R_L is the load resistance.

In the ideal case the true secondary current can be expressed as

$$I_{2t} = \frac{I_1 N_1}{N_2} . \tag{2.39}$$

Due to the finite gain of the power amplifier, however, the secondary current deviates from the true one. The relative error can be defined as follows:

$$\epsilon_r = \frac{I_2 - I_{2t}}{I_{2t}} . \tag{2.40}$$

Substituting the above expressions of B, U_H, I_2 and I_{2t} into (2.40), the relative error of the transfer factor of the dc current transformer is:

$$\epsilon_r = -\frac{d_a}{\mu_0} \cdot \frac{d}{R_H I} \cdot \frac{R_L + R_2}{N_2 A} . \tag{2.41}$$

In practice approximately 0.1% uncertainty can be achieved. At low current levels, however, the error increases because of the offset voltage of the Hall generator. At higher frequencies the winding capacitances cause an increasing error.

Dc current transformers are manufactured up to rated primary currents of 50 kA (Lem, 1986).

2.3.10 Measuring Amplifiers

Measuring amplifiers are used for the measurement of low-level signals and if necessary for isolation.

Inverting amplifier

The inverting amplifier, composed of the operational amplifier, input resistor and feedback resistor is illustrated in Fig. 2.54.

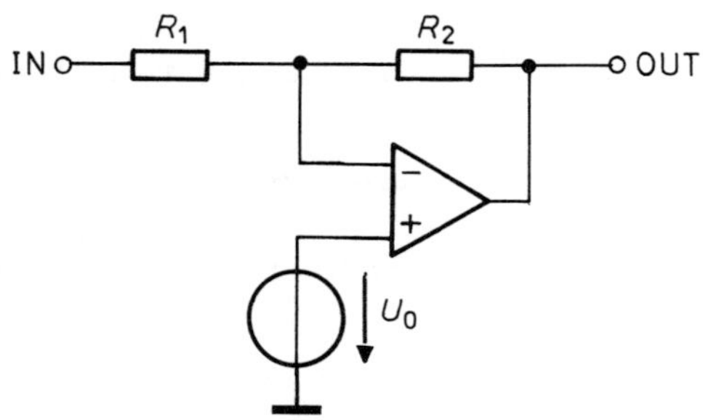

Fig. 2.54 Inverting amplifier

The gain of the ideal inverting amplifier is

$$H = -\frac{R_2}{R_1} \,. \tag{2.42}$$

The input offset voltage U_0 causes an output offset, thus

$$U_{\text{out}} = -\frac{R_2}{R_1} U_{\text{in}} + \left(1 + \frac{R_2}{R_1}\right) U_0 \,. \tag{2.43}$$

The input offset voltage changes with temperature, time and power supply voltage. These parameters can be very critical for low-voltage measurements.

At higher frequencies the transfer function deviates from the ideal one. Supposing that the transfer function of the operational amplifier is $-A_0/(1 + j2\pi fT)$, the following equation holds:

$$H = -\frac{R_2}{R_1} \cdot \frac{1}{1 + \left(1 + \frac{R_2}{R_1}\right) \frac{1 + j2\pi fT}{A_0}} \,. \tag{2.44}$$

The phase error of the inverting amplifier increases with the frequency:

$$\delta \approx -\left(1 + \frac{R_2}{R_1}\right) \frac{f}{f_1} \,, \tag{2.45}$$

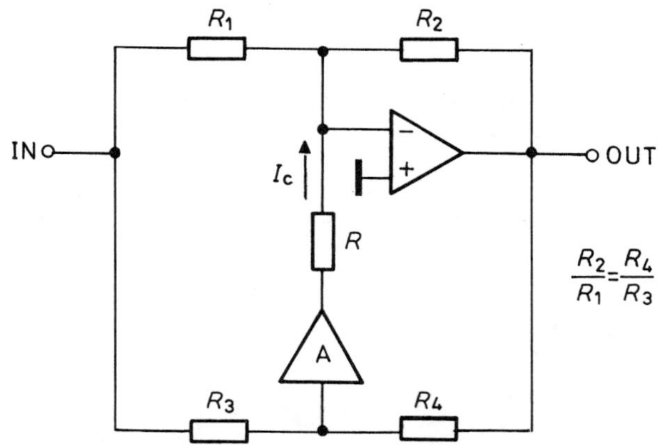

Fig. 2.55 Error compensation of the inverting amplifier

where $f_1 = A_0/(2\pi T)$ is the unity-gain frequency bandwidth of the operational amplifier.

The phase error can be compensated by means of a second amplifier (see Fig. 2.55).

The input voltage of amplifier A is proportional to the phase error of the inverting amplifier, thus the resultant compensating current I_c reduces the phase error.

Noninverting amplifier

Noninverting amplifiers have a very high input impedance.

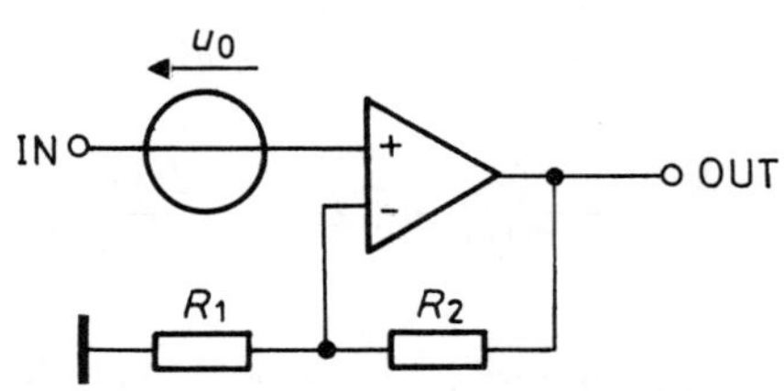

Fig. 2.56 Noninverting amplifier

The gain of the ideal noninverting amplifier is

$$H = 1 + \frac{R_2}{R_1}. \tag{2.46}$$

The offset and common-mode voltages cause errors, thus

$$U_{out} = \left(1 + \frac{R_2}{R_1}\right)\left(U_{in} + U_0 + \frac{U_{in}}{CMRR}\right), \tag{2.47}$$

where $CMRR$ is the common-mode rejection ratio of the operational amplifier.

Differential amplifier

Differential amplifiers are used for the measurement of floating voltage sources as shown in Fig. 2.57.

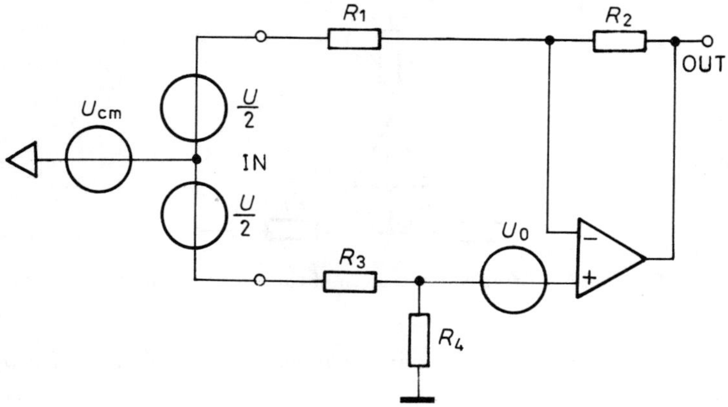

Fig. 2.57 Differential amplifier

In the case of perfect common-mode rejection, the output voltage is given by

$$U_{\text{out}} = -\frac{R_2}{R_1}U\,, \tag{2.48}$$

if $R_2/R_1 = R_4/R_3$.

Resistance tolerances, however, reduce the common-mode rejection. The output voltage of the nonideal differential amplifier is given by

$$U_{\text{out}} = -\frac{R_2}{R_1}\left(U - U_0\left(1 + \frac{R_1}{R_2}\right) + \frac{U_{\text{cm}}}{CMRR} + U_{\text{cm}}\frac{R_1}{R_1 + R_2}(\epsilon_{1\text{r}} - \epsilon_{2\text{r}} + \epsilon_{4\text{r}} - \epsilon_{3\text{r}})\right)\,, \tag{2.49}$$

where U_0 is the input offset voltage and ϵ_{ir} is the relative tolerance of the resistance R_{i}.

Instrumentation amplifier

Instrumentation amplifiers (IA) are closed loop, differential input gain blocks exhibiting high input impedance and high common-mode rejection (CMR). They are generally used in applications where extracting and accurately amplifying low level differential signals floating on high common-mode voltages is very important.

An instrumentation amplifier, composed of two noninverting and one differential amplifiers, is shown in Fig. 2.58.

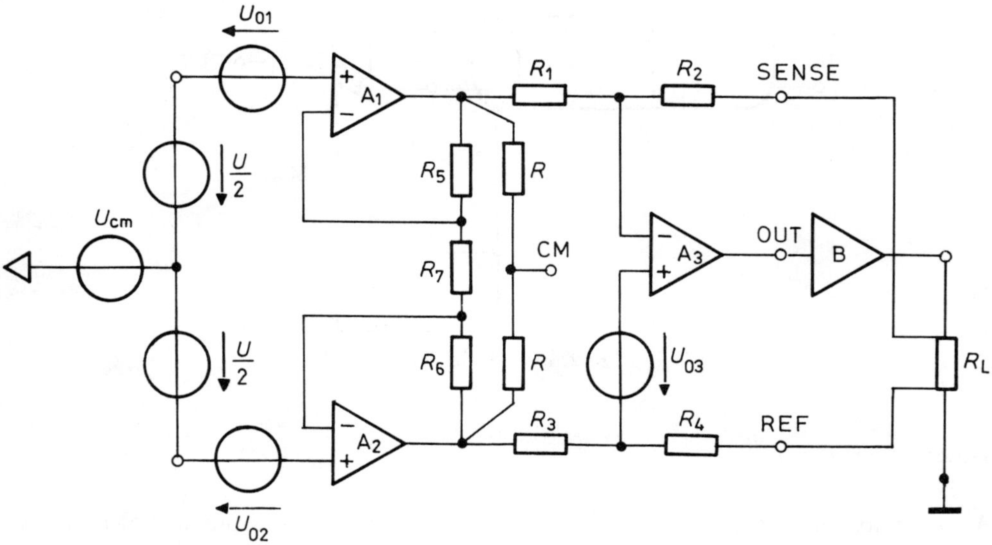

Fig. 2.58 Instrumentation amplifier

The gain of the ideal IA is given by

$$H = -\left(1 + 2\frac{R_5}{R_7}\right)\frac{R_2}{R_1}, \tag{2.50}$$

if $R_5 = R_6$ and $R_2/R_1 = R_4/R_3$.

The output voltage of the nonideal IA is

$$U_{\text{out}} = H \cdot \left(U + U_{01} - U_{02} + \frac{U_{\text{cm}}}{CMRR1} - \frac{U_{\text{cm}}}{CMRR2} + \frac{U_{\text{cm}}}{1 + 2\frac{R_5}{R_7}}\right.$$

$$\left.\times\left(\frac{1}{CMRR3} + \frac{R_1}{R_1 + R_2}(\epsilon_{1\text{r}} - \epsilon_{2\text{r}} + \epsilon_{4\text{r}} - \epsilon_{3\text{r}})\right)\right) + \left(1 + \frac{R_2}{R_1}\right)U_{03}\,. \tag{2.51}$$

As can be seen, the asymmetry of the input stage dominates, and the error of the differential stage is reduced by the gain factor of the input stage.

Output-sense and reference terminals allow greater accuracy in connecting the four-terminal load and an optional buffer amplifier (B). Using the terminal CM, degradation of *CMR* due to the distributed capacitance on the input lines can be reduced, as shown in Fig. 2.59.

The unity-gain voltage follower drives the shield of the cable, thus minimizing the voltage-difference between the input lines and the shield.

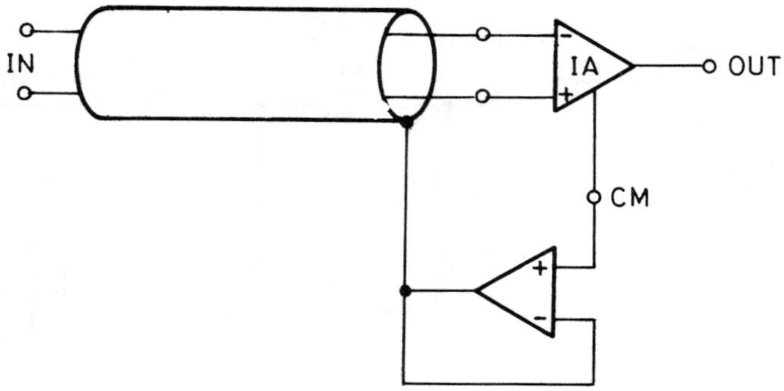

Fig. 2.59 Shield driving

Isolation amplifier

An isolation amplifier provides ohmic isolation between the input signal and the output of the amplifier. It consists of an input operational amplifier or instrumentation amplifier, followed by a unity-gain isolation stage. Fig. 2.60 shows a typical isolation amplifier.

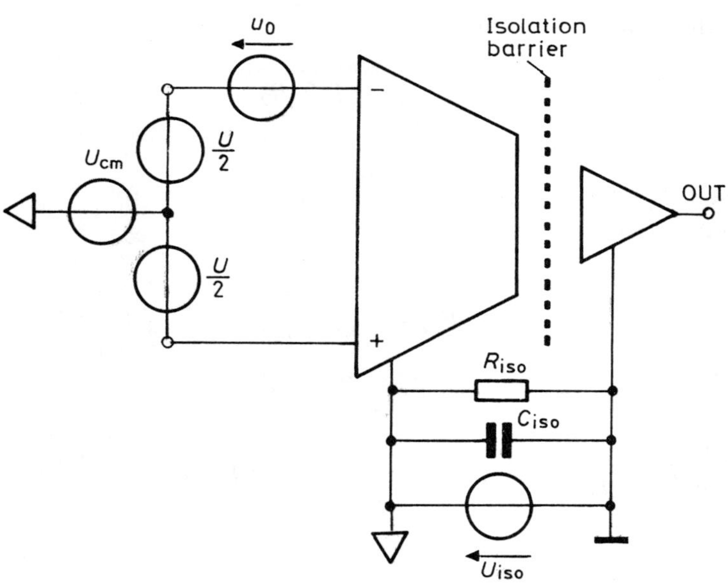

Fig. 2.60 Isolation amplifier

The output voltage of the isolation amplifier is as follows:

$$U_{\text{out}} = H \cdot \left(U + U_0 + \frac{U_{\text{cm}}}{CMRR} + \frac{U_{\text{iso}}}{IMRR} \right), \tag{2.52}$$

where U_{iso} is the isolation-mode voltage across the isolation barrier, *IMRR* is the isolation-mode rejection ratio, U_{cm} is the common-mode voltage and *CMRR* is the common-mode rejection ratio.

With the isolation amplifier, as low as $0.1-1\%$ uncertainty can be achieved. These amplifiers are used for measuring low-level signals floating on high common-mode voltages, and/or for breaking ground loops. They provide isolation protection for electronic equipment.

Chopper amplifier

Chopper amplifiers are used for microvolt-range measurements. The direct input voltage is converted into an alternating voltage, amplified by an ac amplifier and converted back into a dc voltage as shown in Fig. 2.61.

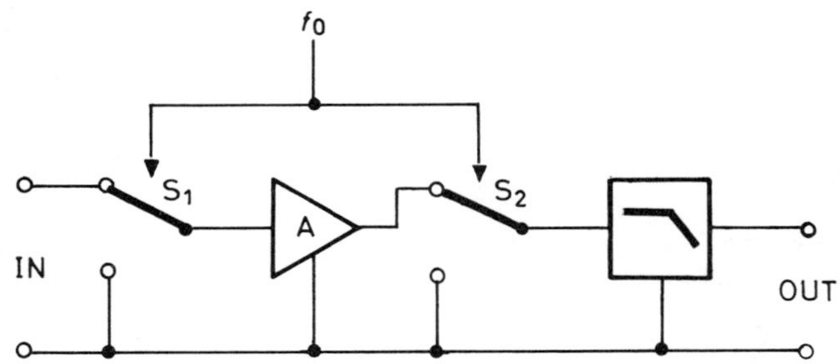

Fig. 2.61 Chopper amplifier

Switches S_1 and S_2 are controlled by the same signal of frequency f_0. They are used as choppers for modulation and demodulation. The lowpass filter provides a dc output voltage proportional to the input signal.

Chopper amplifiers help to avoid the drift problems of direct coupled amplifiers.

There are chopper-stabilized operational amplifiers that have a high dc-ac performance. Chopper stabilization eliminates both the initial offset and changes in the offset due to time, temperature and common-mode voltage.

2.3.11 Voltage-to-Current Converters

Voltage measurement can be carried out by means of an ammeter, using a voltage-to-current converter.

Resistance voltage-to-current converters are shown in Fig. 2.62.

The error caused by the internal impedance Z_A of the ammeter can be reduced by means of an operational amplifier used in active converters. Multi-input converters are used for the weighted summation of voltages.

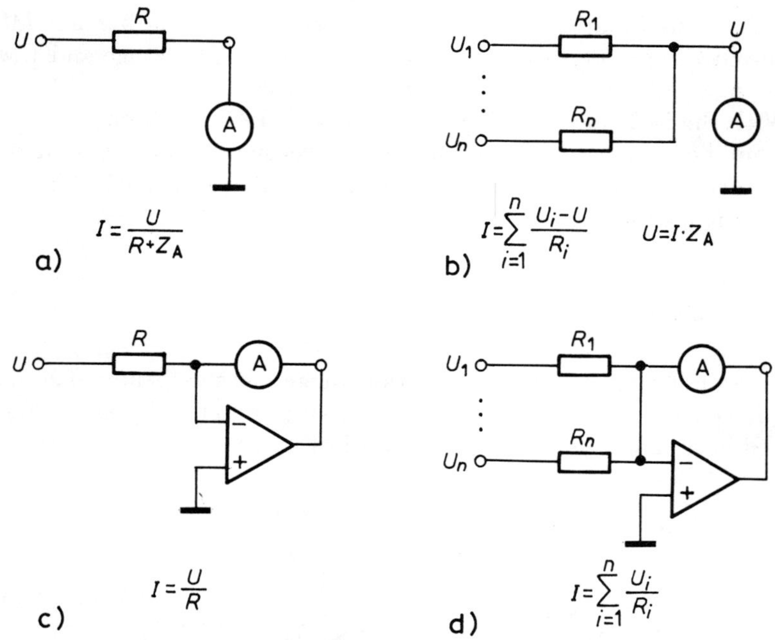

Fig. 2.62 Resistance voltage-to-current converters *a)* passive *b)* passive, multi-input *c)* active *d)* active, multi-input

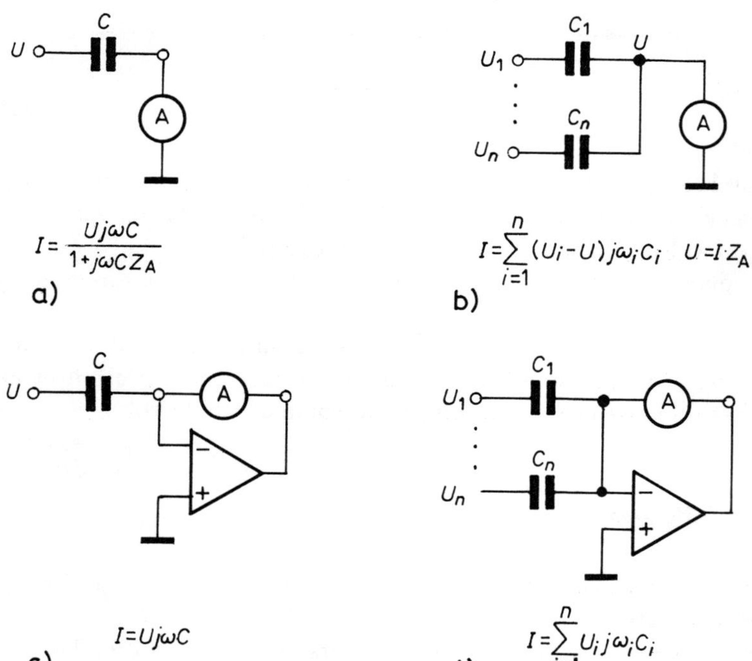

Fig. 2.63 Capacitance voltage-to-current converters *a)* passive *b)* passive, multi-input *c)* active *d)* active, multi-input

Capacitance voltage-to-current converters are shown in Fig. 2.63. These converters can be used for ac measurements only.

The loading error of the ammeter can be eliminated by means of an operational amplifier. To avoid the instability of active converters, special frequency compensation is required. Weighted summation of the voltages can also be realized.

The resistance and capacitance voltage-to-current converters above can be used for high-voltage measurements as well.

2.3.12 Current-to-Voltage Converters

Current measurement can be carried out by means of a voltmeter using a current-to-voltage converter.

Resistance current-to-voltage converters are shown in Fig. 2.64.

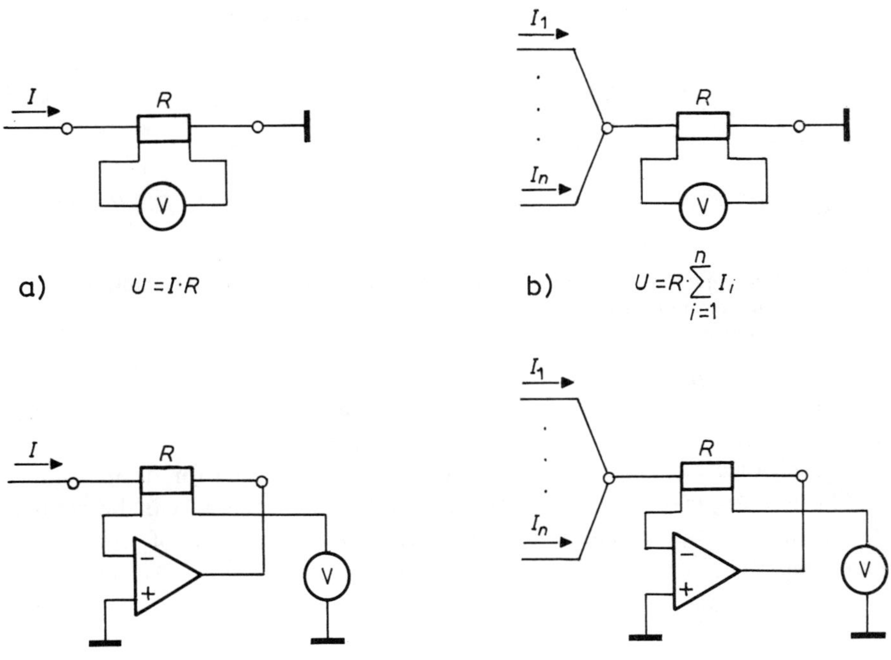

Fig. 2.64 Resistance current-to-voltage converters *a)* passive *b)* passive, multi-input *c)* active *d)* active, multi-input

The loading error caused by the four-terminal resistor R can be reduced by means of an operational amplifier, used in active converters. Multi-input converters are used for the summation of currents.

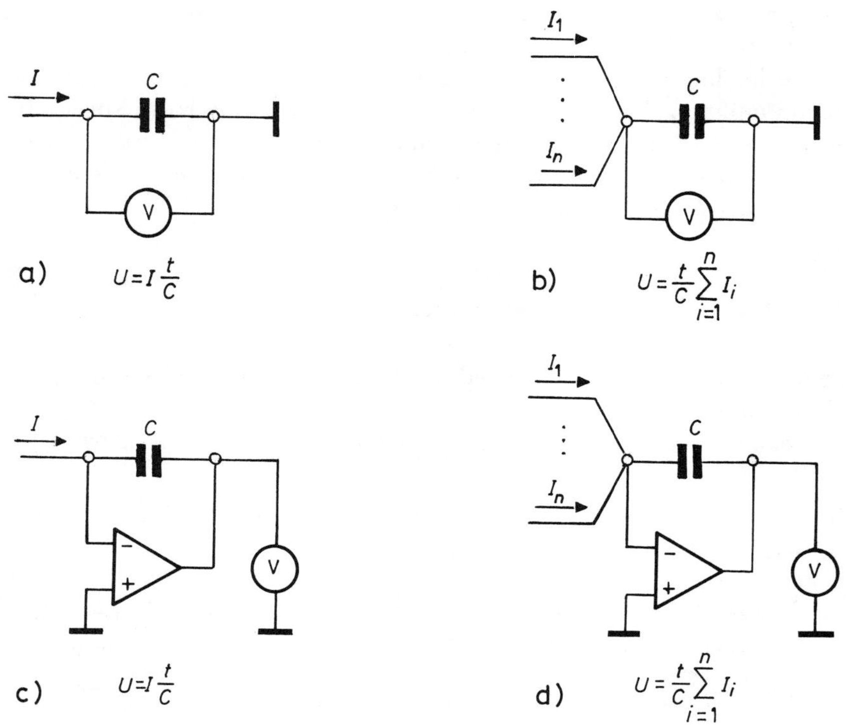

Fig. 2.65 Capacitance current-to-voltage converters *a)* passive *b)* passive, multi-input *c)* active *d)* active, multi-input

Capacitance current-to-voltage converters are given in Fig. 2.65. Due to the integrating technique very low currents can be measured.

Capacitance current-to-voltage converters can be used for dc measurements only. The loading error caused by the charged capacitance C can be reduced by means of an operational amplifier. However, the bias current of the operational amplifier causes an error.

2.3.13 Rectifiers

Rectifiers are used in ac-dc converters. Half-wave, full-wave, phase-sensitive and peak rectifiers will be treated here.

Half-wave and full-wave rectifiers

A simple passive half-wave rectifier is given in Fig. 2.66. The nonlinear transfer characteristic can be approached by straight lines as shown in the figure.

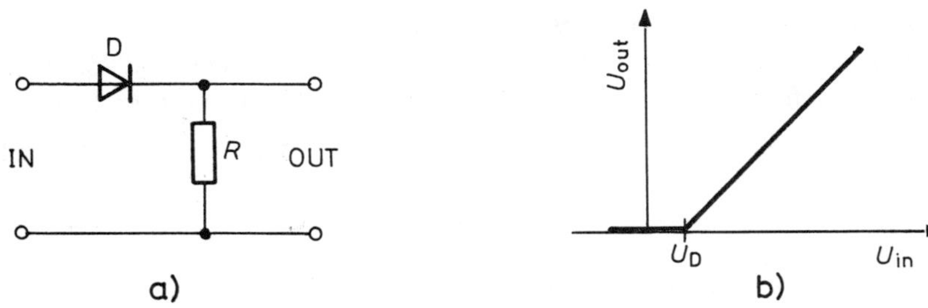

Fig. 2.66 Half-wave rectifier *a)* circuit *b)* transfer characteristic

The lower the voltage to be rectified, the higher is the error due to the forward voltage of the diode.

A passive full wave rectifier is shown in Fig. 2.67. The breakpoints are at $\pm 2U_d$.

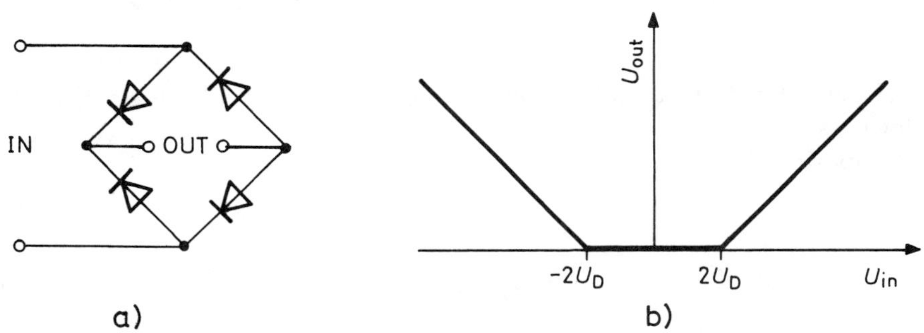

Fig. 2.67 Full-wave rectifier *a)* circuit *b)* transfer characteristic

An improved half-wave rectifier is given in Fig. 2.68. The breakpoint is now at $-U_d/A$, where A is the gain factor of the operational amplifier at the operating frequency. Resistor R_3 in series with the diode prevents the overdriving of the operational amplifiers for positive input voltages.

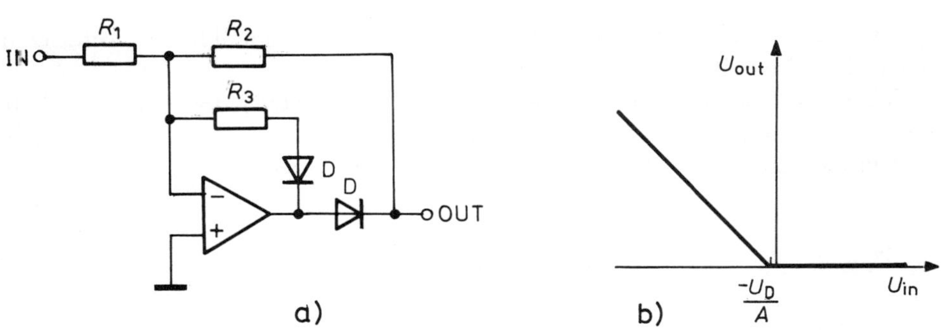

Fig. 2.68 Active half-wave rectifier *a)* circuit *b)* transfer characteristic

Half-wave and full-wave rectifiers are used for the measurement of the mean absolute value of the signal. The half-wave rectifier can be used for this purpose only if the waveform is symmetric with respect to the zero voltage level.

Phase-sensitive rectifier

The scheme of a phase-sensitive rectifier is given in Fig. 2.69.

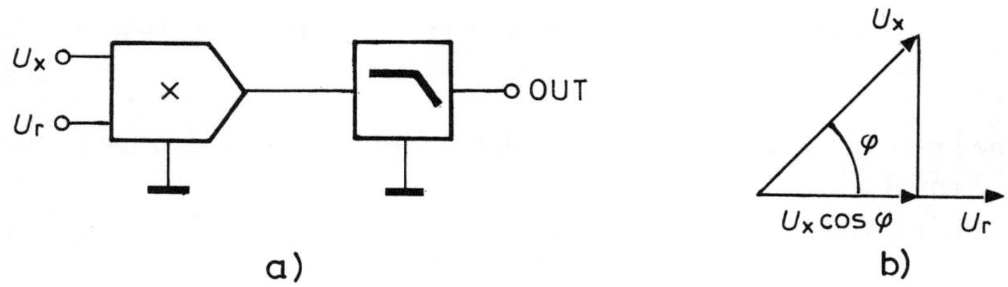

Fig. 2.69 Phase-sensitive rectifier *a)* block diagram *b)* phasors

The input signal U_x and the reference signal U_r are multiplied, then filtered by a lowpass filter.

Supposing that

$$u_x(t) = \sqrt{2}U_x \sin(\omega t + \varphi) \tag{2.53}$$

and

$$u_r(t) = \sqrt{2}U_r \sin \omega t\,, \tag{2.54}$$

the output voltage of the multiplier is given by

$$u(t) = U_x U_r \big(\cos\varphi - \cos(2\omega t + \varphi)\big)\,. \tag{2.55}$$

The lowpass filter produces the dc component, thus the output voltage is

$$U_{out} = U_r U_x \cos\varphi\,. \tag{2.56}$$

As can be seen, the output voltage depends on the phase difference of the two input signals.

The phase-sensitive rectifier produces a signal proportional to the vector components of U_x, thus it can be used in vector voltmeters.

Multiplication can be carried out by means of a controlled switch. Fig. 2.70a shows a switching phase-sensitive rectifier.

Switch S is controlled by the following switching signal:

$$s_r(t) = \frac{4}{\pi}\sum_{i=0}^{\infty}\frac{1}{2i+1}\sin(2i+1)\omega t\,. \tag{2.57}$$

Supposing that

$$u_x(t) = \sqrt{2}U_x \sin(\omega t + \varphi)\,, \tag{2.58}$$

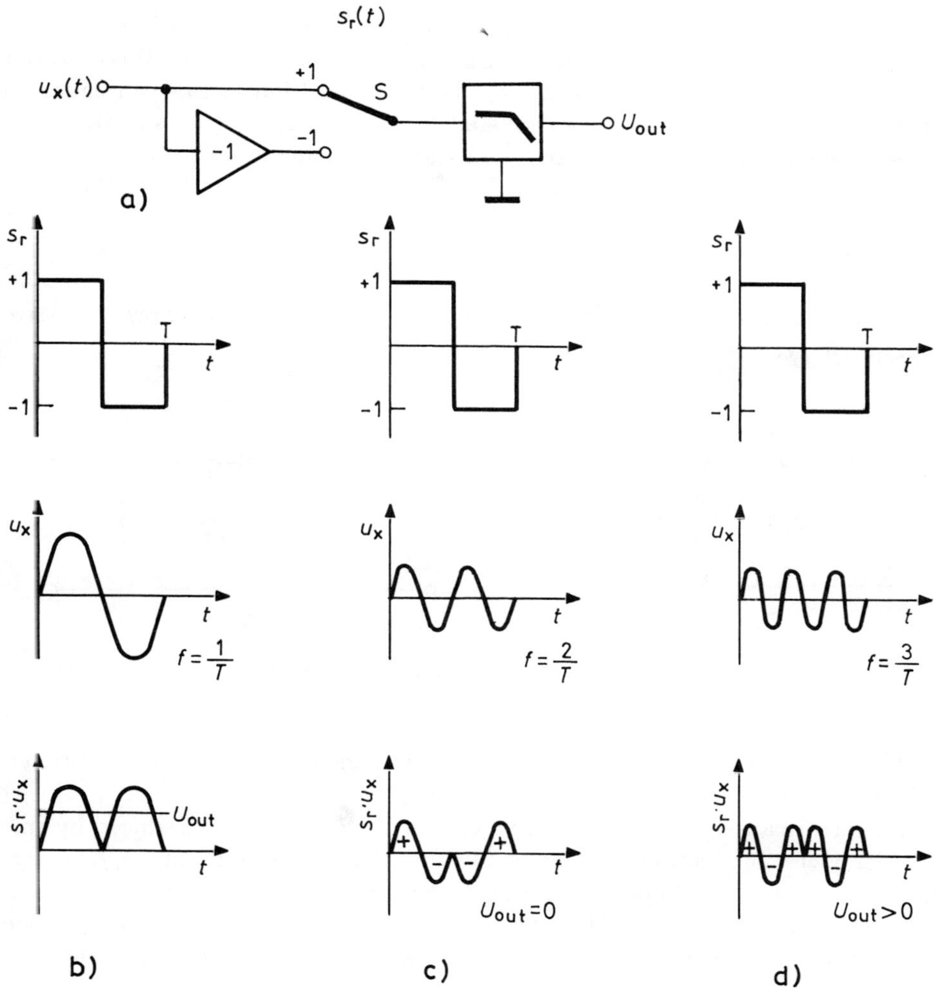

Fig. 2.70 Switching phase-sensitive rectifier with different input signals *a)* block diagram *b)* fundamental harmonic *c)* second harmonic *d)* third harmonic

the expected value of $s_r u_x$ is as follows:

$$\mathcal{E}\Big\{u_x(t)s_r(t)\Big\} = \mathcal{E}\Big\{\frac{4\sqrt{2}}{\pi}U_x \sin(\omega t + \varphi)\sin\omega t\Big\} = \frac{2\sqrt{2}}{\pi}U_x \cos\varphi\,. \tag{2.59}$$

The output voltage of the switching phase-sensitive rectifier is given by

$$U_{out} = \frac{2\sqrt{2}}{\pi}U_x \cos\varphi\,. \tag{2.60}$$

Fig. 2.70b. shows the signals of the phase-sensitive rectifier for $\varphi = 0$.

Higher-order harmonics of the input signal U_x are sources of error. The expected value of the even harmonics during $T/2$ is always zero, so these do not disturb the measurement (see Fig. 2.70c). Only the odd harmonics can cause an error, therefore these have to be suppressed. Fig. 2.70d shows the signals of the phase-sensitive rectifier at the presence of a third harmonic.

Peak rectifier

Peak rectifiers measure the peak value of the input signal. Fig. 2.71 shows a passive and an active peak rectifier.

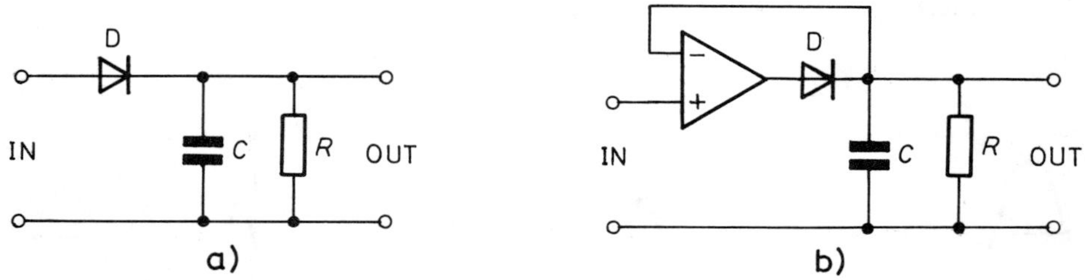

Fig. 2.71 Peak rectifiers *a)* passive *b)* active

The peak voltage is obtained by periodic charging of the capacitor C through the diode D.

The error caused by the forward voltage of the diode can be reduced by employing an operational amplifier. The bandwidth of active peak rectifiers, however, is limited because of the finite bandwidth of the operational amplifier.

Discharging of capacitor C through resistor R causes an error in the peak value measurement, but it makes it possible to follow the changes of the peak value.

2.3.14 Rms Converters

Rms converters are used for determining the root-mean-square value of a signal. Fig. 2.72 shows an rms-to-dc converter based on differential thermal converter.

A thermal converter consists of a resistor and a thermocouple. Thermocouples T_1 and T_2 in Fig. 2.72 measure the temperature of resistors R_1 and R_2, respectively. The resistor R_1 is heated by the voltage to be measured. The amplifier A keeps the temperature of R_2 equal to the temperature of R_1, thus

$$U_{out} = U_{eff}, \tag{2.61}$$

if $R_1 = R_2$.

Using bipolar transistors as temperature sensors much higher sensitivity can be achieved.

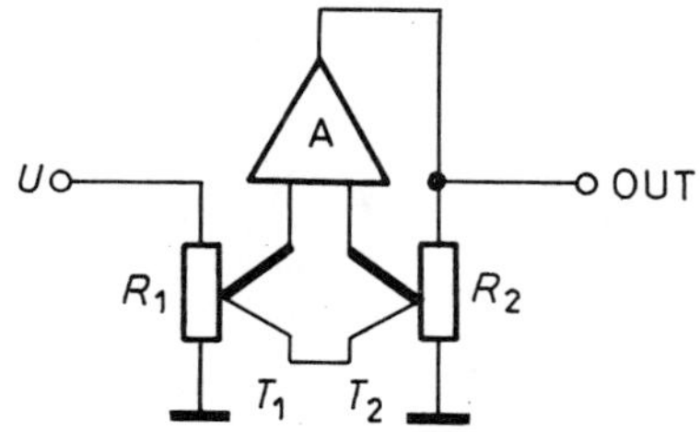

Fig. 2.72 Rms-to-dc converter based on differential thermal converter

The uncertainty of thermal rms converters for industrial purposes can be 0.05% up to 100 kHz. Thermal voltage converters can be used as ac-dc voltage transfer standards as well. The uncertainty of the ac-dc voltage transfer can be 0.002–0.01% between 10 Hz and 100 kHz.

Rms-to-dc conversion can be carried out by means of a multiplier and an integrator as shown in Fig. 2.73.

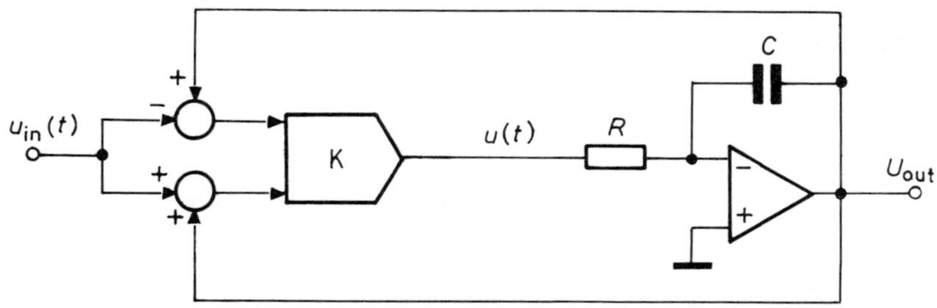

Fig. 2.73 Rms-to-dc converter

The ac component of $u(t)$ is filtered by the integrator. In the balanced state the expected value of $u(t)$ is zero. The output voltage of the multiplier is given by

$$u(t) = K\left(U_{\text{out}}^2 - u_{\text{in}}^2(t)\right). \tag{2.62}$$

The expected value of $u(t)$ is as follows:

$$\mathcal{E}\left\{u(t)\right\} = KU_{\text{out}}^2 - K\mathcal{E}\left\{u_{\text{in}}^2(t)\right\}. \tag{2.63}$$

Supposing that the input signal is periodic and

$$u_{\text{in}}(t) = \sum_{i=1}^{\infty} u_{\text{i}}(t) = \sum_{i=1}^{\infty} \sqrt{2}U_{\text{i}} \sin(i\omega t + \varphi_{\text{i}}), \tag{2.64}$$

the following equation can be written:

$$u_{\text{in}}^2(t) = \sum_{i=1}^{\infty}\left(\sum_{k=1}^{\infty} u_{\text{i}}(t)u_{\text{k}}(t)\right). \tag{2.65}$$

The expected value of the terms $u_i u_k$ is zero for $i \neq k$, thus

$$\mathcal{E}\Big\{u_{\text{in}}^2(t)\Big\} = \mathcal{E}\Big\{\sum_{i=1}^{\infty} u_{\text{i}}^2(t)\Big\}. \tag{2.66}$$

Since $u_{\text{i}}(t)$ is sinusoidal, the expected value is as follows:

$$\mathcal{E}\Big\{u_{\text{in}}^2(t)\Big\} = \mathcal{E}\Big\{\sum_{i=1}^{\infty} U_{\text{i}}^2\Big(1 - \cos 2(i\omega t + \varphi i)\Big)\Big\} = \sum_{i=1}^{\infty} U_{\text{i}}^2. \tag{2.67}$$

In the balanced state the expected value of $u(t)$ is zero, that is

$$U_{\text{out}}^2 - \sum_{i=1}^{\infty} U_{\text{i}}^2 = 0. \tag{2.68}$$

As it can be seen from this equation, the dc output voltage of the converter is equal to the rms value of the input voltage:

$$U_{\text{out}} = \sqrt{\sum_{i=1}^{\infty} U_{\text{i}}^2}. \tag{2.69}$$

2.3.15 Electromechanical Converters

The operation of electromechanical ammeters and voltmeters is based on electromechanical conversion (Golding and Widdis, 1963). The deflecting torque produced by the current or voltage to be measured is balanced by the controlling torque of the springs. In the balanced state the deflection is proportional to the measurand.

The permanent-magnet type moving-coil instrument can be used for dc measurements only, because its deflection is proportional to the mean value of the signal.

The dynamometer type moving-coil instruments are applicable either for dc or ac rms measurements. These types of instrument can be used as transfer standards as well.

Moving-iron instruments are intended for the measurement of dc values and the rms value of ac quantities.

Electrostatic voltmeters are used either for dc or ac rms measurements of very high voltages.

Measuring instruments based on electromechanical conversion are discussed in Subsection 2.5.1.

2.3.16 Sample and Hold Amplifiers

Sample and hold amplifiers (S/H) are used for the measurement of the instantaneous value of a signal.

A simplified schematic diagram of an S/H amplifier is shown in Fig. 2.74.

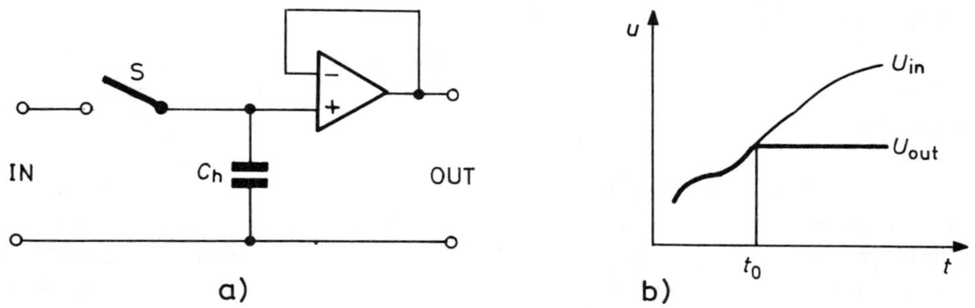

Fig. 2.74 Sample and hold amplifier *a)* circuit *b)* input and output signals

During sample (or track) mode the switch is on, thus the output follows the input voltage. At the moment t_0 the switch is opened and the hold capacitor C_h stores the last value that is the instantaneous value $u_{in}(t_0)$.

2.3.17 Analog-to-Digital Converters

An analog-to-digital converter (ADC) converts the analog input voltage into its equivalent digital form as shown in Fig. 2.75.

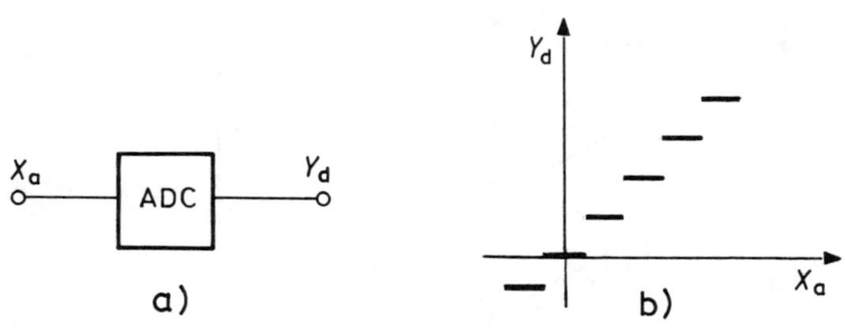

Fig. 2.75 ADC transfer function *a)* symbolic representation *b)* transfer characteristic

Sampling and quantization are usually performed in the same block, and they often cannot be separated. The A/D converter is generally preceded by a sample-and-hold unit, in this case timing is determined arbitrarily, but without the sample-hold the A/D conversion principle determines the timing of the samples. There are three important possibilities of this timing:

- sampling at predefined (usually equidistant) instants (parallel ADC-s)

- sampling at amplitude dependent instants (successive approximation ADC-s)
- integrating sampling (dual-slope and sigma-delta modulation ADC-s).

During the past few decades lots of A/D converter types have been developed. In practice, however, parallel, successive approximation, dual-slope and Σ-Δ ADCs are commonly used (Burr-Brown, 1990).

Parallel ADC

A parallel n-bit ADC (flash converter) is shown in Fig. 2.76. The input voltage to be converted is compared to reference voltages $U_{r1} - U_{rm}$ simultaneously. The comparator unit (CU) consists of $m = 2^n - 1$ voltage comparators. The outputs of the comparator unit are connected to the logic unit (LU) which generates the digital output.

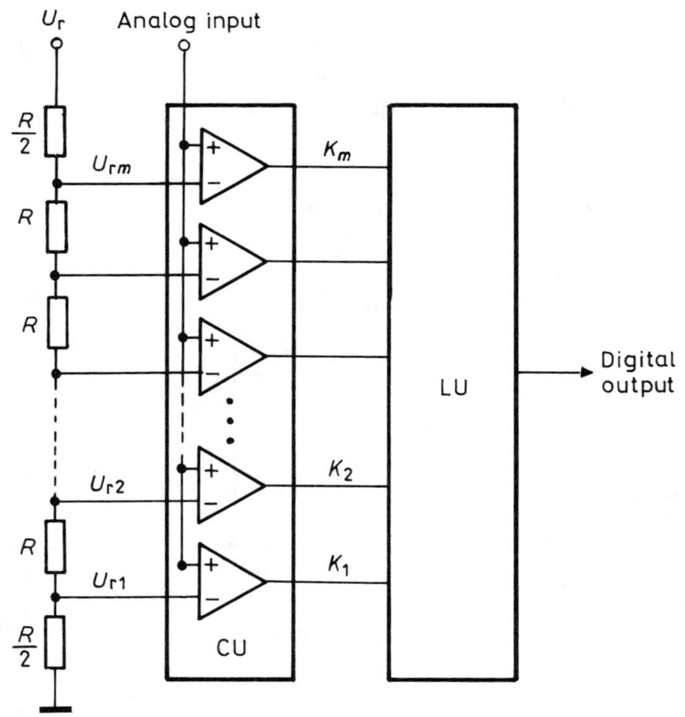

Fig. 2.76 Parallel ADC

The advantage of the parallel ADC is its very high speed, limited only by the switching time of the comparators and gates to 10–100 ns. Unfortunately, the resolution of a parallel ADC is also limited by the number of available comparators.

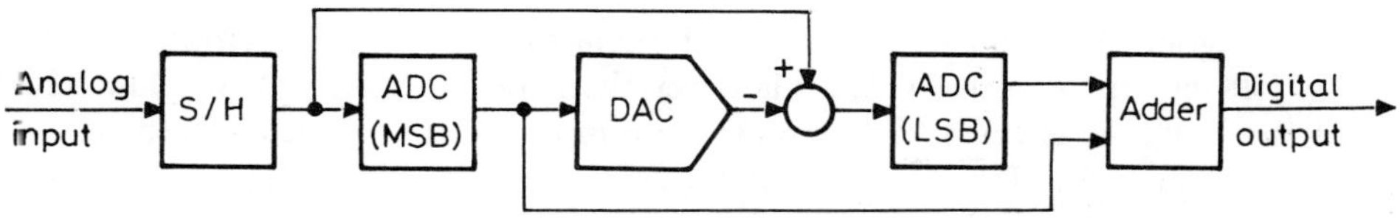

Fig. 2.77 Two-step subranging ADC

Using a two-step subranging ADC, the resolution can be improved. The two-step subranging ADC shown in Fig. 2.77 consists of a sample/hold amplifier, two parallel ADC-s, a DAC and an Adder.

The input signal is sampled and converted to digital form with a coarse ADC, then it is converted back to analog form with a high-accuracy DAC. The output voltage of the DAC is subtracted from the S/H output and the amplified difference is converted to digital form with a second coarse ADC. Finally, the Adder combines the digital output from the first ADC (MSB) with the digital output from the second ADC (LSB).

Using the technique described above, a sampling rate of 10 MHz and a resolution of 12 bit can be achieved.

Successive approximation ADC

Figure 2.78a shows a successive approximation ADC, consisting of comparator (C), digital-to-analog converter (DAC), successive approximation register (SAR) and control logic (CL).

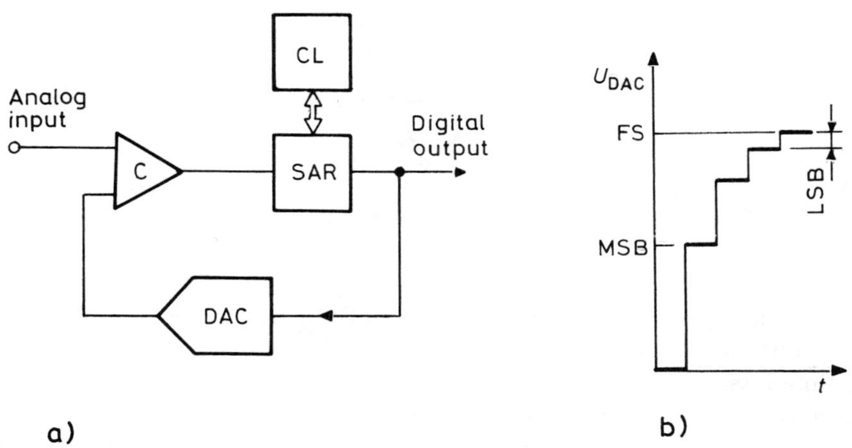

Fig. 2.78 Successive approximation ADC *a)* block diagram *b)* DAC output

The input voltage is compared to the output voltage of the DAC by the comparator. First, the most significant bit (MSB) is switched on. If the input voltage is higher than the DAC output, it remains on and an attempt with the next bit follows. If the input voltage is lower than the DAC output, it is switched off and the next bit is tried. The process continues until the least significant bit (LSB). Using an n-bit DAC, the input signal is converted into digital form in n steps. Fig. 2.78b shows the change of the DAC output of a 4-bit ADC when a full-scale (FS) voltage is applied to its input.

The typical conversion time of a 16-bit successive approximation ADC is 17 μs.

Dual-Slope ADC

The block diagram of a dual-slope ADC is given in Fig. 2.79.

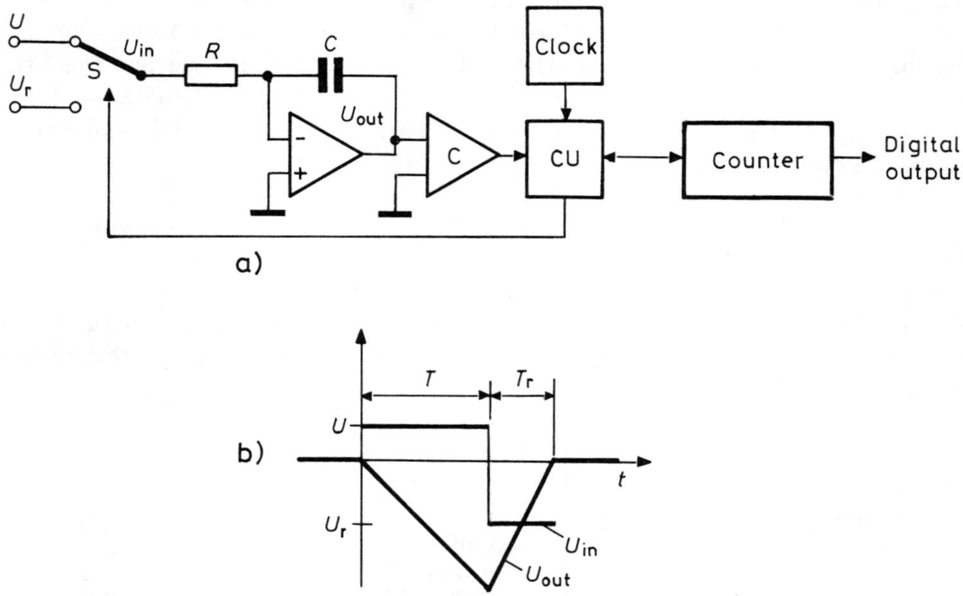

Fig. 2.79 Dual-slope ADC *a)* block diagram *b)* waveforms

The dual-slope ADC consists of a switch S, an integrator, a comparator C, a control unit CU, a clock, and a counter. First, the input voltage U is integrated during time interval T. In the second phase the reference voltage is connected to the integrator. The comparator is used to determine the time interval required for the output voltage to reach zero again. The following equation can be written for the first and second phase:

$$\frac{1}{RC}\int_0^T U\,\mathrm{d}t = \frac{1}{RC}\int_0^{T_\mathrm{r}} U_\mathrm{r}\,\mathrm{d}t\,. \tag{2.70}$$

If U is a constant direct voltage, the result is given by

$$U = U_r \frac{T_r}{T} \,. \tag{2.71}$$

The voltage to be measured is proportional to the ratio of T_r and T which can be measured accurately by means of the counter. If the voltage to be measured changes with time, the integrator produces the mean value taken over the integrating time T.

Dual-slope ADCs are widely used in digital voltmeters, where high accuracy, high resolution and a high normal-mode rejection ratio are required.

Sigma-delta modulation ADC

Sigma-delta modulation ADC-s make use of the principle described in Subsection 6.1.3. The signal is several times oversampled, and the resulting bit series is filtered by an appropriate digital filter (Fig. 2.80).

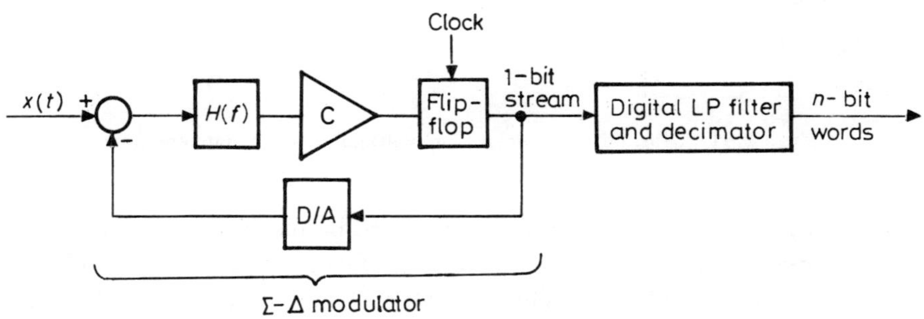

Fig. 2.80 Sigma-delta modulation ADC

Since only a single comparator is used for quantization, there is no element which would significantly deteriorate the linearity: even 18-20 bit ADC-s can be easily realized using this principle. A further advantage is that because of oversampling, no very selective anti-aliasing filter is necessary; at the input it is enough to apply a simple RC filter, and the rest is done by the digital subsampling one. The disadvantages are that because of the closed loop and the internal lowpass filter or integrator, this A/D converter cannot be multiplexed; further, because of the output digital filter the A/D converter has a significant delay (in the order of several hundred μs in the case of ADC-s with 40-80 kHz output rate).

In principle the result of the ***stochastic-ergodic converter*** could be used like the binary signal of the sigma-delta modulator. However, because of the relatively high variance (the noise power is not shifted towards high frequencies) and the realization difficulties of an appropriate (high-frequency, independent) uniform dither, this type has negligible importance.

2.4 TYPES OF METER

There are different types of meter for measuring dc and ac quantities. Their ability to reject superimposed noises has to be investigated. These signals are called normal-mode noise signals.

2.4.1 Nonintegrating dc Meters

Nonintegrating dc meters are not able to reject normal-mode noise signals. Fig. 2.81 shows the model of the source to be measured.

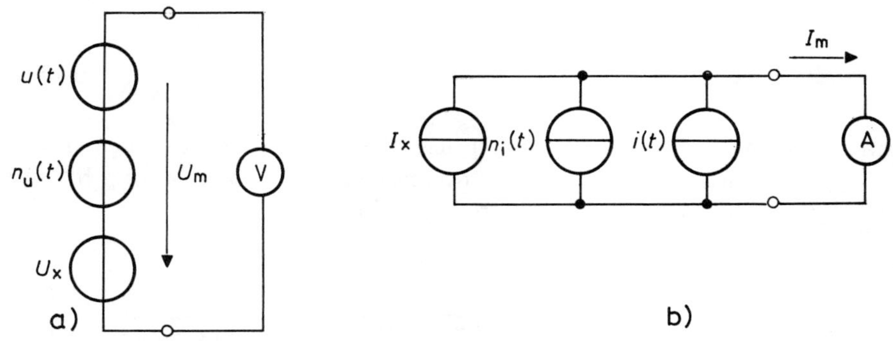

Fig. 2.81 Normal-mode noises *a)* voltage noise *b)* current noise

U_x and I_x are the dc voltage and current to be measured, respectively. $n_u(t)$ and $n_i(t)$ represent the thermal noises, $u(t)$ and $i(t)$ represent the periodic components of the superimposed noise. The frequency of these periodic signals is typically an integer multiple of the line frequency.

Nonintegrating dc meters measure the instantaneous value, thus the measured values are as follows:

$$U_m = U_x + n_u(t_0) + u(t_0) \tag{2.72}$$

and

$$I_m = I_x + n_i(t_0) + i(t_0)\,, \tag{2.73}$$

where t_0 is the time instant of sampling.

The relative errors in voltage and current measurement are

$$\epsilon_u = \frac{n_u(t_0) + u(t_0)}{U_x} \tag{2.74}$$

and

$$\epsilon_i = \frac{n_i(t_0) + i(t_0)}{I_x}\,, \tag{2.75}$$

respectively.

As can be seen, the relative error changes with t_0.

Nonintegrating dc meters use, for example, parallel or successive approximation ADCs.

2.4.2 Integrating dc Meters

Integrating dc meters are used for the rejection of superimposed noise signals. The reading of these meters is given by

$$X_{\mathrm{m}} = \frac{1}{T_{\mathrm{i}}} \int_0^{T_{\mathrm{i}}} x_{\mathrm{in}}(t)\,\mathrm{d}t\,, \tag{2.76}$$

where T_{i} is the integration interval, and $x_{\mathrm{in}}(t)$ is the input signal of the meter.

Averaging can be carried out by the inertia of a permanent-magnet moving-coil type instrument, by a lowpass filter or by a controlled integrator.

The mean value can also be measured by digital methods, according to the following equation:

$$X_{\mathrm{m}} = \frac{1}{n} \sum_{i=1}^{n} x_{\mathrm{m}}(i)\,, \tag{2.77}$$

where n is the number of samples and $x_{\mathrm{m}}(i)$ is the ith sampled value.

The thermal noise $n(t)$ to be rejected can be characterized by a Gaussian probability density function

$$p(n) = \frac{1}{\sigma\sqrt{2\pi}} e^{-n^2/2\sigma^2}\,, \tag{2.78}$$

where σ is the rms value of the noise signal and the mean value is zero.

Because of the finite integration time the reading of the meter differs from zero, if $x_{\mathrm{in}}(t) = n(t) \neq 0$.

Integrating dc meters suppress also the periodic component of the noise signal. Supposing that

$$x_{\mathrm{in}}(t) = X_{\mathrm{p}} \sin \omega t\,, \tag{2.79}$$

the reading of the meter is given by

$$X_{\mathrm{m}} = \frac{1}{T_{\mathrm{i}}} \int_0^{T_{\mathrm{i}}} X_{\mathrm{p}} \sin \omega t\,\mathrm{d}t\,. \tag{2.80}$$

The result can be expressed as follows:

$$X_{\mathrm{m}} = \frac{X_{\mathrm{p}}}{2\pi} \cdot \frac{T}{T_{\mathrm{i}}} \cdot \left(1 - \cos 2\pi \frac{T_{\mathrm{i}}}{T}\right), \tag{2.81}$$

where $T = 2\pi/\omega$. The normalized error is then

$$\frac{X_{\mathrm{m}}}{X_{\mathrm{p}}} = \frac{1}{2\pi} \cdot \frac{T}{T_{\mathrm{i}}} \cdot \left(1 - \cos 2\pi \frac{T_{\mathrm{i}}}{T}\right), \tag{2.82}$$

illustrated in Fig. 2.82.

The error caused by the sinusoidal noise is zero if T_{i}/T is an integer number, and the local maximum decreases if T_{i}/T increases.

The above integrating technique can be carried out for example by means of the well known dual-slope ADC. Thermoelectric voltages, however, can not be eliminated by integration.

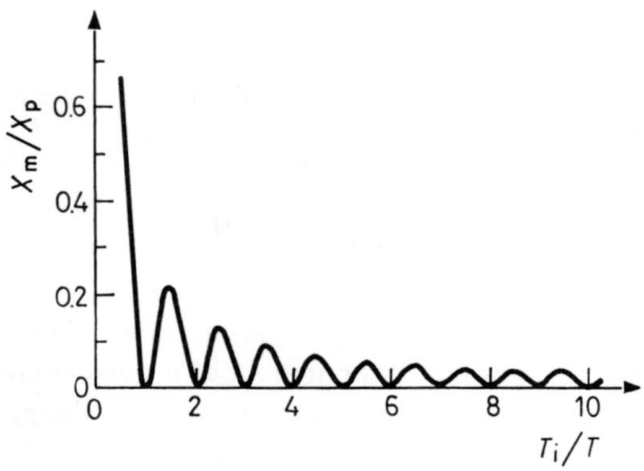

Fig. 2.82 Normalized error vs. relative integrating time

2.4.3 Average-responding Meters

The reading of an average-responding meter for periodic signals is given by

$$X_{\mathrm{m}} = \frac{\pi}{2\sqrt{2}} \cdot \frac{1}{T} \int_0^T |x_{\mathrm{in}}(t)| \,\mathrm{d}t\,, \tag{2.83}$$

where the constant is the form factor of the sine wave:

$$k_{\mathrm{f}} = \frac{\text{rms value}}{\text{average absolute value}}\,. \tag{2.84}$$

For a sine wave:

$$k_{\mathrm{f}} = \frac{\pi}{2\sqrt{2}} \approx 1.11\,. \tag{2.85}$$

X_{m} is equal to the rms value for a sine wave (see (2.83)).

Average-responding meters use full-wave rectifiers followed by a lowpass filter. The accuracy depends on the harmonic content of the signal. In the case of a fundamental signal and a single harmonic the waveform is as follows:

$$x(t) = X_{\mathrm{p}}\big(\sin\omega t + k\sin(i\omega t + \varphi)\big)\,, \tag{2.86}$$

where k is the relative amplitude of the ith-order harmonic.

The average absolute value can be expressed by the following equation:

$$X_{\mathrm{av}} = \frac{1}{T}\int_{t_1}^{t_2} x(t)\,\mathrm{d}t - \frac{1}{T}\int_{t_2}^{t_1+T} x(t)\,\mathrm{d}t\,, \tag{2.87}$$

where t_1 and t_2 are the zero crossing points of $x(t)$:

$$\sin\omega t + k\sin(i\omega t + \varphi) = 0\,, \quad \text{if} \quad t = t_1 \text{ or } t = t_2\,. \tag{2.88}$$

It should be noted that highly distorted signals may have more zero crossings than the first harmonic. In this case, (2.87) must be modified.

The relative error can be defined as follows:

$$\epsilon_r = \frac{k_f X_{av} - X_{eff}}{X_{eff}}, \tag{2.89}$$

where k_f is the form factor of the sine wave.

Supposing a single second harmonic ($i=2$) and $\varphi=0$, the zero-crossing points are at $t_1=0$ and $t_2=T/2$, as shown in Fig. 2.83. The relative amplitude k, however, has to be smaller than 0.5.

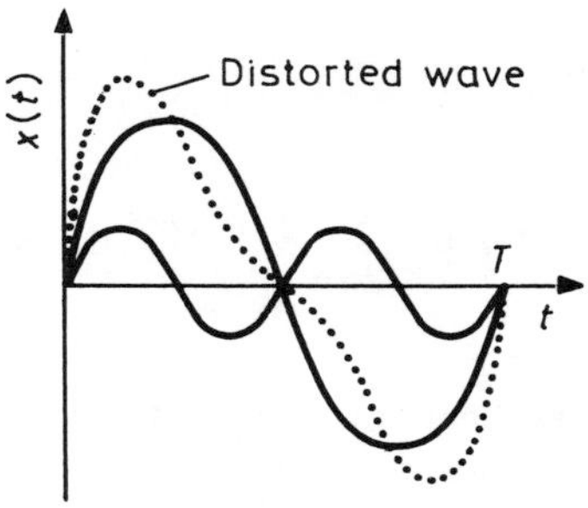

Fig. 2.83 In-phase second harmonic

The figure shows that the average absolute value is not influenced by the second harmonic, and is given by

$$X_{av} = \frac{2}{\pi} X_p . \tag{2.90}$$

The rms value, however, depends on the second harmonic:

$$X_{eff} = \frac{X_p}{\sqrt{2}} \sqrt{1+k^2} . \tag{2.91}$$

Substituting (2.90) and (2.91) in (2.89) yields

$$\epsilon_r = \frac{1}{\sqrt{1+k^2}} - 1 \approx -\frac{k^2}{2} . \tag{2.92}$$

10% second harmonic distortion causes approximately 0.5% error.

When more than one harmonic is present, the error calculation becomes much more complicated.

Using a switching-mode phase-sensitive (synchronous) rectifier, the error, caused by the even harmonics, can be fully eliminated.

The reading of an average-responding meter to gaussian noise is as follows:

$$X_{\mathrm{m}} = \frac{\pi}{2\sqrt{2}} \int_{-\infty}^{\infty} |n| p(n)\, \mathrm{d}n\,. \tag{2.93}$$

If $p(n)$ is symmetrical about zero,

$$X_{\mathrm{m}} = \frac{\pi}{\sqrt{2}} \int_{0}^{\infty} n p(n) \mathrm{d}n\,. \tag{2.94}$$

Substituting (2.78) into (2.94) gives

$$X_{\mathrm{m}} = \frac{\pi}{\sqrt{2}} \int_{0}^{\infty} n \frac{1}{\sqrt{2\pi}\sigma} e^{-n^2/2\sigma^2}\, \mathrm{d}n = \frac{\sqrt{\pi}}{2}\sigma\,. \tag{2.95}$$

It is assumed that no overload occurs on the peaks of the noise signal.

In Table 2.1 average absolute values of typical signals and the readings of average-responding meters are listed.

Table 2.1 Readings of average-responding meters

Signal	Average absolute value	Reading	Form factor
Sine	$\frac{2}{\pi} X_{\mathrm{p}}$	$\frac{1}{\sqrt{2}} X_{\mathrm{p}}$	$\frac{\pi}{2\sqrt{2}}$
Triangular	$\frac{1}{2} X_{\mathrm{p}}$	$\frac{\pi}{4\sqrt{2}} X_{\mathrm{p}}$	$\frac{2}{\sqrt{3}}$
Square	X_{p}	$\frac{\pi}{2\sqrt{2}} X_{\mathrm{p}}$	1
Gaussian noise	$\sqrt{\frac{2}{\pi}}\sigma$	$\frac{\sqrt{\pi}}{2}\sigma$	$\sqrt{\frac{\pi}{2}}$

2.4.4 Peak-responding Meters

In practice, positive peak, negative peak and peak-to-peak values have to be measured. There are dc-coupled and ac-coupled peak rectifiers.

The reading of a peak-responding meter is given by

$$X_{\mathrm{m}} = \frac{X_{\mathrm{p}}}{\sqrt{2}}\,, \tag{2.96}$$

where X_p is the peak value, and $\sqrt{2}$ is the crest factor of the sine wave. The crest factor is defined as follows:

$$k_c = \frac{\text{peak value}}{\text{rms value}}. \tag{2.97}$$

As usual, the reading is equal to the rms value for the sine wave.

Obviously, the harmonic distortion of the input signal causes an error in the reading of the peak-responding meter.

The advantage of a peak-responding meter is that the peak rectifier can be placed into the probe when no ac preamplification is required. The measuring leads are used for dc signal transmission only, thus the peak-responding voltmeter can be used up to hundreds of megahertz.

The disadvantage is the limited sensitivity because of the nonlinear diode characteristics.

In Table 2.2 peak values of typical signals and the readings of peak-responding meters are shown.

Table 2.2 Readings of peak-responding meters

Signal	Peak value	Reading	Crest factor
Sine	X_p	$\frac{X_p}{\sqrt{2}}$	$\sqrt{2}$
Triangular	X_p	$\frac{X_p}{\sqrt{2}}$	$\sqrt{3}$
Square	X_p	$\frac{X_p}{\sqrt{2}}$	1

2.4.5 Rms-responding Meters

Rms-responding meters provide a reading for periodic signals as follows:

$$X_m = \sqrt{\frac{1}{T}\int_0^T x^2(t)\,dt} \tag{2.98}$$

The methods of rms-to-dc conversion are treated in Subsection 2.3.14.

An rms voltmeter with high crest-factor rating must contain amplifiers with sufficient dynamic range to avoid nonlinear operation. On the other hand, the maximum crest factor is determined by the bandwidth of the amplifier. Obviously, the higher the crest factor of a signal the wider is the frequency spectrum.

Superimposed noise signals composed of thermal and periodic noises cause an error in rms measurement. Supposing that the input voltage is

$$u_{in}(t) = u_x(t) + n(t) + u(t), \tag{2.99}$$

the reading of the rms-responding meter is as follows:

$$U_{\mathrm{m}} = \sqrt{U_{\mathrm{xeff}}^2 + \sigma^2 + U_{\mathrm{eff}}^2}\,. \tag{2.100}$$

The relative error can be defined as follows:

$$\epsilon_{\mathrm{r}} = \frac{U_{\mathrm{m}} - U_{\mathrm{xeff}}}{U_{\mathrm{xeff}}}\,. \tag{2.101}$$

Substituting (2.100) in (2.101) gives

$$\epsilon_{\mathrm{r}} = \sqrt{1 + \left(\frac{\sigma}{U_{\mathrm{xeff}}}\right)^2 + \left(\frac{U_{\mathrm{eff}}}{U_{\mathrm{xeff}}}\right)^2} - 1\,. \tag{2.102}$$

For small relative errors the following approximation can be used:

$$\epsilon_{\mathrm{r}} \approx \frac{1}{2}\left(\frac{\sigma}{U_{\mathrm{xeff}}}\right)^2 + \frac{1}{2}\left(\frac{U_{\mathrm{eff}}}{U_{\mathrm{xeff}}}\right)^2\,. \tag{2.103}$$

As can be seen, low-level rms measurements can be erroneous in the presence of noise. Using selective voltmeters this problem may be overcome.

Table 2.3 shows the rms values of typical signals and the readings of rms-responding meters.

Table 2.3 Readings of rms-responding meters

Signal	Rms value and reading
Sine	$\frac{X_{\mathrm{p}}}{\sqrt{2}}$
Triangular	$\frac{X_{\mathrm{p}}}{\sqrt{3}}$
Square	X_{p}
Gaussian noise	σ

2.5 MEASURING INSTRUMENTS

In this section measurement methods and instruments for dc and ac quantities will be introduced. These instruments can be composed from the building blocks presented and analyzed in Section 2.3.

2.5.1 Electromechanical Instruments

The permanent-magnet type moving-coil instrument can be used for dc measurement. Such instruments can be constructed for measurement ranges of $1\,\mu$A to 0.5 A and 1 mV to 600 V.

For range extension, shunts and series resistors are widely used as shown in Figs. 2.64 and 2.62. The uncertainty can be 0.1%, and the power consumption is low. The internal resistance is typically $20\,\text{k}\Omega/\text{V}$, thus in this case the full-scale power consumption is only 0.5 mW in the 10 V-range. Because of the inertia of the moving coil, superimposed noise signals are rejected.

In order to reach very high sensitivity, moving-coil galvanometers are used. Their resolution can be 10 pA or 10 nV, thus they are used in null detectors.

Electrodynamic instruments are applicable either for dc or ac rms measurements. They can be constructed for measurement ranges of 30 mA to 100 A and 15 V to 600 V. In many cases the coil of the instrument is divided into two parts, thus range extension can be carried out by parallel or series connection of these parts. The uncertainty can be 0.1% at a few VA power consumption. An electrodynamic instrument can also be used as a transfer standard up to hundreds of Hertz.

Moving-iron instruments can be used for dc and ac rms measurements. The measurement ranges of moving-iron instruments for laboratory purposes can be 0.1 A to 10 A and 50 V to 750 V. In special cases, however, the range of hundreds of Amperes can be achieved.

For range extension, tapped coil or series/parallel connection of the coil parts can be used. The uncertainty can be 0.2% at a few VA power consumption from dc up to 100 Hz.

Electrostatic voltmeters are used either for dc or ac rms measurement of voltages. The measurement range can be 10 V to 1 MV. With a light beam as a pointer, 0.2% uncertainty can be achieved. The electrostatic voltmeter which has extremely low power consumption can be used from dc up to 10 kHz.

For more information about electromechanical instruments see Golding and Widdis (1963).

2.5.2 Dc Voltmeters

Analog dc voltmeters are used for the measurement of dc voltages up to 1000 V. A typical analog dc voltmeter is shown in Fig. 2.84.

The voltage range is selected by the switch S using the resistance voltage divider VD. For low-voltage measurement, however, the signal is even amplified by the amplifier A of variable gain.

Superimposed noise signals can overdrive the amplifier. To avoid this overdriving, a low pass filter can be switched on.

The output voltage of the amplifier is proportional to the input voltage and is measured by the permanent-magnet type moving-coil instrument V.

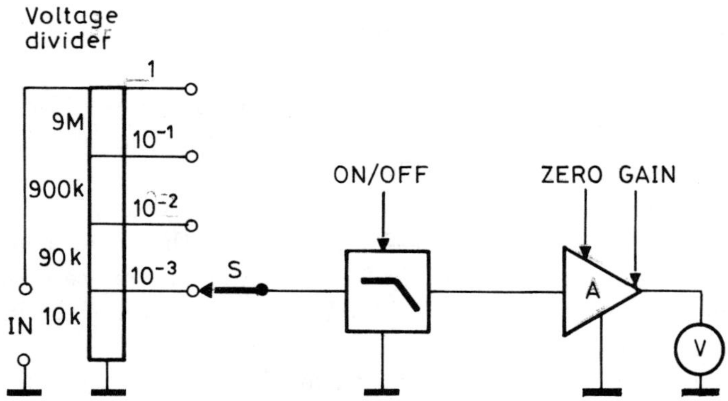

Fig. 2.84 Dc voltmeter

The accuracy of the above electronic meter is limited by the moving-coil instrument, thus the uncertainty can be 0.2–1%. The high input resistance results in a very low power consumption.

For low-voltage measurements, special dc amplifiers are used, which have very low voltage offset and drift. Any remaining offset voltage can be compensated by the zero control provided on the instrument.

2.5.3 Low Frequency Voltmeters

Low frequency voltmeters are used for the measurement of ac voltages between 10 Hz and 1 MHz. The main building blocks are given in Fig. 2.85.

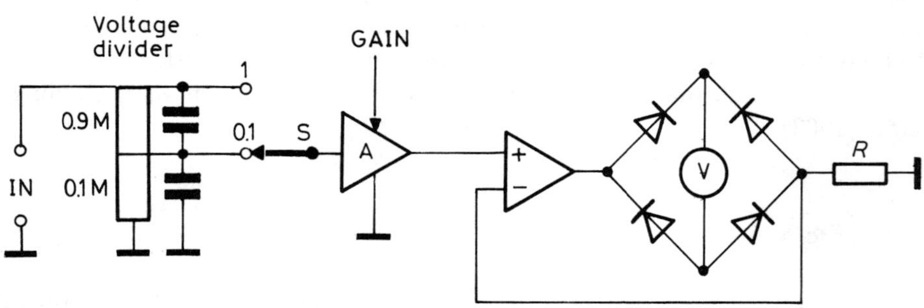

Fig. 2.85 LF voltmeter

The voltage range is selected by the switch S using the compensated-resistance voltage divider VD. Low voltages can be amplified by the ac-coupled amplifier of variable gain. The output voltage of the amplifier is then rectified and measured by the permanent-magnet type moving-coil instrument V. An active full-wave rectifier is used for the reduction of the error caused by the nonlinearity of the diodes.

Low frequency voltmeters are used for the measurement of ac voltages between 1 mV and 300 V. Voltages lower than 1 mV can be measured by means of selective voltmeters which suppress the noise.

At higher frequencies the error increases because of the frequency dependence of the gain and the parasitic impedances of the measuring leads, thus low frequency voltmeters can be used up to 1 MHz only.

2.5.4 High Frequency Voltmeters

High frequency voltmeters are used for measurement of ac voltages up to hundreds of megahertz. Fig. 2.86 shows a simple HF voltmeter.

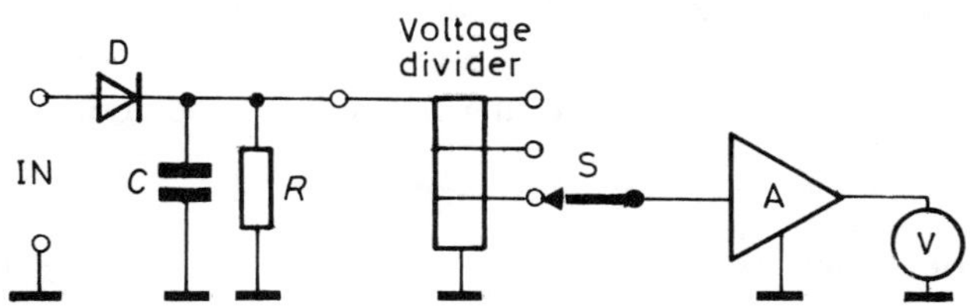

Fig. 2.86 High frequency voltmeter

The input voltage is rectified by the peak rectifier. The measurement range is selected by the switch S using resistance voltage divider VD. The DC amplifier A with high input resistance drives the permanent-magnet type moving-coil instrument.

Because of the nonlinearity of the rectifying diode D, HF voltmeters can be used from 1 V up to 300 V only.

For measurements at very high frequencies the peak rectifier is placed in the probe, thus the measuring leads are used for dc signal transmission only.

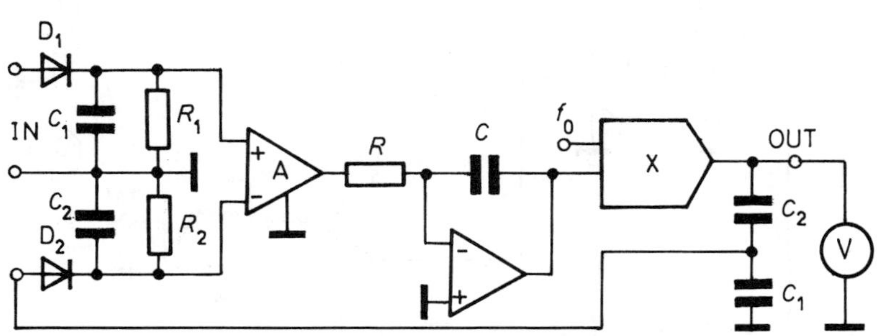

Fig. 2.87 High sensitivity HF voltmeter

For the compensation of the error caused by the diode, a second compensating peak rectifier is used, which has the same characteristics as the measuring one. The differential amplifier A measures the voltage difference between the two rectifier outputs. The amplified error signal is integrated, then multiplied by an ac signal with f_0 frequency.The amplitude modulated signal is then divided by the capacitance voltage divider and fed back to the compensating rectifier (see Fig. 2.87). In the balanced state the output peak voltage is given by

$$U_{\text{pout}} = \left(1 + \frac{C_1}{C_2}\right) U_{\text{pin}} \,. \tag{2.104}$$

The peak value of the input signal is amplified, but the frequency of the output signal is always f_0 and does not depend on the input frequency. f_0 is typically 100 kHz, thus the output signal can be rectified and measured with high accuracy. Using matched pairs of diodes even 1–10 mV can be measured. For measurements at very high frequencies the measuring and compensating rectifiers are placed in the probe and it is arranged that the thermal coupling be good.

2.5.5 Vector Voltmeters

Vector voltmeters are used for the measurement of vector components of complex voltages. Fig. 2.88 shows the principle of an analog vector voltmeter using switching-mode phase-sensitive rectifiers.

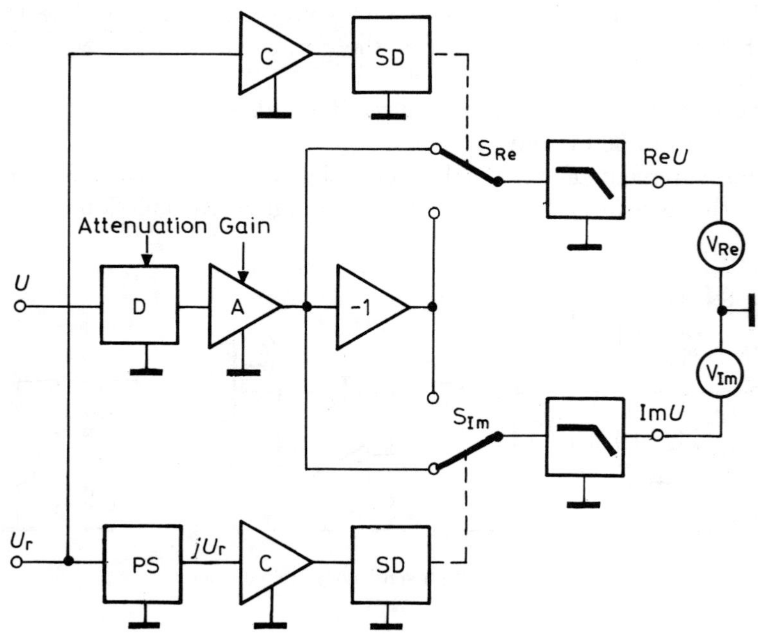

Fig. 2.88 Analog vector voltmeter

The voltage to be measured is attenuated by the variable divider D, or amplified by the variable amplifier A. The conditioned signal is then rectified by two switching-mode phase-sensitive rectifiers. The comparators (C) produce square-wave signals from U_r and jU_r and control the switch drivers (SD). The phase of the reference signal U_r is shifted by the phase shifter PS.

The output voltages of the phase-sensitive rectifiers are proportional to the real and imaginary parts of U, as it was discussed in Subsection 2.3.13. The two vector components are then measured and displayed by the two meters V_r and V_i.

Unfortunately, the analog vector voltmeter is sensitive to the odd harmonics (see Subsection 2.3.13.).

Vector voltmeters can be realised digitally as well (Nagy, 1989). Fig. 2.89 shows the main parts of a digital vector voltmeter.

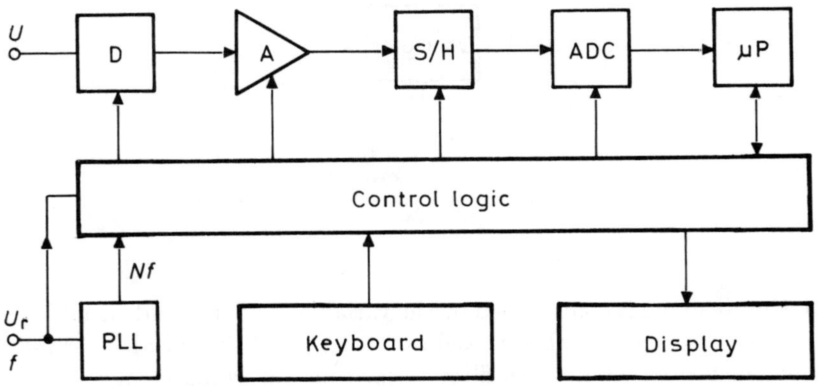

Fig. 2.89 Digital vector voltmeter

The input signal is conditioned by the divider D and amplifier A. Samples are taken from the conditioned signal by the sample/hold amplifier. After analog-to-digital conversion a discrete Fourier transform (DFT) is made by the microprocessor.

Sampling signals are generated from the reference signal U_r by means of a frequency multiplying PLL. The complex amplitude of the first harmonic can be determined by the following equation:

$$U_1 = \frac{2}{N}\sum_{i=0}^{N-1}\left(\cos\frac{2\pi}{N}i + j\sin\frac{2\pi}{N}i\right)u\left(i\frac{T}{N}\right), \tag{2.105}$$

where N is the number of samples during one period, T is the period length of the input signal, $u\left(i\frac{T}{N}\right)$ is the ith sample taken from $u(t)$.

The advantage of the digital vector voltmeter is that the higher-order harmonics are highly rejected.

Vector voltmeters can be used for measurement of complex voltages, currents and impedances.

2.5.6 Selective Voltmeters

Selective voltmeters are used for measurement of very low ac voltages and for frequency spectrum analysis. The principle of the heterodyne selective voltmeter is shown in Fig. 2.90.

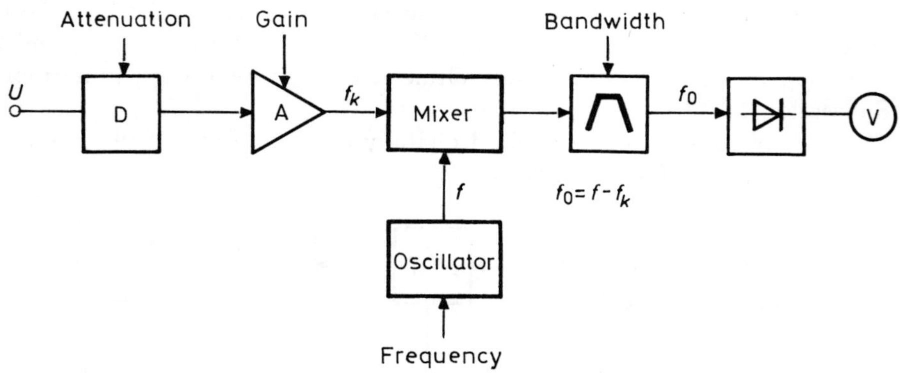

Fig. 2.90 Heterodyne selective voltmeter

The input signal is conditioned by the variable divider D and variable amplifier A. The conditioned signal is then mixed by the signal of the variable oscillator. The centre frequency of the bandpass filter is f_0, thus the component with frequency f_k of the input signal can be measured by setting the oscillator frequency to $f = f_k + f_0$. The bandwidth of the bandpass filter is variable in two or more steps. The advantage of the heterodyne principle is that the bandwidth does not depend on the measurement frequency.

Due to their selectivity, these voltmeters can reject the superimposed noise voltages, thus they are applicable for μV measurements in a wide frequency range.

2.5.7 Dc Compensators

For high-accuracy measurements compensators (dc potentiometers) can be used (Luppold, 1969). These realize a direct comparison of the measurand to a quantity of the same kind having a known value. Moreover, compensators realize a null method of measurement in which the value of the measurand is determined by balancing.

Fig. 2.91 shows the principle of dc compensators. In the balanced state of the voltage compensator the current I_n of the null detector ND is zero, thus $U_x = U_c$. The variable compensating voltage U_c has a known value.

This compensator realizes an ideal voltmeter causing no loading error. In practice, however, the current of the null detector somewhat differs from zero, thus

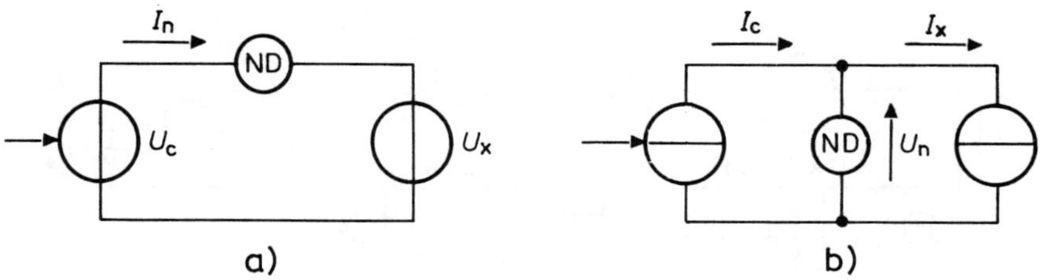

Fig. 2.91 DC compensators *a)* voltage compensator *b)* current compensator

the input resistance can be estimated by

$$R_{in} = \pm \frac{U_x}{I_{nmin}}, \tag{2.106}$$

where I_{nmin} is the current not detectable by the null detector.

The operation of the current compensator is similar to that of the voltage compensator. The current compensator is an ideal ammeter having zero input resistance. Due to the finite sensitivity of the null detector, however, the input resistance differs from zero and can be estimated by

$$R_{in} = \pm \frac{U_{nmin}}{I_x}, \tag{2.107}$$

where U_{nmin} is the voltage which is not detectable by the null detector.

In Fig. 2.92 a simple voltage compensator (dc potentiometer) is shown.

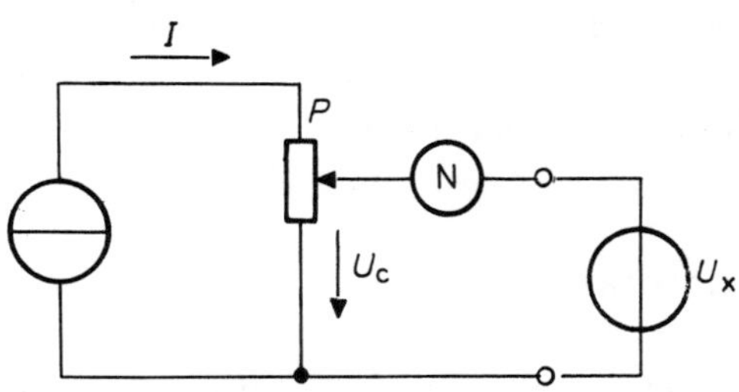

Fig. 2.92 DC potentiometer

The current source produces a stable current of known value. The compensating voltage can be adjusted by the potentiometer P. In balanced state the current flowing through the null detector is zero, and U_c is known from the position of the potentiometer.

In practice, high-accuracy resistance dividers are used. As an example, Fig. 2.93 shows a cascade compensator having four decades.

The operation of the cascade divider is discussed in Subsection 2.3.3 The current I can be determined by a second voltage compensator, composed of the

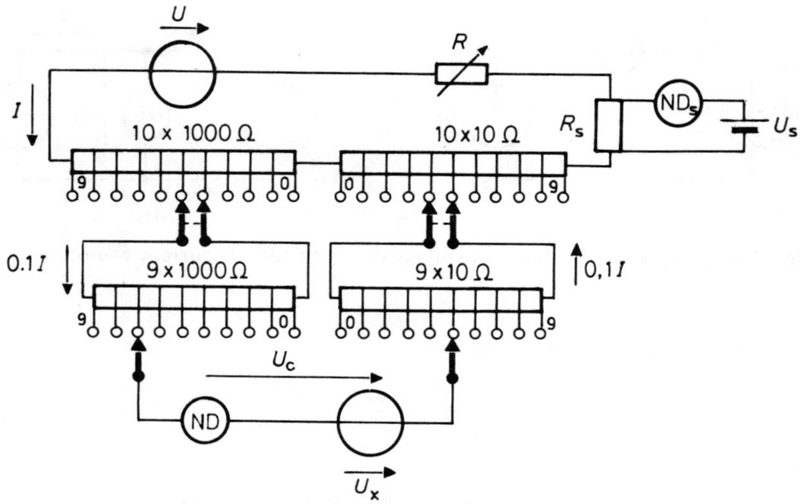

Fig. 2.93 Cascade compensator

standard voltage source U_s (standard cell), standard resistor R_s and null detector ND_s. The current I is adjusted by the variable resistor R. In the balanced state of the second compensator the current is given by U_s/R_s. Typically moving-coil galvanometers are used as null detectors.

The uncertainty of voltage compensators can be 0.01%, up to 10 V. For higher voltages precision resistance dividers (volt boxes) are used for attenuation of the voltage to be measured.

For very low-voltage measurements special voltage compensators (potentiometers) have to be used which eliminate the effect of thermoelectric voltages.

There are many other types of dc compensator which can be operated either manually or automatically (Luppold, 1969).

2.5.8 Ac Compensators

Ac compensators (ac potentiometers) realize a direct comparison of ac quantities, combined with the null method. Using ac compensators, vector components of ac voltages and currents can be determined. Fig. 2.94 shows their principle.

The operation of ac compensators is similar to the operation of dc ones. The compensating voltage or current, however, consists of real and imaginary parts. If the measurand is distorted, the balancing can be carried out at the first harmonic only, thus the null detector of ac compensators is always selective.

Fig. 2.95 shows a simple ac compensator (ac potentiometer).

The primary voltage of the voltage transformers T_1 and T_2 are U_g and jU_g, respectively. The desired phase shift (90°) is produced by the phase shifter PS. The real and imaginary parts of the compensating voltage can be adjusted by potentiometers P_1 and P_2, respectively.

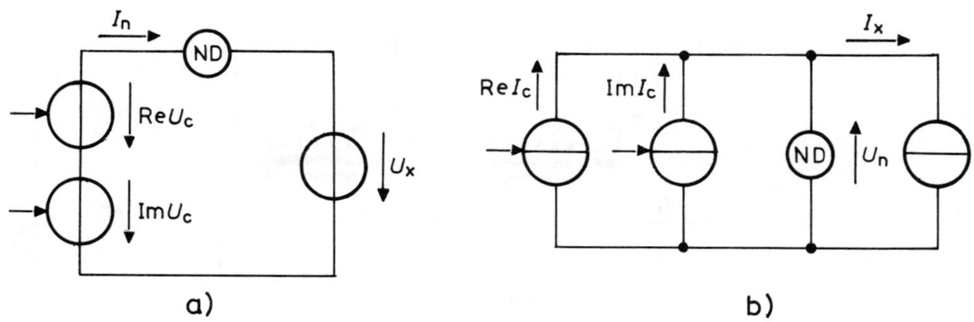

Fig. 2.94 Ac compensators *a)* voltage compensator *b)* current compensator

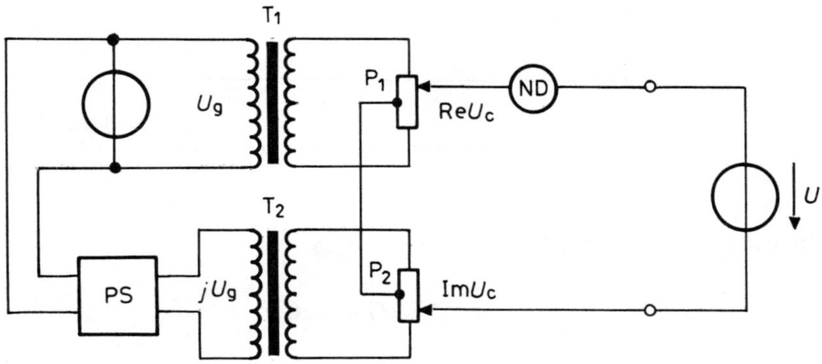

Fig. 2.95 Ac potentiometer

Fig. 2.96 shows a current transformer accuracy test, based on an ac current compensator.

For direct comparison, a standard current transformer SCT is used. The secondary current I_x of the CT differs from the true one,represented by the secondary current I_s of the SCT. The error current is given by

$$I_e = I_x - I_s \,. \tag{2.108}$$

For measurement of the error current, a standard resistor R_s, a differential amplifier A and the adjusting elements C, G are used. In the balanced state, when the voltage across the null detector is zero, the error current can be expressed as follows:

$$I_e = I_s R_s A(G + j\omega C) \,. \tag{2.109}$$

The relative error of the CT is defined by the following equation:

$$\epsilon_r = \frac{I_x - I_s}{I_s} = \frac{I_e}{I_s} \,. \tag{2.110}$$

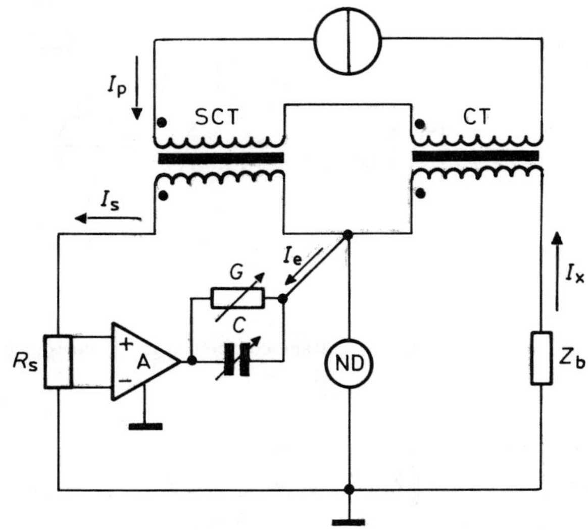

Fig. 2.96 Current transformer accuracy test

Substituting (2.109) into (2.100) gives

$$\epsilon_r = AR_s(G + j\omega C)\,. \tag{2.111}$$

It can be seen that the in-phase component of the relative error is AR_sG and the quadrature component is $AR_s\omega C$.

AC compensators can be operated either manually or automatically. They are used for the measurement of ac currents, ac voltages and impedances.

2.5.9 Digital Multimeters

Digital multimeters convert analog signals to digital form and can measure ac/dc voltage, ac/dc current and resistance. DMMs are widely used in electronic instrumentation where accuracy, high speed, automatic operation and possibility of programming are required.

A digital multimeter consists of a digital voltmeter and signal conditioners as shown in Fig. 2.97.

be controlled either manually or automatically. The selectable modes are as follows:

- dc voltage
- ac voltage
- dc current
- ac current
- resistance.

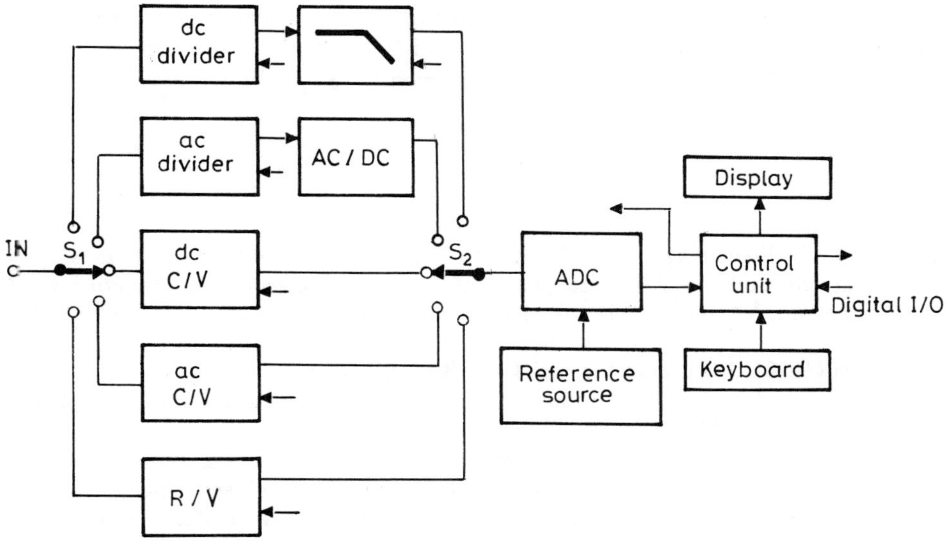

Fig. 2.97 Digital multimeter

The signal-to-noise ratio can be enhanced by means of a low pass filter. The analog-to-digital converter is usually a dual-slope one which guarantees a good normal-mode rejection.

Self-calibration is a typical feature of high-quality multimeters. Self-calibration can be carried out for example by means of a modified input stage as shown in Fig. 2.98.

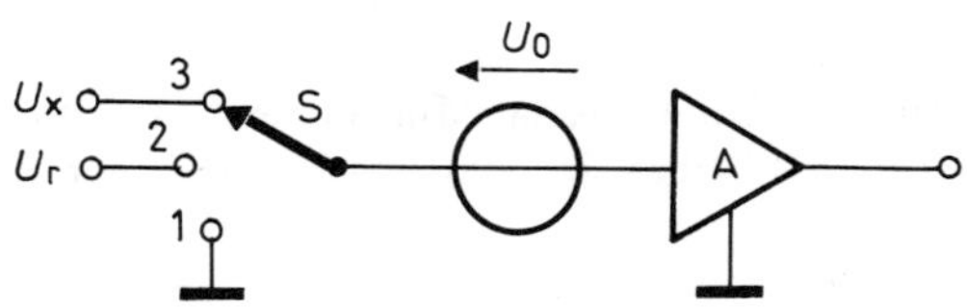

Fig. 2.98 Self-calibration

In switch position 1

$$U_1 = U_0 A , \tag{2.112}$$

will be measured, in position 2

$$U_2 = (U_r + U_0)A , \tag{2.113}$$

and in position 3

$$U_3 = (U_x + U_0)A . \tag{2.114}$$

The offset voltage and the gain factor of the input amplifier can be eliminated as follows:

$$\frac{U_3 - U_1}{U_2 - U_1} = \frac{U_x}{U_r} . \tag{2.115}$$

Digital multimeters have a very broad measurement range up to 10 A and 1000 V with a maximum resolution of 0.1 to 1 nA and 0.1 to 1 μV. Their uncertainty can be 0.5 to 0.002%. For calibration purposes even an 8 1/2-digit digital multimeter was developed (Stever, 1989).

References

Boll, R. (1990): Weichmagnetische Werkstoffe: Einführung in den Magnetismus, VAC-Werkstoffe und ihre Anwendungen (Soft Magnetic Materials: Introduction to the Magnetism, VAC-Materials and their Applications). Berlin, Siemens AG. (In German)

Burr–Brown (1990): Integrated Circuits Data Book Supplement, Vol. 33b.

Golding, E. W. and Widdis, F. C. (1963): Electrical Measurements and Measuring Instruments. London, Sir Isaac Pitman & Sons.

IEEE (1977): IEEE Standard Dictionary of ELectrical and Electronics Terms. Second ed. New York, The Institute of Electrical and ELectronics Engineers, Inc.

Jones, L. D. and Chin, F. A. (1983): Electronic Instruments and Measurements. New York, John Wiley & Sons.

Keithley Instruments (1984): Low Level Measurements. Cleveland, Ohio.

Lang, T. T. (1987): Electronics of Measuring Systems. Chichester, John Wiley & Sons.

Lem Sa (1986): Das LEM Modul. Beschreibung und Anwendungen (The LEM-Module. Specification and Applications). Genf, Schweiz. (In German)

Luppold, D. S. (1969): Precision dc Measurements and Standards. Reading, MA, Addison-Wesley.

Nagy, F. (1989): A New Digital Vector Voltmeter. *Proceedings of the 3rd IMEKO TC4 Symposium on Measurement in Electrical and Electronic Power Systems,* Zurich, Sept. 20–22, 1989. pp. 279–287.

Oliver, B. M. and Cage J. M. (1971): Electronic Measurements and Instrumentation. New York, McGraw-Hill, Inc.

Stever, S. D. (1989): An 8 1/2-Digit Digital Multimeter Capable of 100,000 Reading per Second and Two-Source Calibration. *Hewlett-Packard Journal,* Vol. 40, No. 2, April, pp. 6-7.

Stoll, D. (1976): EMC – Elektromagnetische Verträglichkeit (Electromagnetic Compatibility). Berlin, Elitera-Verlag GmbH. (In German)

Tettex AG (1987): General Catalog. Zurich, Switzerland.

Zinke, O. und Seither, H. (1982): Widerstände, Kondensatoren, Spulen und ihre Werkstoffe (Resistors, Capacitors, Coils and their Materials). Berlin, Springer Verlag. (In German)

Zoltán, I. (1989): Capacitive Error Compensation in Current Transformers. *Proceedings of the 3rd IMEKO TC4 Symposium on Measurement in Electrical and Electronic Power Systems,* Zurich, Sept. 20–22, 1989. pp. 65–70.

3

Measurement of Power and Energy

Gy. Korányi

Editorial introduction

The measurement of electrical power and energy has outstanding importance in energy distribution and accounting, in the regulation of power consumption, in the supervision of electrical machines, in the investigation of effectiveness etc. This chapter deals with the methods and instruments of such measurements. The discussion focuses on the methods applicable in the line frequency range (50 – 60 Hz), but methods are also introduced which can be used at higher frequencies.

3.1 MEASURED QUANTITIES

3.1.1 Definition of Power

The instantaneous value of the power is the product of the instantaneous values of the current and the voltage:

$$p(t) = u(t)i(t)\,. \tag{3.1}$$

The average power can be calculated from the integral of the instantaneous power:

$$P = \frac{1}{T}\int_0^T u(t)i(t)\,\mathrm{d}t \tag{3.2}$$

where T is the period time. The power has a sign, which depends on the direction of the energy flow. The consumed power is positive and the produced power is negative.

Technology of Electrical Measurements, Edited by L. Schnell
© 1993 John Wiley & Sons Ltd.

The power of a dc circuit with voltage U and current I is:

$$P = p(t) = \pm UI\,. \tag{3.3}$$

The dissipated power on the resistance R of the network, using Ohm's law, is:

$$P = I^2R = U^2/R\,. \tag{3.4}$$

In the case of a pure sinusoidal voltage and linear network, the current and voltage are

$$i(t) = \sqrt{2}I\cos\omega t\,; \qquad u(t) = \sqrt{2}U\cos(\omega t + \varphi)\,,$$

where I and U are the effective (rms) values of the current and the voltage, respectively. The instantaneous power of the network is:

$$u(t)i(t) = p(t) = UI\cos\varphi(1 - \cos 2\omega t) - UI\sin\varphi\sin 2\omega t\,. \tag{3.5}$$

The active power is:

$$P = \frac{1}{T}\int_0^T p(t)\,\mathrm{d}t = UI\cos\varphi\,, \tag{3.6}$$

where $\cos\varphi$ is the power factor.

The second part in Eq. (3.5) is related to the reactive power. Its mean value is zero, and its maximum is conventionally called reactive power:

$$Q = UI\sin\varphi\,, \tag{3.7}$$

while the apparent power is defined as:

$$S = UI = \sqrt{P^2 + Q^2}\,. \tag{3.8}$$

Nonsinusoidal periodic signals can be expanded into harmonic components. Active power is only generated by currents and voltages of the same frequency, therefore the active power of periodic voltage and current is the sum of the active powers of the harmonics:

$$P = I_0U_0 + \sum_{k=1}^{\infty} U_kI_k\cos\varphi_k\,. \tag{3.9}$$

Those instruments are applicable for the measurement of power that are able to measure the above sum.

Similarly, for the reactive power, the following relationship can be written:

$$Q = \sum_{k=1}^{\infty} U_kI_k\sin\varphi_k\,. \tag{3.10}$$

The apparent power is

$$S = UI = \sqrt{\sum_{k=0}^{\infty} U_k^2} \sqrt{\sum_{k=0}^{\infty} I_k^2}\,, \tag{3.11}$$

that is,

$$S = \sqrt{P^2 + Q^2 + D^2}\,, \tag{3.12}$$

where D is the distortion power, calculated from the summation of the inner products of current and voltage harmonics with different frequencies.

According to another concept, the reactive power is

$$P_\mathrm{r} = \sqrt{Q^2 + D^2}\,, \tag{3.13}$$

as shown in Enslin and Wyk (1988), that is,

$$S^2 = P^2 + P_\mathrm{r}^2\,. \tag{3.14}$$

The definition of the power factor is:

$$\cos\varphi = \frac{P}{S} = \frac{\frac{1}{T}\int_0^T u(t)i(t)\,\mathrm{d}t}{\sqrt{\frac{1}{T}\int_0^T u^2(t)\mathrm{d}t}\sqrt{\frac{1}{T}\int_0^T i^2(t)\,\mathrm{d}t}}\,. \tag{3.15}$$

In a multiphase network the power of the generator or of the load is the sum of the powers of the single phases:

$$P = \sum_{i=1}^{n} P_i\,, \tag{3.16}$$

where P_i is the power of the ith phase and n is the number of phases. In practice, the most important is the three-phase system ($n=3$). In a symmetrical system the relationships

$$P = 3U_\mathrm{p}I_\mathrm{p}\cos\varphi = 3I_\mathrm{p}^2R = 3GU_\mathrm{p}^2 = \sqrt{3}U_\mathrm{l}I_\mathrm{l}\cos\varphi \tag{3.17}$$

give the power of the system. U_p, I_p are the phase voltage and current, and U_l, I_l are the line voltage and current (rms values).

Practical measurement techniques define the relationships for nonsinusoidal multiphase systems similarly to Eqs. (3.9)–(3.16).

It is possible to determine the power also from the complex effective values of the voltage and the current ($\mathbf{U}$, $\mathbf{I}$). If the complex conjugate of $\mathbf{I}$ is denoted by $\mathbf{I}^*$, the complex power is

$$\mathbf{S} = \mathbf{U}\mathbf{I}^* = UIe^{j\varphi}\,, \tag{3.18}$$

where U and I are effective values, and φ is the phase angle between $\mathbf{U}$ and $\mathbf{I}$. Using Euler's formula

$$\mathbf{S} = UI(\cos\varphi + j\sin\varphi) = P + jQ\,, \tag{3.19}$$

and

$$P = \mathrm{Re}\mathbf{S}, \qquad Q = \mathrm{Im}\mathbf{S}, \qquad S = |\mathbf{S}|\,. \tag{3.20}$$

3.1.2 Definition of Energy

The electrical energy (work) is the integral of the power vs. time:

$$W = \int_0^{t_1} u(t)i(t)\,\mathrm{d}t = \int_0^{t_1} p(t)\,\mathrm{d}t\,, \tag{3.21}$$

where t_1 is the duration of measurement. This definition is valid in general, and it can be determined for every kind of power defined previously.

3.2 INSTRUMENTS OF POWER MEASUREMENT

3.2.1 Electromechanical Instruments

The electromechanical instrument most often used for the measurement of electrical power is the electrodynamic instrument (either iron-cored or ironless), discussed in Subsection 2.5.1.

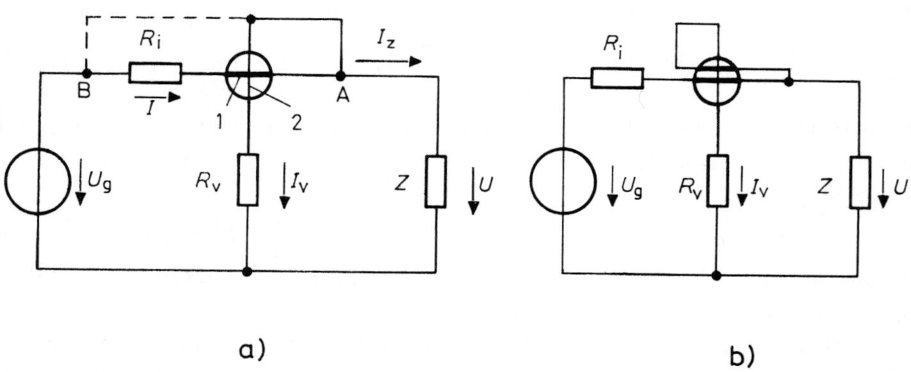

Fig. 3.1 Arrangement of the electrodynamic instrument *a)* basic type, *b)* special construction

Fig. 3.1 shows the circuit diagram of the instrument. The excitation coil (1) is usually called the current coil and the moving coil (2) is the voltage coil. R_i denotes the internal resistance of the current coil and R_v that of the voltage coil plus the series resistance. The deflection torque of the instrument can be derived from the energy stored by the electromagnetic field of the coils. It can easily be shown that the deflection angle (α) of the movement is in the case of ac currents:

$$\alpha = kII_\mathrm{v}\cos\varphi\,,$$

where I and I_v are the rms values of the currents flowing in the coils, φ is the phase angle between I and I_v, and k is a constant, dependent on the construction. Since

$I_v = U/R_v$, the deflection is

$$\alpha = \frac{k}{R_v} IU \cos\varphi = \frac{k}{R_v} P, \tag{3.22}$$

thus it is proportional to the active power P. If the voltage coil is connected to point A (Fig. 3.1), the instrument measures the sum of the powers on the impedance Z and on the voltage coil of the wattmeter. So the power P_Z on the impedance Z is

$$P_Z = P - \frac{U^2}{R_v}. \tag{3.23}$$

If a connection is made to point B, the correction must be calculated with respect to the power consumption of the current coil:

$$P_Z = P - I^2 R_i. \tag{3.24}$$

Fig. 3.1b illustrates a special construction of the electrodynamic instrument, where the measured value is automatically corrected for the power consumption of the voltage coil. The stator has an additional fixed coil which carries a current equal to that of the voltage coil and develops a correcting torque (Drechsler, 1975).

Wattmeters have separate voltage and current pole pairs for connection. When connecting these instruments, attention must be paid to the nearly equipotential connection of the voltage and current coils (Fig. 3.1), so that no large voltage difference between the coils can appear.

Because of the characteristics of the cosine function, the nature of the load impedance (whether it is capacitive or inductive) cannot be determined on the basis of the deflection, but the sign of the deflection will change if the direction of the energy flow changes. Multirange wattmeters are generally equipped with a switch that work in the voltage branch (e. g. it changes the polarity of the winding), thereby making possible the use of the instrument for both directions of the energy flow.

Ironless electrodynamic instruments are manufactured generally for laboratory use in the accuracy classes of 0.1 and 0.2. Iron-cored instruments are available as panelmeters or portable instruments with accuracy classes up to 0.5. In the case of currents larger than 5–50 A, or voltages higher than 600 V, the use of measuring transformers becomes necessary.

3.2.2 Electronic Power Meters

Electronic power meters realize the multiplication and averaging of the voltage and current, belonging to the power to be measured, by means of electronic devices. Electronic devices may have an analog, digital or hybrid construction. This section describes first the analog and later the digital systems.

Power meter with analog multiplier

Fig. 3.2 shows the scheme of an analog wattmeter. In the connection arrangement according to Fig. 3.2a, the analog multiplier, the transfer factor of which is K, multiplies the voltage $u_x(t)$ across the impedance Z and the voltage $u_y(t) = i(t)R$. The output voltage of the multiplier is

$$u_1(t) = K u_x(t) u_y(t) = K R u_x(t) i(t)\,,$$

and the output voltage of the lowpass filter is

$$U_2 = KR\frac{1}{T}\int_0^T u_x(t)i(t)\,\mathrm{d}t = cP\,, \tag{3.25}$$

thus, it is proportional to the power on the impedance Z. If the input range of the analog multiplier is too low, the current and the voltage of the impedance Z are coupled to the multiplier through measuring transformers (Fig 3.2b).

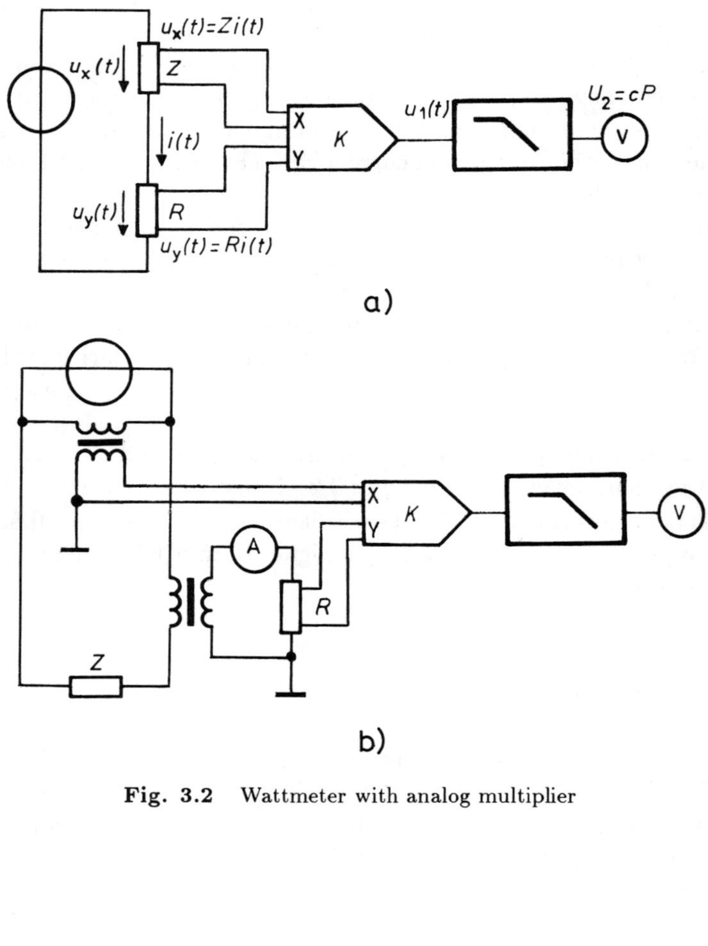

Fig. 3.2 Wattmeter with analog multiplier

Power meter with time division multiplier

As a multiplying circuit of the analog wattmeter, a time division multiplier (TDM) can be used. Figure 3.3 shows the connection diagram of the multiplier.

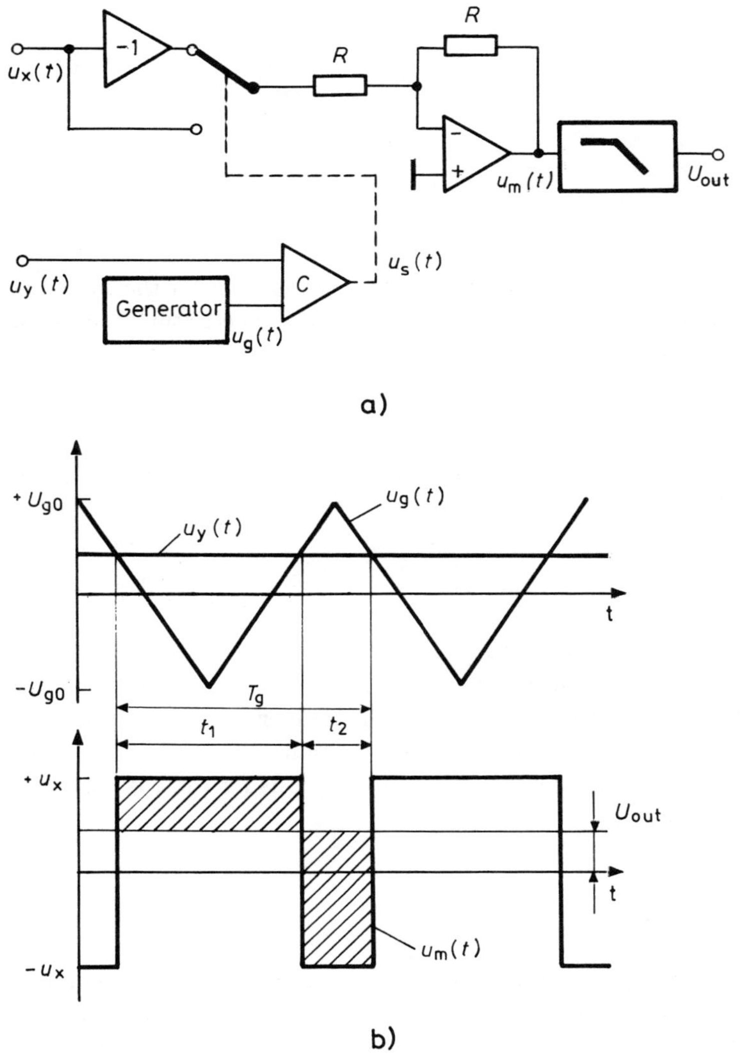

Fig. 3.3 TDM wattmeter *a)* simplified block diagram *b)* time diagram

When using the time division multiplication (TDM) method, a square wave of constant period time $T_g = t_1 + t_2$ is generated, whose amplitude and duty cycle are determined by $u_x(t)$ and $u_y(t)$, respectively (see Fig. 3.3b). The duty cycle is set by the comparison of the voltages $u_g(t)$ and $u_y(t)$. The voltage $u_g(t)$ has a triangular waveform. If T_g is much smaller than the period time of the signal to be measured, the voltages $u_x(t)$ and $u_y(t)$ can be regarded as constant during time T_g. If

$$u_g(t) = \frac{4U_{g0}}{T_g}t \qquad \text{when} \qquad 0 \le t \le \frac{T_g}{4},$$

then

$$t_2 = 2\left(\frac{T_g}{4} - \frac{u_y T_g}{4U_{g0}}\right),$$

and

$$t_1 - t_2 = \frac{T_g}{U_{g0}}u_y .$$

The output voltage $u_m(t)$ of the inverting amplifier is for the time duration t_1 equal to $+u_x$, and for the time duration t_2 to $-u_x$. The output voltage of the lowpass filter is the mean value of $u_m(t)$:

$$U_{out} = \frac{1}{T_g}\int_0^{T_g} u_m(t)\mathrm{d}t = K'\left(\int_0^{t_1} u_x \mathrm{d}t + \int_{t_1}^{t_1+t_2} (-u_x)\mathrm{d}t\right) = K'u_x(t_1 - t_2) = K u_x u_y .$$

In the general case:

$$U_{out} = < K u_x(t) u_y(t) > = KP . \tag{3.26}$$

The high frequency limit is determined by the filter and it must be smaller than half of the frequency of the signal $u_g(t)$. The frequency limit is typically f_c =200 Hz–20 kHz, maybe up to 100 kHz. The uncertainty is 0.01–0.2%. Since it has many advantages, there is a great number of variations for the basic connection as shown by Filipski (1990).

Digital power meter

A possible simplified connection diagram of the digital power meter is shown in Fig. 3.4, where the microprocessor can be replaced by other digital hardware, but in most cases a circuit with microprocessor is used (Hill and Anderson, 1981).

The voltage and current waveforms are sampled at a rate f_s and the integral in Eq. (3.6) is evaluated numerically by summing the discrete sample products over the measurement interval. The discrete equivalent to Eq. (3.6), which is used to calculate the average power, is then

$$P = \frac{1}{N}\sum_{k=1}^{N} u_k i_k , \tag{3.27}$$

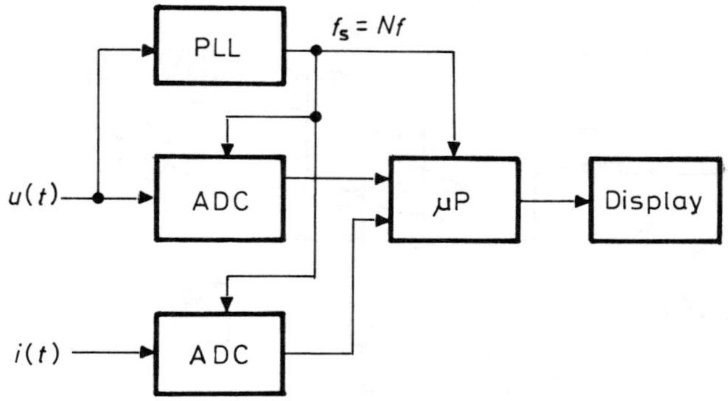

Fig. 3.4 Block diagram of digital wattmeter

where N sample pairs are taken over an integer number of cycles of the voltage waveform. It has been shown that a minimum of 3 sample pairs is required to give an exact computation of power in the case of sinusoidal waveforms, and that these samples may be taken over several cycles provided that the ratio $2M/N$ is not integer, where M represents the number of cycles. In practice, more than 3 samples are needed because of the presence of harmonics.

In the application of digital wattmeters the crucial components are the A/D converters, especially the A/D converter for the current. Suppose that the amplitude of the current (I) changes from 5% to 250% of its nominal value. Let the accuracy of the power measurement be 0.5%, and suppose that $\cos\varphi = 1$. Since the whole domain of the maximum current must be covered by the A/D converter, the following inequality must hold:

$$2.5 \times 2 \times \sqrt{2} I < 2^B q_i \,, \tag{3.28}$$

where B is the number of bits and q_i is the quantum size. The requirement of accuracy can be met if

$$b_i < 1\,\text{LSB} = q_i < 0.005 \times 0.05 I \,. \tag{3.29}$$

From Eqs. (3.28) and (3.29)

$$B > 14.8 \approx 15\,\text{bits}\,.$$

As it has been proved by Kollár (1985), this result is too pessimistic, and if a dither is applied, an 11-bit A/D converter is already appropriate to achieve the above accuracy.

3.2.3 Wattmeter with Hall Device

The Hall device is a thin semiconductor plate. The voltage $u_H(t)$ is proportional to the product of two time-dependent quantities:

$$u_H(t) = c_H i(t) B(t) \,, \tag{3.30}$$

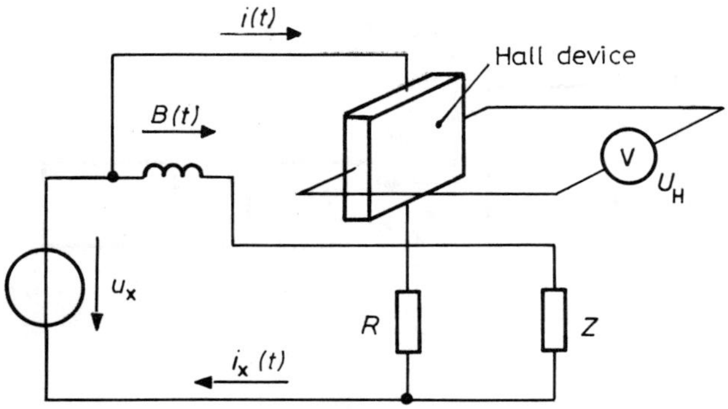

Fig. 3.5 Power measurement by Hall device

where c_H is the Hall constant, B is the magnetic induction and i is the current through the device as shown in Fig. 3.5. The effective power P, to which the deflection of the instrument V is proportional, can be expressed by $u_X(t) = ai(t)$ and $i_X(t)=bB(t)$, where a and b are proportionality factors:

$$P = \frac{1}{T}\int_0^T u_X(t) i_X(t)\,\mathrm{d}t = ab\frac{1}{T}\int_0^T i(t)B(t)\,\mathrm{d}t = \frac{ab}{c_H}U_H\,, \tag{3.31}$$

where U_H is the average of the voltage $u_H(t)$.

The usual construction of the power meter with Hall multiplier can be seen in Fig. 3.5. The magnetic field is proportional to the current to be measured and the optimum of i can be set by the resistance R. In the GHz frequency range the Hall device is placed directly into the waveguide, sensing its magnetic field.

Considering the changes of the Hall constant as a function of time and temperature, the measuring arrangement often has to be calibrated. It can be used in the frequency range from dc to GHz.

3.2.4 Power Measurement by Means of Thermocouples

The basic connection is shown in Fig. 3.6. A and B denote two thermocouples with identical and ideal characteristics ($U=kI^2$). Through the filament of thermocouple A the current $i_A(t)$ is proportional to the current $i(t)$ and the voltage $u(t)$:

$$i_A(t) = ai(t) + bu(t)\,, \tag{3.32}$$

where a and b are proportionality factors. Similarly, the current through thermocouple B is

$$i_B(t) = ai(t) - bu(t)\,. \tag{3.33}$$

It can easily be proved that the instantaneous power can be expressed by

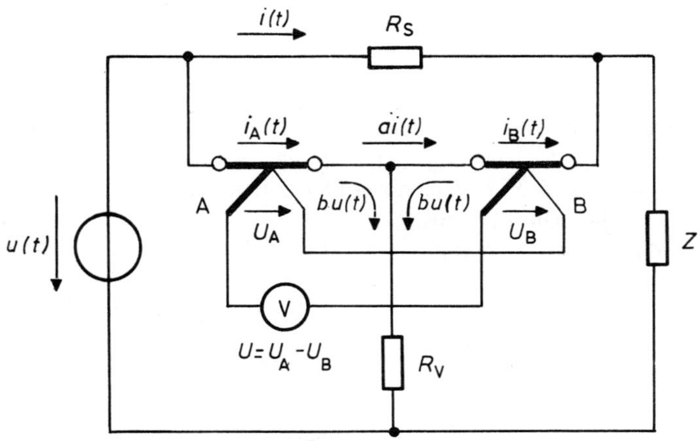

Fig. 3.6 Power measurement with thermocouple

$$u(t)i(t) = \frac{1}{4ab}\left(i_{\mathrm{A}}^2(t) - i_{\mathrm{B}}^2(t)\right). \tag{3.34}$$

Calculating the average:

$$P = \frac{1}{T}\int_0^T u(t)i(t)\,\mathrm{d}t = \frac{1}{4ab}\left(\frac{1}{T}\int_0^T i_{\mathrm{A}}^2(t)\mathrm{d}t - \frac{1}{T}\int_0^T i_{\mathrm{B}}^2(t)\mathrm{d}t\right) = \frac{1}{4ab}(I_{\mathrm{A}}^2 - I_{\mathrm{B}}^2). \tag{3.35}$$

Due to the currents with effective values I_{A} and I_{B}, the thermocouples A and B give the dc voltages U_{A} and U_{B}. Considering the properties of the thermocouples:

$$U_{\mathrm{A}} = kI_{\mathrm{A}}^2; \qquad U_{\mathrm{B}} = kI_{\mathrm{B}}^2,$$

the voltage difference $U_{\mathrm{A}} - U_{\mathrm{B}}$ is proportional to the power P:

$$P = \frac{1}{4abk}(U_{\mathrm{A}} - U_{\mathrm{B}}). \tag{3.36}$$

The power at the resistances R_{S} and R_{V} means systematic error that can be corrected. Instruments working on the previously discussed principle can be used up to the MHz frequencies with an uncertainty of 1%.

Thermal converters have neither perfectly quadratic nor identical characteristics and this leads to inaccuracies. The error can to a great extent be reduced by ensuring identical temperature for the thermocouples. This is done by equalizing the heating powers of the ac currents $i_{\mathrm{A}}(t)$ and $i_{\mathrm{B}}(t)$ by a dc current I, added to the smaller current. It means that the output voltage will be zero. If $I_{\mathrm{A}} > I_{\mathrm{B}}$,

$$< i_{\mathrm{A}}^2 > \; = \; < (i_{\mathrm{B}} + I)^2 > \; = \; < i_{\mathrm{B}}^2 > \; + I^2 .$$

Expressing I^2:

$$I^2 = I_A^2 - I_B^2 = \; < 4abu(t)i(t) > \; = 4abP\,. \tag{3.37}$$

It seems that the added current, I^2, is proportional to the power to be measured.

Using the arrangement of Fig. 3.7, no correction is necessary, and, applying the coupled thermocouples, the measurement accuracy can also be improved as shown in Kusters and Cox (1980). The output voltages of the summing circuits A1 and A2 are:

$$u_{A1} = k\big(u(t) + U_{dc}\big)\,,$$
$$u_{A2} = k\big(u_i(t) + U_{out}\big)\,.$$

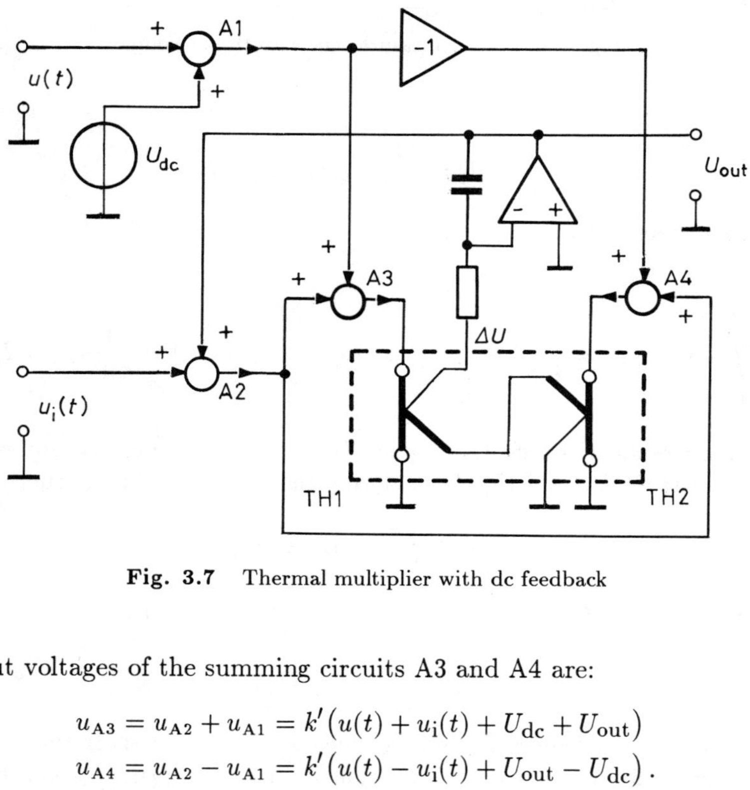

Fig. 3.7 Thermal multiplier with dc feedback

The output voltages of the summing circuits A3 and A4 are:

$$\begin{aligned} u_{A3} &= u_{A2} + u_{A1} = k'\big(u(t) + u_i(t) + U_{dc} + U_{out}\big) \\ u_{A4} &= u_{A2} - u_{A1} = k'\big(u(t) - u_i(t) + U_{out} - U_{dc}\big)\,. \end{aligned} \tag{3.38}$$

These voltages arrive at the differentially connected thermocouples TH1 and TH2. The output voltage (ΔU) goes to the integrator. Its output voltage (U_{out}) is changing in such a way that the two thermovoltages are equal ($\Delta U \to 0$), that is

$$< u_{A3}^2 > \; = \; < u_{A4}^2 > \,. \tag{3.39}$$

After substitution and rearrangement:

$$u(t)u_i(t) + U_{dc}U_{out}\,, \qquad P = cU_{dc}U_{out} = kU_{out}\,, \tag{3.40}$$

where $P = UI\cos\varphi$ is the active power to be measured, and $-cU_{dc}U_{out}$ is proportional to a dc power. So this arrangement realizes an ac/dc transfer of power. If U_{dc} is known, the voltage U_{out} is proportional to the power. The minimal uncertainty is about 10 ppm. The frequency range is limited by the summing circuits (< 1 kHz). The realization of the thermocouples with appropriate characteristics is rather expensive and it can be used economically with precision instruments only.

3.3 MEASUREMENT OF ELECTRICAL POWER

Electrical power measurements have significance at industrial frequencies, radio frequencies and in the microwave frequency range. This chapter mainly discusses the methods applicable at industrial frequencies and the methods that are applicable at other frequencies will be merely referred to.

One of the main problems of power and energy measurement is that the measurements have to be carried out in networks where the loads are nonlinear. Let us think, for example, of electronic power controlled systems, fluorescent lights, ac/dc and dc/ac converters, or switching type power supplies. Because of these reasons the best way to describe the power is to use Eq. (3.9).

To demonstrate the above described problem, let us consider Fig. 3.8.

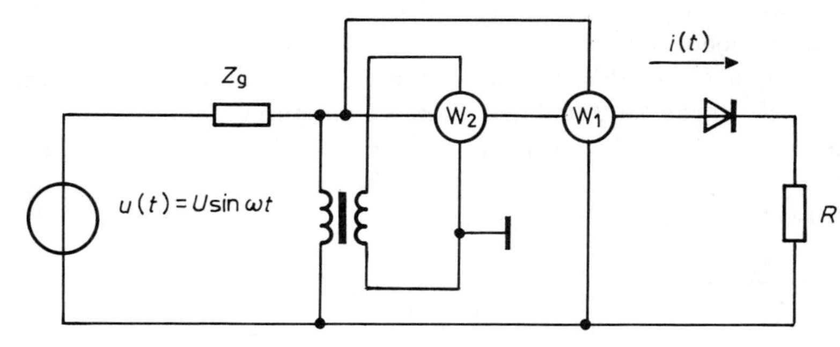

Fig. 3.8 Power measurement in a nonlinear circuit

The applied voltage (U) is a pure sinusoidal voltage with the angular frequency ω, and the current waveform is a half-wave rectified sinewave. The current of the nonlinear circuit causes a voltage drop on both the generator's internal impedance Z_g and the load resistance R. These voltage drops have a dc component. The voltage transformer separates the dc component of the voltage, therefore the deflections of the two wattmeters (W_1, W_2) are different. The deflection of W_1 corresponds to Eq. (3.9), while the deflection of W_2 is proportional to

$$\sum_{k=1}^{\infty} U_k I_k \cos\varphi_k \,.$$

From the foregoing considerations it follows that when measuring power and energy, not only the measuring instruments and methods are to be considered at the evaluation of the measured data but also other characteristics of the network as well. It is reasonable to observe the voltage and the current of the network e. g. by means of an oscilloscope or spectrum analyzer. The measurement accuracy (see the measurement errors to be dealt with later) characterizes the measurement method and instruments for the case only when an ideal network is measured (pure sinusoidal voltage, linear load).

3.3.1 Direct Measurement of Power

The word 'direct' means that the instrument has two inputs (a current and a voltage input) and the output is proportional to the power. The output can be deflection, voltage or some other quantity. The basic connection is shown in Fig. 3.1.

Wattmeters have separate voltage and current connectors. When connecting these instruments, attention must be paid to the nearly equipotential connection of the current input and one end of the voltage inputs.

The output of the instruments is proportional to the effective power, but the load of the inputs is determined by the maximum values of the current and the voltage, independently from the value of the power factor ($\cos\varphi$). The wattmeter indicates zero if U and I are large but $\cos\varphi$ is equal to zero. It is reasonable therefore to supplement the wattmeter connections with at least an ammeter, thereby avoiding the overload of the inputs. The connection, extended by an ammeter and a voltmeter, has the additional advantage that the other characteristic power quantities (Q, S, φ) can be calculated from the measured data.

If the input ranges of wattmeters are too low, they can be extended mainly by instrument transformers. A detailed description of instrument transformers is given in Subsections 2.3.7., 2.3.8. and 2.3.9.

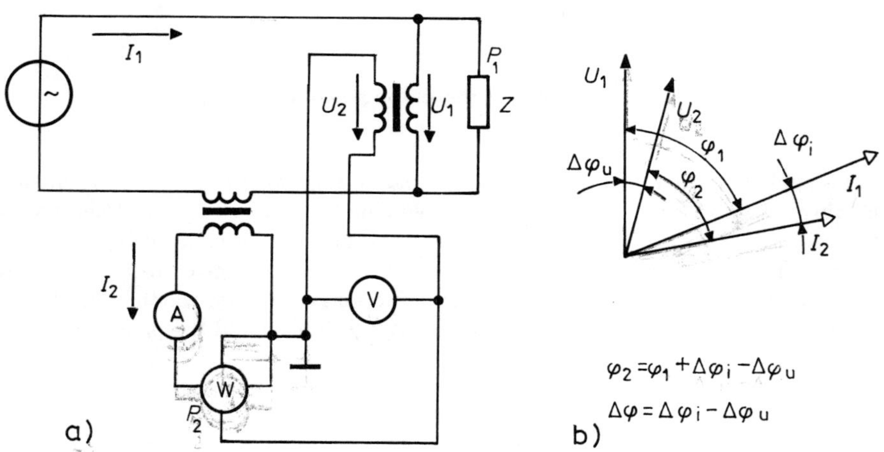

Fig. 3.9 Power measurement with transformers *a)* network *b)* phasor diagram

In Fig. 3.9a the measurement of the power on an impedance Z with the use of instrument transformers can be seen. In the above mentioned subsections amplitude and phase errors of instrument transformers were described. The power on the impedance Z is

$$P_1 = U_1 I_1 \cos\varphi_1 ,$$

but the wattmeter measures the power

$$P_2 = U_2 I_2 \cos\varphi_2 ,$$

even if the systematic errors due to the consumption of the coils are neglected. The relationship of the two powers is:

$$P_1 = \frac{P_2}{a_{\mathrm{u}} a_{\mathrm{i}}} ,$$

where a_{u} and a_{i} are the transfer factors of the voltage and the current transformers, respectively. The relative error caused by the nonideal nature of the instrument transformers is, according to Example 1.4c:

$$\frac{\Delta P_1}{P_1} = \epsilon_2 + \epsilon_{\mathrm{u}} + \epsilon_{\mathrm{i}} - \Delta\varphi \tan\varphi_2 , \tag{3.41}$$

where $\epsilon_2 = \Delta P_2 / P_2$ is the relative error, originating from the inaccuracy of the power meter, measuring P_2; ϵ_{u} and ϵ_{i} are the ratio errors of the voltage and current transformers, respectively; the meaning of $\Delta\varphi$ is demonstrated in Fig. 3.9b. The last term in the error expression indicates that the phase errors of the instrument transformers introduce a larger error in the power measurement the closer the phase angle φ_2 is to $\pi/2$.

Multiphase systems are most frequently four-wire three-phase (with neutral wire) or three-wire three-phase (without neutral wire) systems. Theoretically, the number of wattmeters required to measure correctly the power of multiphase systems is by one less than the number of wires in the system. This statement involves no limitation regarding the waveform, frequency or degree of symmetry in the phase currents. In a four-wire, three-phase system the current coil must be connected in line, either directly or through a current transformer, but the voltage coil may be connected in several different ways. If the neutral wire is available, the connection can be made between the respective line and the neutral wire (Fig. 3.10a).

When the neutral point is not available, a similar set of conditions can be set up by utilizing Y-connected circuits consisting of the three identical impedances of the voltage inputs (Fig. 3.10b).

The three-phase power is:

$$P = P_1 + P_2 + P_3 = U_1 I_1 \cos\varphi_1 + U_2 I_2 \cos\varphi_2 + U_3 I_3 \cos\varphi_3 . \tag{3.42}$$

Generally the summing is made by the instruments (e. g. in the case of electromechanical instruments by a common shaft) and the indication is the three-phase power.

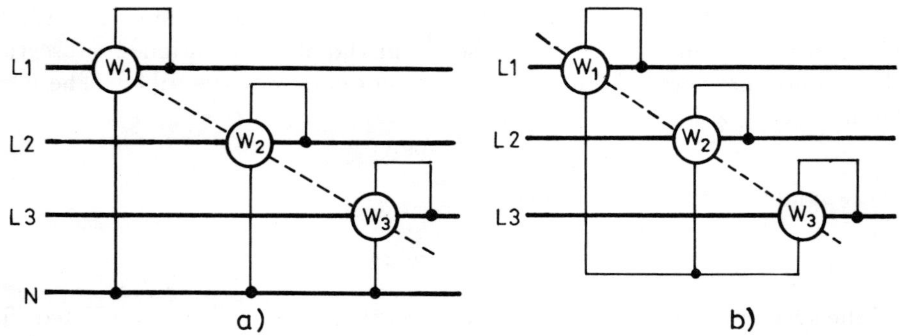

Fig. 3.10 Power measurement in four-wire three-phase systems *a)* with neutral wire *b)* without neutral wire

In three-wire, three-phase systems the connection shown in Fig. 3.11a is used (the so called Aron method), which makes it possible to measure the three-phase power with two wattmeters. To prove this, let us write the complex power of the three phases (see Subsection 3.1.1):

$$\mathbf{S} = \mathbf{U}_1\mathbf{I}_1^* + \mathbf{U}_2\mathbf{I}_2^* + \mathbf{U}_3\mathbf{I}_3^* . \tag{3.43}$$

In three-wire systems without neutral wire, the following equations are valid for the line currents:

$$\mathbf{I}_1 + \mathbf{I}_2 + \mathbf{I}_3 = 0 \qquad \text{and} \qquad \mathbf{I}_1^* + \mathbf{I}_2^* + \mathbf{I}_3^* = 0 . \tag{3.44}$$

Using the second equation

$$\mathbf{I}_3^* = -\mathbf{I}_1^* - \mathbf{I}_2^* ,$$

$$\mathbf{S} = (\mathbf{U}_1 - \mathbf{U}_3)\mathbf{I}_1^* + (\mathbf{U}_2 - \mathbf{U}_3)\mathbf{I}_2^* = \mathbf{U}_{13}\mathbf{I}_1^* + \mathbf{U}_{23}\mathbf{I}_2^* = \mathbf{S}_{\mathrm{I}} + \mathbf{S}_{\mathrm{II}} . \tag{3.45}$$

The two power meters measure the real part of the complex power:

$$\mathrm{Re}\mathbf{S} = P = P_{\mathrm{I}} + P_{\mathrm{II}} . \tag{3.46}$$

Thus, the sum of the two indications gives the three-phase power.

If the loads of the phases are symmetrical,

$$P_{\mathrm{I}} = UI\cos(30^\circ - \varphi) ,$$

and

$$P_{\mathrm{II}} = UI\cos(30^\circ + \varphi) .$$

The three-phase active power is their sum:

$$P_{\mathrm{I}} + P_{\mathrm{II}} = UI\Big(\cos(\varphi - 30^\circ) + \cos(\varphi + 30^\circ)\Big) = \sqrt{3}UI\cos\varphi = P , \tag{3.47}$$

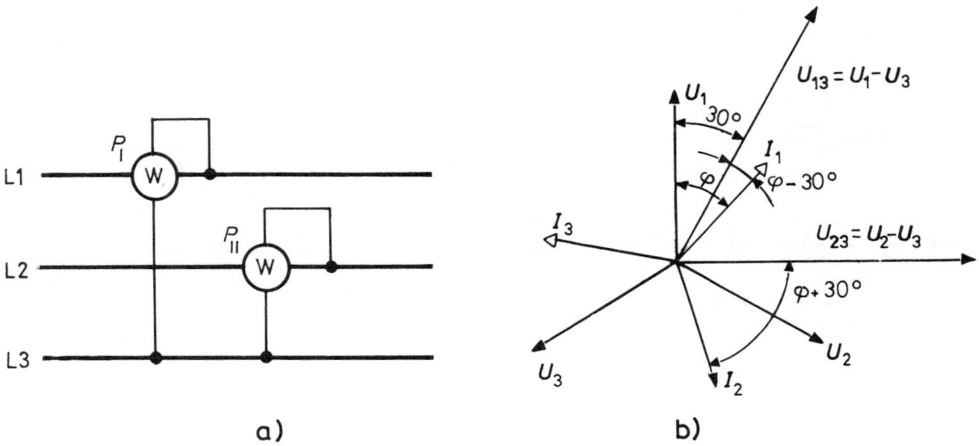

Fig. 3.11 Power measurement in three-wire, three-phase systems *a)* the network *b)* phasor diagram

and their difference is the reactive power:

$$P_{\text{I}} - P_{\text{II}} = UI\Big(\cos(\varphi + 30°) - \cos(\varphi - 30°)\Big) = UI \sin\varphi = Q/\sqrt{3}\,, \tag{3.48}$$

and

$$\tan\varphi = \sqrt{3}\frac{P_{\text{II}} - P_{\text{I}}}{P_{\text{II}} + P_{\text{I}}}\,. \tag{3.49}$$

3.3.2 Indirect Methods of Power Measurement

A method is indirect if the active power must be calculated from the readings of the instruments.

The three-voltmeter method for the measurement of power on an impedance Z can be used if a known resistance R is used. The measuring circuit and the phasor diagram are shown in Fig. 3.12. If the consumption of the voltmeter is negligible, the power can be derived by means of the phasor diagram as follows:

$$U^2 = U_{\text{R}}^2 + U_{\text{Z}}^2 + 2RIU_Z \cos\varphi_{\text{Z}}\,. \tag{3.50}$$

The active power is:

$$P_{\text{Z}} = IU_{\text{Z}} \cos\varphi_{\text{Z}} = \frac{U^2 - (U_{\text{R}}^2 + U_{\text{Z}}^2)}{2R}\,. \tag{3.51}$$

The apparent power is:

$$S = \frac{U_{\text{Z}}U_{\text{R}}}{R}\,, \tag{3.52}$$

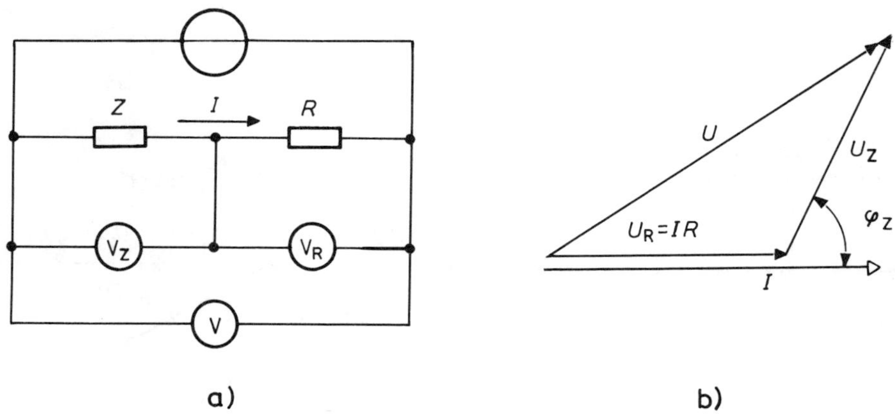

Fig. 3.12 Power measurement with three voltmeters *a)* measuring circuit *b)* phasor diagram

and the power factor is:

$$\cos\varphi_Z = \frac{U^2 - U_R^2 - U_Z^2}{2U_Z U_R}. \tag{3.53}$$

The accuracy of this method is limited by the voltage measurement. In the case of true rms meters it can be used generally. Analyzing the relationship (3.51), it can be seen that the choice $U_Z \approx U_R$ means that $|Z| \approx R$, and the error is minimum. The error approaches infinity in the case of $\varphi \to \pi/2$.

If the power dissipated on a known resistance R has to be measured, it can be determined from the voltage measured across the resistance, according to the relationship:

$$P_R = \frac{U^2}{R}. \tag{3.54}$$

If the power on a resistance, across which a dc voltage can be measured, has to be determined, one of the connections on Fig. 3.13 can be applied considering the achievable minimum of the systematic error caused by the consumption of the measuring instrument.

3.3.3 Measurement of Reactive Power

The definition of the reactive power is only precise for pure sinusoidal and symmetrical networks, therefore, in general, it can be calculated by Eq. (3.8):

$$Q = \sqrt{S^2 - P^2} = \sqrt{(UI)^2 - P^2}, \tag{3.55}$$

from the data of the voltage, current, and power measurements. In the case of sinusoidal, linear networks the definition formula is given by Eq. (3.7):

$$Q = UI\sin\varphi = UI\cos(\varphi \pm 90°). \tag{3.56}$$

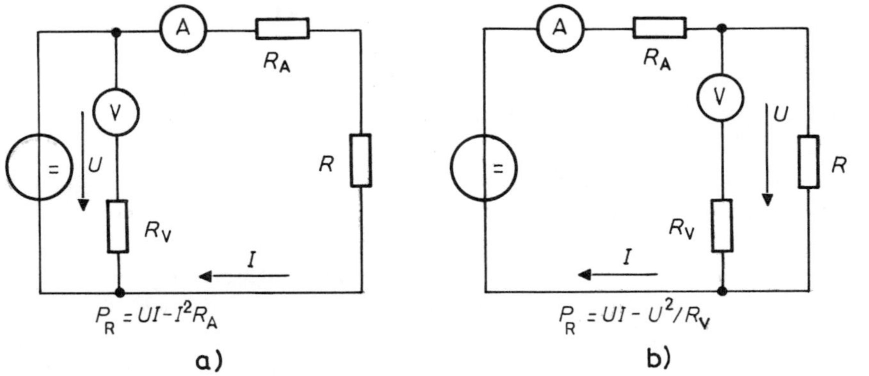

Fig. 3.13 Power measurement with V-A instruments *a)* with unbiased current measurement *b)* with unbiased voltage measurement

The reactive power can be measured with the same instruments and measurement arrangements as the active power. The networks realizing the phase shift of 90° can be of different kinds. For example, in the case of electromechanical wattmeters (Fig. 3.1), instead of the series resistance (R_V) an inductance is used in the circuit of the voltage coil to convert the voltage into current.

In three-phase symmetrical networks two voltages with a phase-shift of 90° can always be found (Fig. 3.14b). The measuring connections, constructed on this basis, are shown in Fig. 3.14a. In the three-phase case the relationship

$$Q = \frac{W_1 + W_2 + W_3}{\sqrt{3}}, \tag{3.57}$$

can be used for the calculation of the reactive power in accordance with Eq. (3.12).

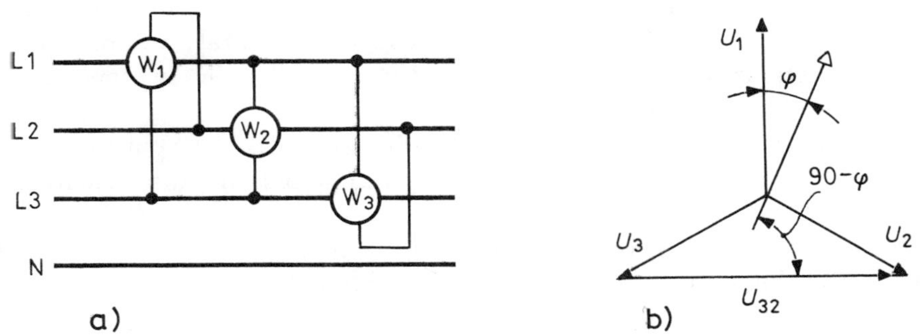

Fig. 3.14 Reactive power measurement in three-phase systems *a)* arrangement *b)* phasor diagram

3.4 MEASUREMENT OF ELECTRICAL ENERGY

3.4.1 Energy Measuring Instruments

According to Eq. (3.21), the electrical energy is given by the time integral of the instantaneous power. Therefore the energy measuring devices are such power meters that are able to integrate the power, too.

Energy measurement can most easily be made with those power meters, the output of which is an electrical signal, proportional to the power. Such are the power meters with thermocouple, the Hall device or the electronic power meters. Here the integration can be made by a voltage-to-frequency converter and counter (Fig. 3.15). At industrial frequency the electronic power meter with Time-Division Multiplier (TDM, Fig. 3.3) has the greatest significance. The output of the VFC converter drives the mechanical counter by means of a stepping-motor. The minimum measurement uncertainty is about 0.05%. It is generally used with external current and voltage transformers in large systems.

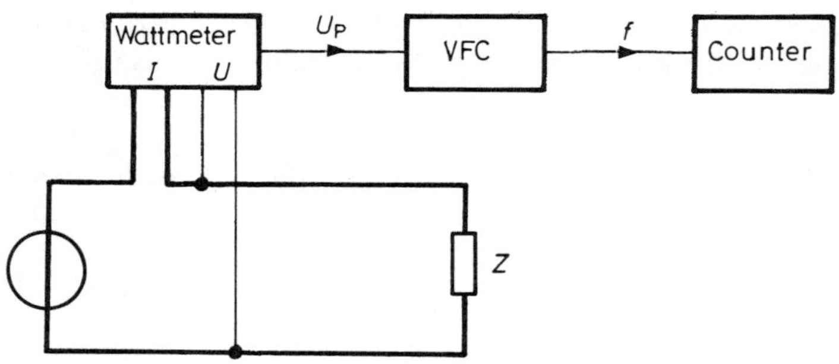

Fig. 3.15 Energy measuring device

The electromechanical watt-hour meters can be regarded as electrical machines. In the dc watt-hour meter the load current flows through the excitation coil, while into the armature, via the commutator, a current proportional to the voltage is led. On the shaft of the motor an eddy-current brake is mounted. The number of rotations is counted by a counter, the count of which is proportional to the consumed electrical energy.

In ac networks the induction type (Ferraris system) watt-hour meters, that are essentially two-phase asynchronous motors with aluminium disc armature, are used (Fig. 3.16).

In the instrument there are two excited magnetic circuits which are closely interleaved by the use of a composite core carrying both the voltage and current coils. The separation between the poles ensures a close interaction between the fluxes $\phi_1(I_1)$ and $\phi_2(I_2)$ and the induced currents I_3 and I_4, and results in an efficient torque-producing system. The flux ϕ_1 is assumed to be in phase with I_1; the induced e.m.f. lags 90° on ϕ_1 and circulates the disc current I_3. I_4 has

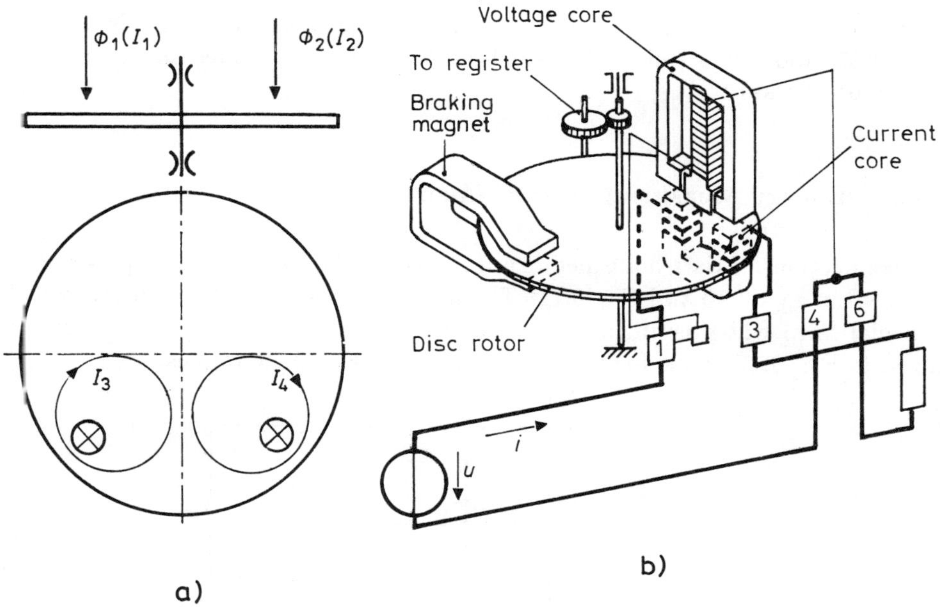

Fig. 3.16 Induction type watt-hour meter *a)* operating principle *b)* diagrammatic representation

a corresponding relationship to ϕ_2. Important requirements are the symmetrical structure and the independent development of the eddy currents (I_3, I_4). If it is true, the value of the rotating torque is

$$M_{\mathrm{r}} = c\omega I_1 I_2 \sin\beta ,$$

where β is the phase angle between I_1 and I_2. If $I_1 \sin\beta = k_1 U \cos\varphi$ (the coil I_1 has a large number of windings) and $I_2 = k_2 I$, it can be written:

$$M_{\mathrm{r}} = c\omega k_2 k_1 U I \cos\varphi = kP .$$

In the induction type kilowatt-hour meter, similarly to the dc energy meter, a brake-magnet is used. On its effect a retaining torque develops which is proportional to $\mathrm{d}\alpha/\mathrm{d}t$, the angular speed, thereby resulting in continuous rotation of the disc. In the time interval (t_1, t_2) the change of the count (n) of the counter is proportional to the electrical work:

$$n = \frac{1}{2\pi} \int_{t_1}^{t_2} a \frac{\mathrm{d}\alpha}{\mathrm{d}t} \mathrm{d}t = \int_{t_1}^{t_2} bP \,\mathrm{d}t = bW , \tag{3.58}$$

where a and b are proportionality factors.

In Eq. (3.58), describing the operation of the induction type watt-hour meters, a lot was neglected. For this reason, to ensure the correct operation, correcting windings and magnetic shunts are used. The current coil is not sensitive to overloads. The presently manufactured types can tolerate overloads as high as 600–700% (about 60 A) in the current branch without exceeding the error limits of the accuracy class.

3.4.2 Energy Measuring Arrangements

The connections of watt-hour meters are similar to the connections of power meters as shown in Fig. 3.10 and Fig. 3.11. The input arrangement of the instrument has a special form which is an international standard. For single-phase systems it is shown in Fig. 3.17, while for the three-phase case Fig. 3.18 applies.

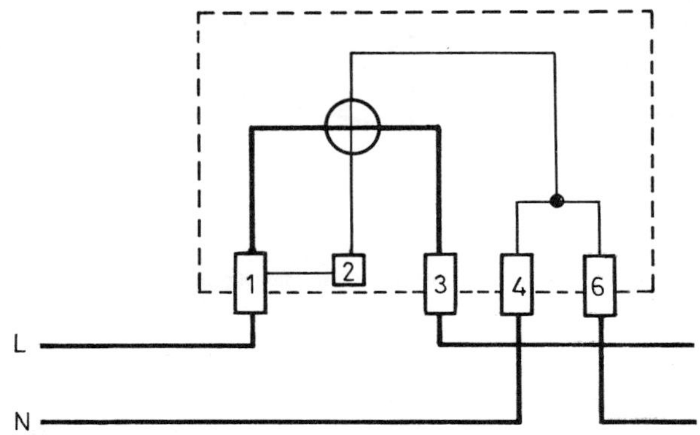

Fig. 3.17 Connection of a single-phase watt-hour meter

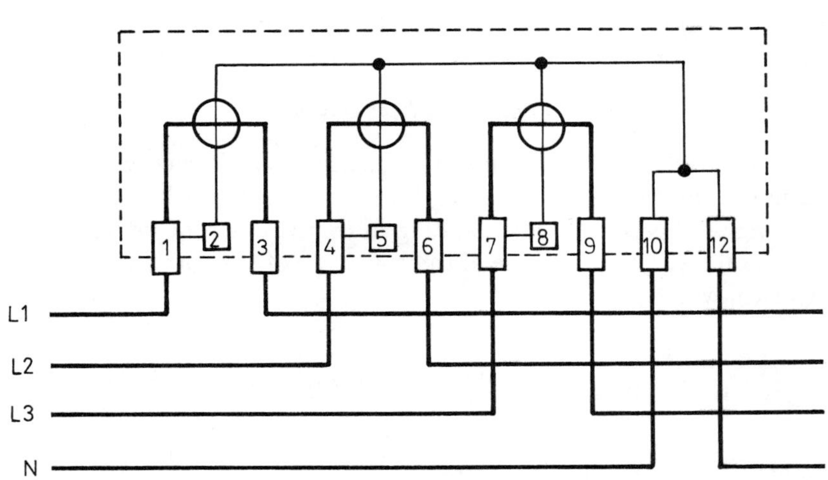

Fig. 3.18 Connection of a four-wire three-phase watt-hour meter

The error of kilowatt-hour meters is always calculated by dividing by the measured value because it has no 'full-scale' deflection. The error is calculated as

$$\frac{\text{actual reading} - \text{true Wh}}{\text{true Wh}} \times 100\% .$$

The error depends on the loading conditions. The class of accuracy of commercial watt-hour meters is 0.5%, 1% or 2%. Recommendations for watt-hour meters are to be found in IEC (1988). The application-specific consumption meters (maximum value meters, meters with double tariff, etc.) are constructed without changes in the basic measuring connections.

References

Drechsler, R. (1975): Messung elektrischer Energie (Measurement of Electrical Energy). Berlin, VEB Verlag Technik.

Enslin, H. R. and van Wyk, J. D. (1988): Measurement and Compensation of Fictitious Power Under Nonsinusoidal Voltage and Current Conditions. *IEEE Trans. on Instrumentation and Measurement*, Vol. IM–37, No. 3, pp. 403–408.

Filipski, P. S. (1990): A TDM Wattmeter with 0.5 MHz Carrier Frequency. *IEEE Trans. on Instrumentation and Measurement*, Vol. IM–39, No. 1, pp. 15–18.

Hill, J. J. and Alderson, W. E. (1981): Design of a Microprocessor Based Digital Wattmeter. *IEEE Trans. on Industrial Electronics and Control Instrumentation*, Vol. 28, No. 3, pp. 180-184, Aug. 1981.

IEC (1988): IEC Standard Publication 521. 2nd edition. Genève, BCCEI.

Kollár, I. (1985): Errors Introduced by Digital Power/Energy Measurements. 10th IMEKO World Congress, Praha, April 22–26, 1985. *Proceedings: Acta IMEKO 1985*, Vol. V, pp 343–349.

Kusters, N. L. and Cox, L. G. (1980): The Development of an Automatic-Reversing Differential Thermal Wattmeter. *IEEE Trans. on Instrumentation and Measurement*, Vol. IM–29. No. 4, pp. 426–431.

Sydenham, P. H. ed. (1982): Handbook of Measurement Science I-II. New York, John Wiley & Sons.

4

Measurement of Impedances

P. Osváth

Editorial introduction

The methods of impedance measurement are inherently more complex than the methods of voltage and current measurement, because the impedance is the ratio of these two quantities, and because impedance measurement requires modelling of a physical object.

Bridge circuits are the classical methods for impedance measurement. During the past 40–50 years high-permeability ferromagnetic materials have been increasingly used in these circuits, in the form of inductive dividers, ratio transformers and current comparators. Using these components it has been possible to increase measurement accuracy by 1–2 orders of magnitude.

This chapter deals with impedance measurement circuits, the effect of measurement leads and the errors due to different stray impedances. Methods to eliminate these effects are also presented.

4.1 PRINCIPLES OF IMPEDANCE MEASUREMENT

The concept of impedance refers to a special model of physical objects. It is usually assumed that when a sinusoidal voltage is applied across an impedance, the current will also be sinusoidal with the same frequency. The ratio of the voltage and current amplitudes gives the absolute value of the impedance at the given frequency, and the phase shift measurable from the current to the voltage gives the phase of it.

There are three basic types of impedance: resistance, capacitance and inductance, with phases of 0°, –90° and +90°, respectively. Resistors, capacitors and inductors can be reasonably well modelled in a certain frequency band by 'clean' impedances, as

$$Z = R, \qquad Z = \frac{1}{j\omega C}, \qquad \text{and} \qquad Z = j\omega L,$$

respectively.

Technology of Electrical Measurements, Edited by L. Schnell
© 1993 John Wiley & Sons Ltd.

However, when using the above models for the description of physical objects, it often turns out that they are not capable to describe the voltage-current relationship with a sufficient accuracy for different frequencies. The reason is that in real objects rather complex physical phenomena occur, and models can only account for the selected properties (see Chapter 1).

As was discussed in Subsection 2.3.2, more realistic models of resistors, capacitors and inductors consist of a few 'clean' impedances. Such a model is valid in a much wider frequency band than the simplest one, and more or less well describes the effects of a few physical phenomena associated with the geometry of the leads and the behaviour of electromagnetic field in materials. Such models can easily describe phase shifts between $\pm 90°$, changing with the frequency.

Impedance measuring instruments usually measure elements of a two-component (parallel or serial) model of an impedance, at a selected frequency.

For a good modelling of electrical two-poles, models of higher complexity may often be necessary. Such models can be measured by the use of impedance analyzers (see Section 4.5).

It must be emphasized that measurement can be understood only within the frame of the modelling procedure. Neglecting significant phenomena will result in erroneous measurement. Nonlinearities (for example the ones due to saturation) can cause very different measured parameter values at different excitation levels; measurement at different levels is also the means for the detection of nonlinearities. Even at a few hundred volts, partial discharges can occur in the insulating materials, resulting in erroneous measured impedance values. Polarisation causes an increase of the capacitance in certain frequency bands, while its effect may be negligible in a given frequency range.

Therefore, each measurement must be preceded by careful consideration of the validity of the model, and during measurements some validity checks should also be performed. Moreover, it is more correct to talk about the measurement of the *parameters of a model*, and not about the measurement of, let us say, a 'real capacitor'.

Having obtained a physically reasonable model, the determination of the parameter values may follow. This step is called measurement in the everyday speech.

The straightforward way of measuring the value of an impedance is to measure the voltage on its terminal and the current flowing through it, and to divide the voltage by the current. In the practice of high-accuracy measurements, voltage and current measurements are accomplished by comparisons. The reference for comparison is the voltage on a known impedance or the current flowing through it.

4.1.1 Measurement by Voltage Comparison

A current I_1 is forced through the unknown impedance Z_X, and another current I_2 through the standard impedance Z_N, as shown in Fig. 4.1.

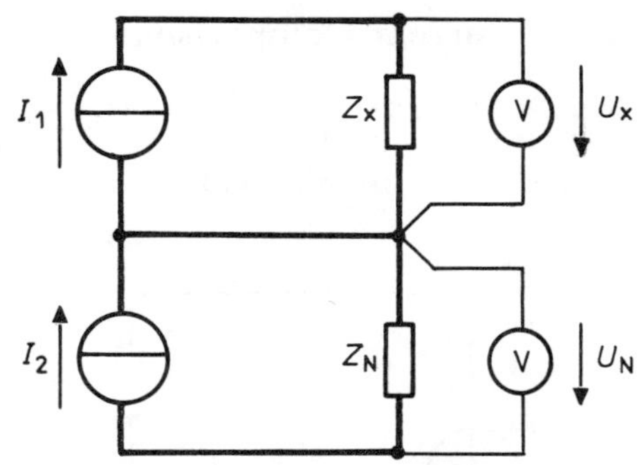

Fig. 4.1 Measurement by voltage comparison

From the voltage U_X, measured across Z_X, and the voltage U_N, measured across Z_N, and from the ratio of the currents I_1 and I_2, the value of Z_X is calculable:

$$Z_X = \frac{I_2 U_X}{I_1 U_N} Z_N . \tag{4.1}$$

It can be seen from (4.1) that it is enough to generate currents of a known ratio instead of two currents of known values. A simple case is shown in Fig. 4.2, where $I_1 = I_2$, that is, the two impedances are connected in series. The calculated value of Z_X is

$$Z_X = \frac{U_X}{U_N} Z_N . \tag{4.2}$$

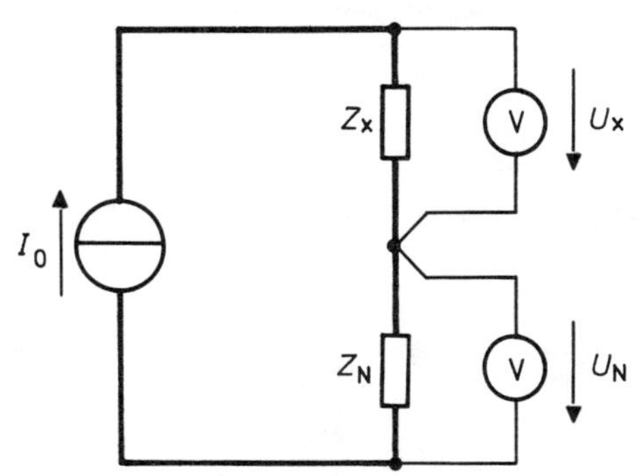

Fig. 4.2 Simplified voltage comparison method

Note that it is not necessary to know the values of the voltages U_X and U_N, only their ratio is important. Therefore, when using the principle of voltage comparison, complex voltage ratios have to be measured. This solution will be analyzed in Subsection 4.3.2.

4.1.2 Measurement by Current Comparison

A voltage of value U_1 is applied at the terminals of the impedance Z_X, and another voltage of value U_2 is applied at the terminals of Z_N, as shown in Fig. 4.3.

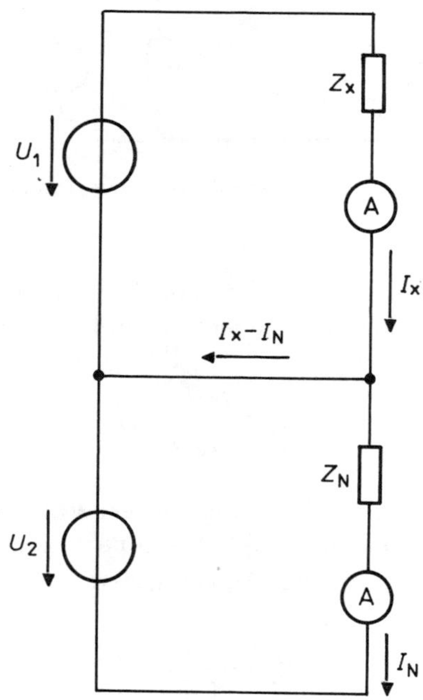

Fig. 4.3 Measurement by current comparison

Z_X is calculable from the currents flowing through Z_X and Z_N and from the known voltage values:

$$Z_X = \frac{U_1 I_N}{U_2 I_X} Z_N . \tag{4.3}$$

Only the ratio of the applied voltages has to be known, and only the ratio of the currents has to be measured. In a simpler case $U_1 = U_2$, that is, the two impedances are connected in parallel, and only the current ratio is to be measured (Fig. 4.4):

$$Z_X = \frac{I_N}{I_X} Z_N .$$

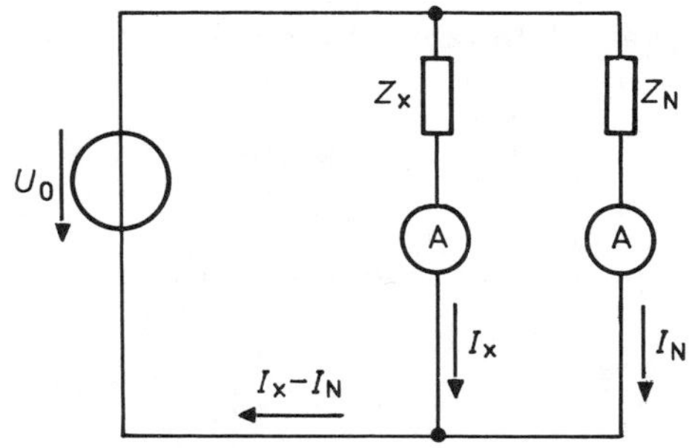

Fig. 4.4 Simplified current comparison method

4.2 QUALITATIVE PROPERTIES OF IMPEDANCE MEASUREMENT CIRCUITS

4.2.1 Errors

Some aspects of the errors, relevant to the following considerations, will be treated here, a more detailed discussion is to be found in Subsections 1.4–1.7.

Definitions of errors

Let us consider the definition of the error in connection with an impedance. The error is the difference of the measured value Z_m, and the true value Z_t:

$$\Delta Z = Z_m - Z_t ,$$

where $Z_m = R_m + jX_m$ and $Z_t = R_t + jX_t$.

Since the impedances are complex, the error ΔZ will also be complex:

$$\Delta Z = \Delta R + j\Delta X . \tag{4.4}$$

Impedances can be expressed by their absolute value and phase angle:

$$Z_m = |Z_m| e^{j\varphi_m} \quad \text{and} \quad Z_t = |Z_t| e^{j\varphi_t} . \tag{4.5}$$

Using this form, the error of the absolute value and the phase of the impedance are

$$\Delta|Z| = |Z_m| - |Z_t| \tag{4.6}$$

and

$$\delta = \varphi_m - \varphi_t . \tag{4.7}$$

The relative error of the absolute value of the impedance is

$$\epsilon_{\mathrm{r}} = \frac{\Delta|Z|}{|Z_{\mathrm{t}}|} \approx \frac{\Delta|Z|}{|Z_{\mathrm{m}}|}\,.$$

The relative value of the phase error is not used because the phase itself is a relative parameter.

Types of error

Absolute errors, relative errors and errors constant along the scale or the measurement range can be defined.

- The dimension of the absolute error is the same as that of the parameter to be measured (see Eq. 4.4), and the attribute *absolute* indicates only that not the relative error is meant. This kind of error must not be confused with the absolute value $|\Delta Z|$.
  Let us consider the error definition $\Delta R = R_{\mathrm{m}} - R_{\mathrm{t}}$. If the value and the sign of ΔR are known, the measured value can be corrected (at least in theory):

$$\widehat{R}_{\mathrm{t}} = R_{\mathrm{m}} - \Delta R\,, \tag{4.8}$$

  where $\widehat{R}_{\mathrm{t}}$ is the estimator of the true value, and ΔR is the correction. In this case $\widehat{R}_{\mathrm{t}}$ is a so-called *point estimator.*
  If a random error is faced, ΔR is not a given number, but an interval $\pm\Delta R$. In this case the expression $R_{\mathrm{m}} \pm \Delta R$ denotes a region containing the true value R_{t} with a given probability, and $\pm\Delta R$ is an *interval estimator.*
- The relative error is the ratio of the absolute error and the true value, which can be approximated by the ratio of the absolute error and the measured value:

$$\epsilon_{\mathrm{r}} = \frac{\Delta R}{R_t} \approx \frac{\Delta R}{R_{\mathrm{m}}}\,. \tag{4.9}$$

  The relative error is usually expressed in per cent, per thousand or parts per million (ppm). This error is due to e. g. the uncertainties of the standard impedances.
- In the case of the constant absolute error along the scale (i. e. in the measurement range), the value ΔR (the absolute error) is constant within the entire measurement range. Such an error can be caused by the finite sensitivity of a null detector or by other effects, introducing a dead zone (e. g. noise).
  The relative error, calculated as $\Delta R/R_{\mathrm{m}}$, increases when R_{m} decreases. Thus, the relative error may appear as a constant $\Delta R/R_{\mathrm{m}}$ along the scale, but it may also appear as a constant ΔR, which means that the relative error is changing with changing R_{m}.
  For a well specified instrument both types of error are given.

Example 4.1

A milliohmmeter measures up to $1\,\Omega$. The catalogue data of the instrument are:

Constant error along the scale: $\Delta R = 0.1\,\mathrm{m}\Omega$
Relative error, with respect to the measured value: 1%
Error limit: $\epsilon = \pm(0.1\,\mathrm{m}\Omega + 1\%)$.

When a resistance of $1\,\Omega$ is measured:

$$\epsilon \approx \pm 10.1\,\mathrm{m}\Omega\,,$$

$$\epsilon_{\mathrm{r}} \approx \pm 1\%\,.$$

When $1\,\mathrm{m}\Omega$ is measured with the same instrument:

$$\epsilon \approx \pm 0.11\,\mathrm{m}\Omega\,,$$

$$\epsilon_{\mathrm{r}} \approx \pm 10\%\,.$$

Frequency dependence of errors

An instrument, measuring complex impedances, is expected to give correct values at all measurement frequencies. If the main parameters of the impedances (R, L, C) have to be measured, these results are expected to be independent of frequency. However, all measuring circuits give frequency dependent results. Some circuits can work only at a given frequency (resonant circuits), others are theoretically frequency independent, but in practice they have frequency dependent errors.

The main sources of frequency dependent error are the standard impedances i. e. the comparison units in the measurement circuit. A theoretically 'pure' resistor has stray inductances and capacitances, and the realized inductances and capacitances have secondary effects, too (see Chapter 2). If these effects are not known, the measured value of the impedance, computed from the nominal values of the standards, will be erroneous. In a classical bridge circuit, the effect of stray impedances of the standards can be reduced by a special compensation technology (Tóth, 1966).

The measured value of Z_{x} is generally a function of the values of standard impedances and their stray impedances:

$$Z_{\mathrm{x}} = Z_{\mathrm{x}}(Z_{\mathrm{N1}}, \ldots, Z_{\mathrm{stray1}}, \ldots)\,. \tag{4.10}$$

A possibility for decreasing frequency dependence is the maximally flat compensation. The derivative of Z_{x} with respect to the frequency f is:

$$\frac{\partial Z_{\mathrm{x}}}{\partial f} = a_0 + a_1 f + a_2 f^2 + \ldots\,. \tag{4.11}$$

By setting $a_0 \ldots a_k = 0$, the maximally flat compensation is realized.

Other sources of frequency dependence are located in the structure of the measuring circuit. The connection leads between the standard impedances and ratio circuitry have stray inductances and capacitances, and the leads or cables connecting the impedance Z_x to the measuring instrument are sources of errors, too. The effect of stray impedances of internal interconnections will be treated in Section 4.3. The effect of connecting leads is the subject of the following part of this section.

The effects of connecting leads on the error

The impedance to be measured is connected to the measuring instrument by leads. Zoltán (1977) analyzed the problems arising in detail. The simplest form of connection is called *two-wire measurement.*

In Fig. 4.5 the impedance Z_x is connected to the measuring instrument by a coaxial cable. The measurement is performed using a voltmeter and an ammeter.

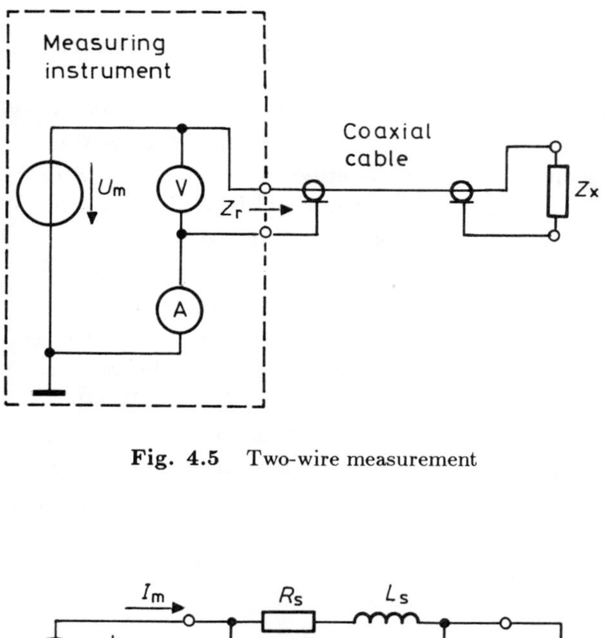

Fig. 4.5 Two-wire measurement

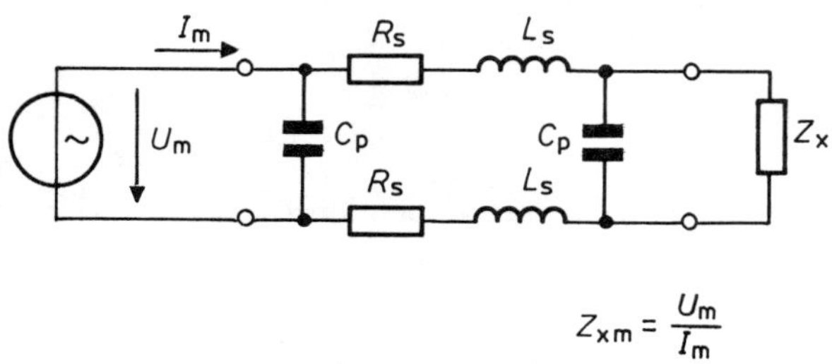

Fig. 4.6 Lumped parameter equivalent of the two-wire measurement

The instrument measures the resultant impedance Z_r of the ensemble of Z_x and the coaxial cable. The coaxial cable is described by a lumped parameter circuit shown in Fig. 4.6. The capacitors C_p each represent half of the cable capacitance, the inductances L_s half of the cable inductance, and the resistances R_s half of the sum of wire and shield (screen) resistances.

It is simple to derive the lumped parameter equivalent of a cable from the transmission parameters of it. The resulting impedance of the cable and the Z_x impedance:

$$Z_r = \frac{Z_x \cosh(\gamma l) + Z_0 \sinh(\gamma l)}{\frac{Z_x}{Z_0} \sinh(\gamma l) + \cosh(\gamma l)},$$

where

$$\gamma = \sqrt{(R+sL)(G+sC)}; \qquad Z_0 = \sqrt{\frac{(R+sL)}{(G+sC)}},$$

and R is the specific resistance of the cable in Ω/m, G is the specific conductance in S/m, L is the specific inductance in H/m and C is the specific capacitance in F/m; l is the length of the cable, γ is the attenuation coefficient, and Z_0 is the wave impedance.

Using the first two members of the Taylor series of γ:

$$Z_r = \frac{Z_x + Z_x\frac{ab}{2} + a}{Z_x b + 1 + \frac{ab}{2}},$$

where

$$a = l(R+sL), \qquad b = l(G+sC).$$

The input impedance of the R-L-C circuit shown in Fig. 4.6 is:

$$Z_r' = \frac{Z_x + Z_x 2\xi\eta + \xi}{Z_x 2\eta(1+\frac{\xi\eta}{2}) + 1 + 2\xi\eta} = \frac{Z_x + Z_x\frac{\xi 4\eta}{2} + \xi}{Z_x\frac{4\eta}{2} + 1 + \frac{\xi 4\eta}{2} + Z_x\xi\eta^2},$$

where

$$\xi = R_s + sL_s, \qquad \eta = sC_p.$$

From the equivalence: $R_s = lR/2$, $L_s = lL/2$, $C_p = lC/2$. The terms which mean the differences between the two expressions are negligible (error<100 ppm) up to 10 MHz.

Example 4.2

A coaxial cable with 50 Ω wave impedance of type RG 58 C/U has the following parameters:

$C = 110$ pF/m; $L = 0.238\,\mu$H/m; $R = 36.2$ mΩ/m.

With these values, for a 1 m long cable:

$C_p = 55$ pF; $L_s = 0.476\,\mu$H; $R_s = 18.1$ mΩ.

The additive term in the denominator reaches the value 1 at 32.5 MHz. This means that the lumped parameter approximation can be used up to 1 MHz, i. e. in the range conforming to the scope of this book.

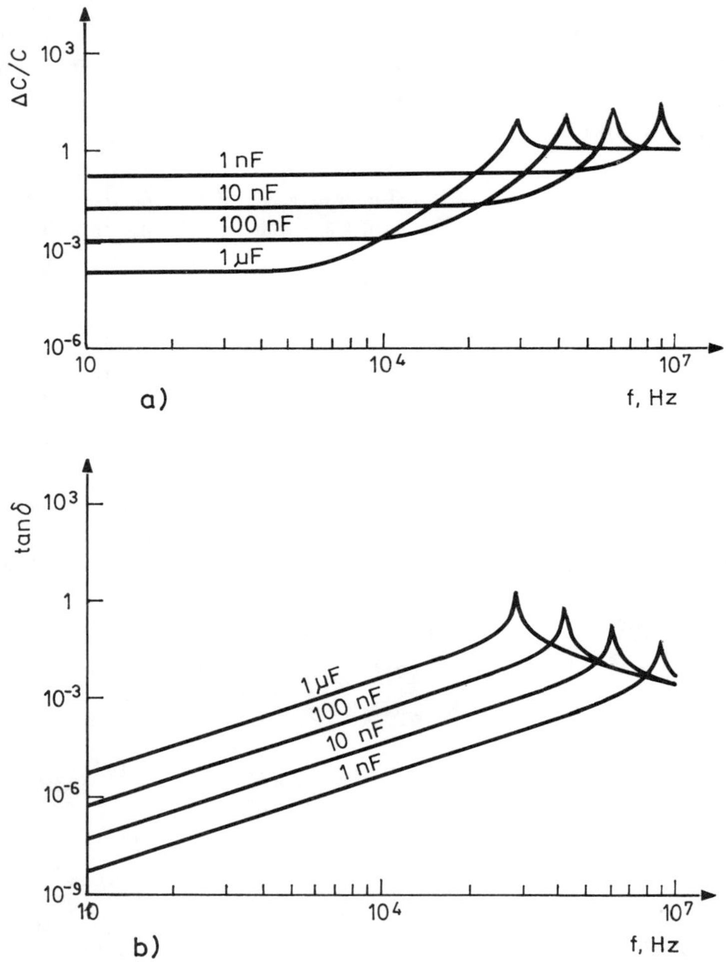

Fig. 4.7 Frequency dependence of the errors of a two-wire measurement *a)* error of the measured capacitance value *b)* error of the measured tanδ value

The frequency dependence of the errors of capacitance and tanδ (loss tangent) measurements is shown in Fig. 4.7.

The errors were calculated for a 1 m long RG 58 C/U measurement cable, for the case of a test object of 1, 10, 100 nF and 1 μF. The value of tanδ of the test object was supposed to be zero. The relative error of the capacitance measurement is constant at frequencies lower than 5 kHz, and it is inversely proportional to the capacitance value. At higher frequencies the inductance of the cable causes an increasing error. The tanδ error increases with increasing frequency and with the capacitance value, owing to the resistance of cable.

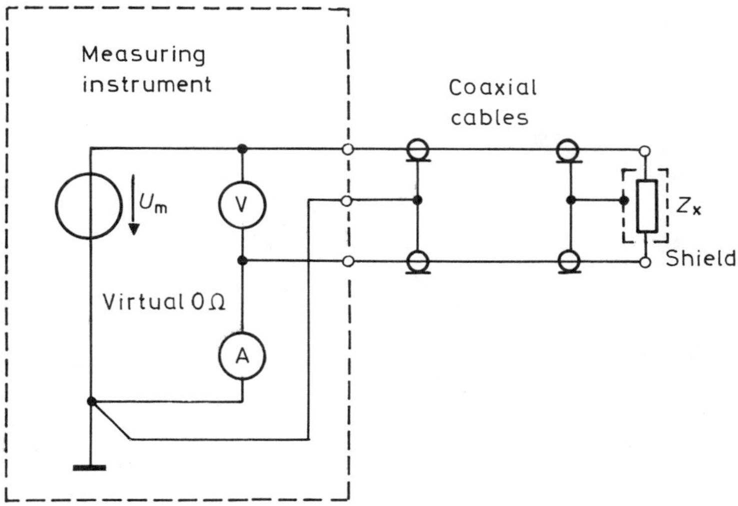

Fig. 4.8 Three-wire measurement

The *three-wire measurement* reduces the effects of cable capacitances. In the circuit shown in Fig. 4.8 the shields of the two connection cables are connected to the neutral point of the measurement circuit.

In Fig. 4.9 the lumped parameter equivalent of three-wire measurement is shown.

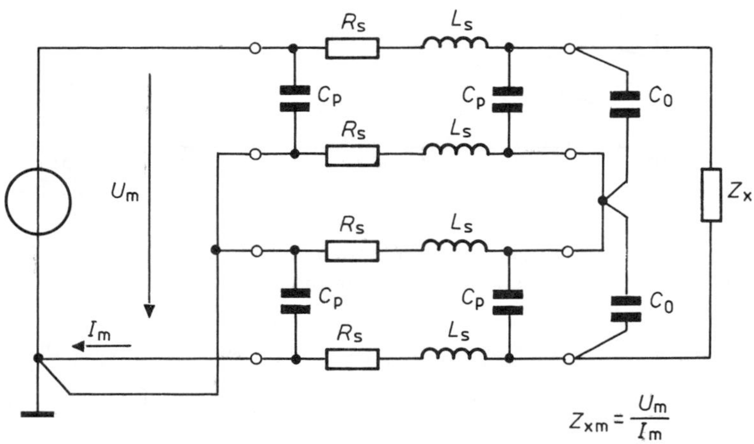

Fig. 4.9 Lumped parameter equivalent of three-wire measurement

The currents flowing through the capacitances C_p bypass the ammeter, and therefore their effect is eliminated. The ohmic resistance of the cable has the same effect as in the case of the two-wire measurement.

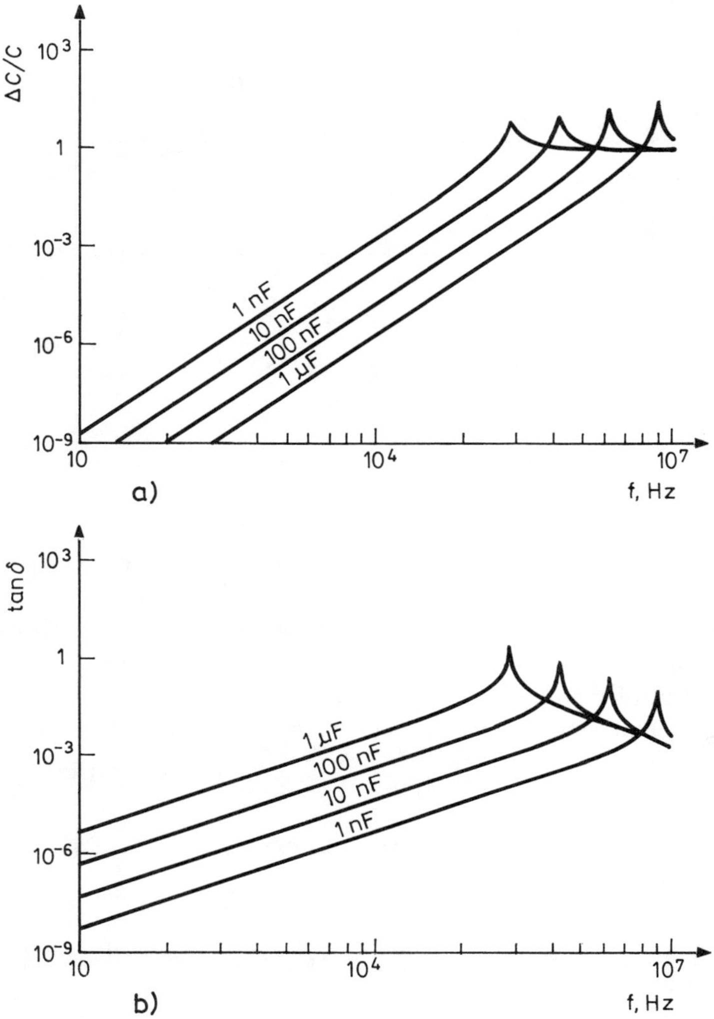

Fig. 4.10 Frequency dependence of the errors of a three-wire measurement *a)* error of the measured capacitance value *b)* error of the measured tanδ value

In Fig. 4.10 the errors of the three-wire measurement for the case of capacitance measurement are shown.

The error of the capacitance measurement at low frequencies is reduced indeed, but the tanδ error is the same as in the case of the two-wire measurement.

In the *five-wire measurement* the effects of capacitances C_p and the effects of R_s and L_s are reduced at the same time. In the circuit shown in Fig. 4.11, two cables carry the current to the current terminals of the impedance to be measured. Two other cables are used to connect the voltmeter to the potential terminals of the impedance to be measured. The shields of the cables are connected to the

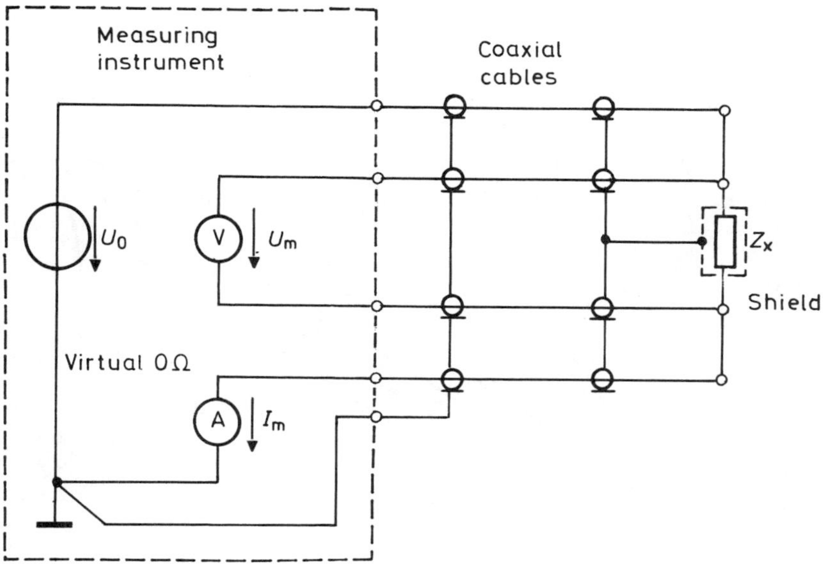

Fig. 4.11 Five-wire measurement

neutral point of the instrument and therefore the currents flowing through the cable capacitances bypass the ammeter. The voltage drops in the current carrying cables are not included in the voltage measurement, therefore they do not cause any error. However, for low values of Z_x, a large signal current flows through the current path, and mutual coupling to the voltage sensing cable will cause an error.

If neither the cables are shielded, nor the guard is connected to the shield of the impedance, the arrangement is called *four-wire measurement.*

The equivalent circuit of the five-wire measurement is shown in Fig. 4.12.

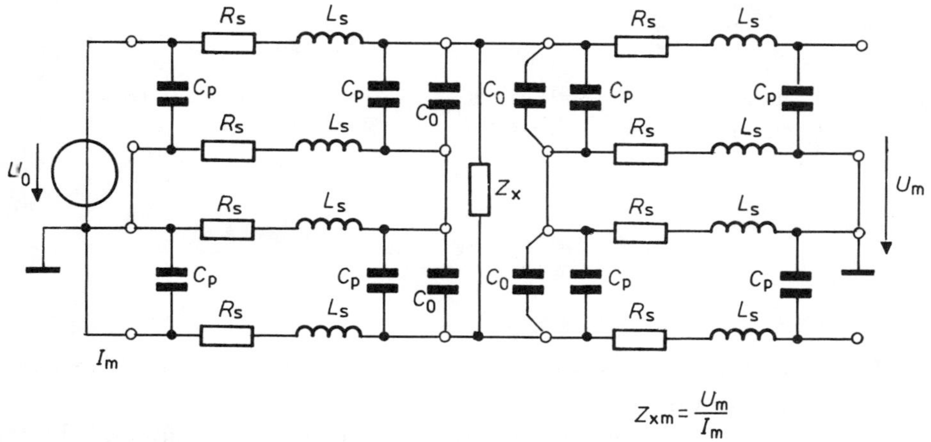

Fig. 4.12 Lumped parameter equivalent of the five-wire measurement

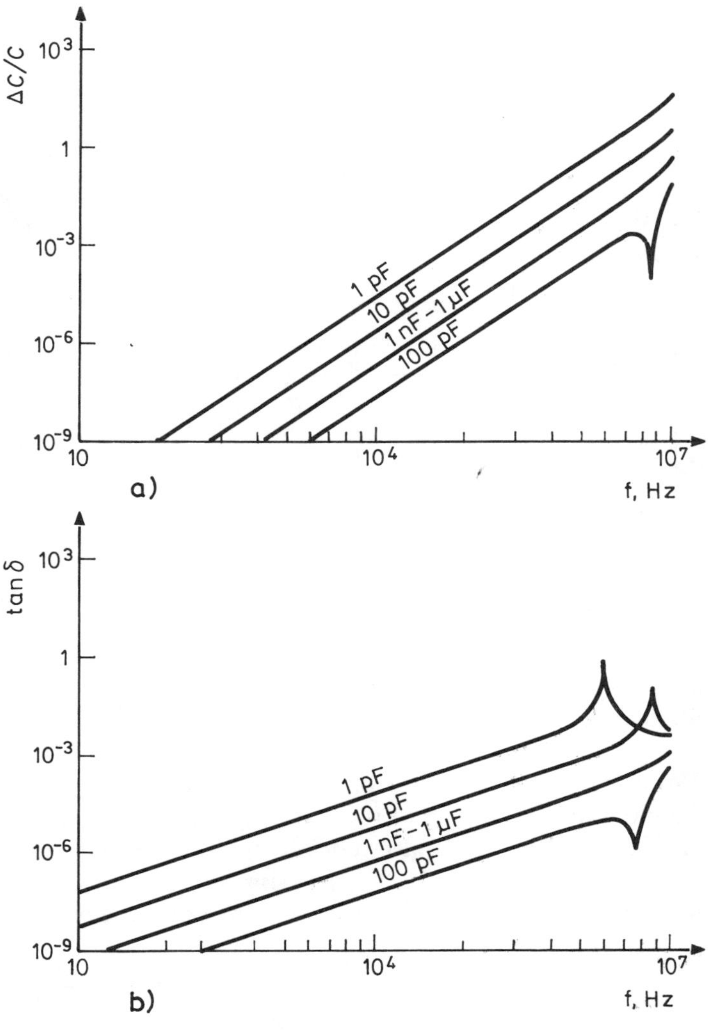

Fig. 4.13 Frequency dependence of the errors of a five-wire measurement *a)* error of the measured capacitance value *b)* error of the measured $\tan\delta$ value

In Fig. 4.13 the errors of a five-wire measurement for the case of capacitance measurement are shown.

The error of the capacitance measurement at low frequencies and the $\tan\delta$ error are reduced. This connection mode is the only correct one, all other simpler connection modes only approximate this, and have a limited application field.

4.2.2 Range of Measurement

An important characteristic of measuring instruments is the measurement range. This is defined as the maximal displayed value in a given setup of the instrument. For example, an analog voltmeter can display no more than 10 V in a given position of the input divider. In this case it is said: the voltmeter is in the 10 V range.

Another parameter is the range of measurement, which can be specified in two ways:

- by the minimal detectable and maximal measurable values. For example, a micro-ohmmeter has a resolution of $0.1\,\mu\Omega$ in the most sensitive position, and the maximal value it can display is $10\,\Omega$. The range of measurement is usually given as $0.1\,\mu\Omega - 10\,\Omega$.
  This specification can be misleading if we are not told what the numbers exactly mean.
- by the smallest and largest values, measurable with the given error. In the example given above, the specified error in each measurement range is 0.1% of the range. The lowest range is $10\,\mu\Omega$, and the highest range is $10\,\Omega$. By this second means of specification the range of measurement is $10\,\mu\Omega - 10\,\Omega$, with 0.1% uncertainty. This is a very realistic means of specification.

The range of measurement is an important parameter in characterizing a method or an instrument. Presently the ranges of measurement, according to the first definition, are in industrial practice:

- for resistances: $10^{-7}\,\Omega - 10^{13}\,\Omega$,
- for capacitances: $10^{-15}\,\text{F} - 0.1\,\text{F}$,
- for inductances: $10^{-8}\,\text{H} - 10^{8}\,\text{H}$.

4.2.3 Speed of Measurement

The methods of impedance measurement are structurally more complex than the methods of voltage and current measurement, because the impedance is a ratio of these two values. For example, the balancing of a bridge circuit is a procedure of several steps. The number of steps, necessary to reach the balanced state, is a measure of the quality of the given method. Modern computerized instruments also need 'steps' to get the results. For example, during a dual-slope ratio measurement a few processing phases (steps) are necessary to reach a correct result.

The speed as a parameter cannot be generally defined because many different processing methods exist. It is often simpler to specify the time up to a correct display, but the speed can also be defined as the number of steps of the procedure of the measurement.

4.3 BASIC TYPES OF IMPEDANCE MEASURING CIRCUIT

In this section the different types of impedance measuring circuit will be analyzed from a general point of view.

4.3.1 Four Arm (Wheatstone) Bridges

Hague (1962) described the history and the terminology of the four arm bridges. Wheatstone (1843) invented the circuit, and Siemens (1892) proposed practical solutions.

Structure of four arm bridges

A four arm bridge is shown in Fig. 4.14.

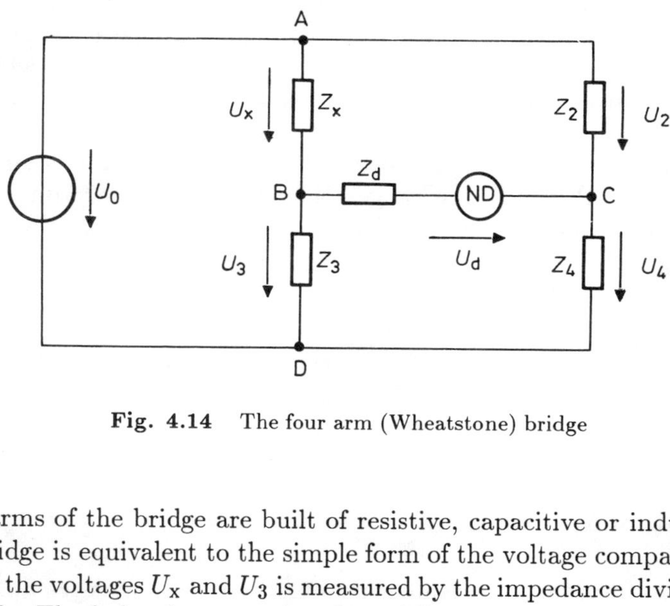

Fig. 4.14 The four arm (Wheatstone) bridge

The arms of the bridge are built of resistive, capacitive or inductive components. A bridge is equivalent to the simple form of the voltage comparison method. The ratio of the voltages U_x and U_3 is measured by the impedance divider composed of Z_2 and Z_4. The balancing procedure is as follows: the ratio and the phase-shift of divider Z_2/Z_4 are adjusted to the values for which the potential difference between the diagonal terminals B–C reaches the zero value. The potential difference between B and C is detected by a null detector (ND). In the balanced state:

$$\frac{Z_x}{Z_3} = \frac{Z_2}{Z_4}, \tag{4.12}$$

or in product form:

$$Z_x Z_4 = Z_2 Z_3 . \tag{4.13}$$

(4.13) can be expressed by the magnitudes and phases of complex impedance vectors:

$$|Z_x|e^{j\varphi_x}|Z_4|e^{j\varphi_4} = |Z_2|e^{j\varphi_2}|Z_3|e^{j\varphi_3}\,. \tag{4.14}$$

Since the impedances have complex values, (4.13) can be split into two equations. If the impedances are expressed in terms of real and imaginary parts, the equations can be solved for the active and reactive (real and imaginary) components of the impedance Z_x. If the impedances are expressed by their absolute values and phases, the balancing properties of the bridge can be deduced:

$$|Z_x| \cdot |Z_4| = |Z_2| \cdot |Z_3| \qquad \text{(amplitude condition)}, \tag{4.15}$$

$$\varphi_x + \varphi_4 = \varphi_2 + \varphi_3 \qquad \text{(phase condition)}. \tag{4.16}$$

Basic errors and sensitivity

The basic relative error is determined by the relative errors of the bridge elements. The relative errors of the absolute value and the phase are:

$$\epsilon_{xr} = \sqrt{\epsilon_{2r}^2 + \epsilon_{3r}^2 + \epsilon_{4r}^2}\,, \tag{4.17}$$

$$\delta_x = \sqrt{\delta_2^2 + \delta_3^2 + \delta_4^2}\,. \tag{4.18}$$

For a more detailed discussion, see Section 1.9).

The limited sensitivity of the null detector is an additional source of error. If a certain change of the value of Z_x does not result in a deviation on the null detector, this possible change must be interpreted as an error. The range between the (theoretical) limits of deviation of Z_x, causing no deflection of the null detector, is called the *domain of insensitivity* or *dead zone.* The origins of limited sensitivity of the null detector are the noises of the amplifiers, or in classical detectors the friction of the display instrument. In dc null detectors the main sources of noise are the *drift* and the so-called *flicker* or $1/f$ noise.

The limit of sensitivity of dc null detectors is presently about 10 nV.

Detectors of ac type are always frequency-selective, therefore the ac bridges are working always at one frequency. Direct filtered amplifiers have typically 100 nV noise in the range of 10 Hz – 100 kHz. The sources of this noise are the Johnson noise of resistors and the distribution noises of semiconductors. The detectors working with frequency transposition (mixers) have higher noise, about 1 μV, coming mainly from the mixers.

The influence of the bridge structure on the dead zone is defined by the bridge sensitivity. The bridge sensitivity S_b is the ratio of the potential difference ΔU_d on the bridge diagonal, in the proximity of zero, to the deviation ΔZ_x causing the above mentioned potential difference.

Let us calculate the value of S_b for a four arm bridge with a voltage-sensitive null detector, having infinitely high input impedance. The voltage measured by the null detector is:

$$U_d = U_0 \frac{Z_2 Z_3 - Z_x Z_4}{(Z_x + Z_3)(Z_2 + Z_4)} \,. \tag{4.19}$$

The sensitivity is:

$$S_b = \left. \frac{\Delta U_d}{\Delta Z_x} \right|_{U_d=0} \approx \frac{dU_d}{dZ_x} = U_0 \frac{Z_4}{(Z_x + Z_3)(Z_2 + Z_4)} \,. \tag{4.20}$$

A higher bridge sensitivity results in a smaller error of measurement for a given sensitivity of the null detector.

To compare the sensitivity of different bridge circuits the term of *'bridge factor'* was introduced into the terminology by Hague (1962), based on definitions of other authors. The *bridge factor* is the ratio of the change of detector voltage, divided by the source voltage of the bridge, to the relative deviation of the impedance Z_x:

$$H = \left. \frac{|\Delta U_d / U_0|}{|\Delta Z_x / Z_x|} \right|_{U_d=0} \,, \tag{4.21}$$

where U_0 is the voltage of the bridge supply.

Let us introduce in addition the term *'bridge ratio' (F)*:

$$F = \frac{Z_x}{Z_3} = \left. \frac{Z_2}{Z_4} \right|_{U_d=0} \,. \tag{4.22}$$

The *bridge factor* can be expressed by the *bridge ratio (F)*:

$$H = \frac{F}{(1+F)^2} \,. \tag{4.23}$$

In the case of a real, positive *bridge ratio* ($\mathrm{Im}\{F\} = 0$, $\mathrm{Re}\{F\} > 0$) the maximum of the absolute value of the *bridge factor* is:

$$|H|_{max} = \frac{1}{4} \qquad \text{for} \qquad F = 1 \,. \tag{4.24}$$

In the case of an imaginary *bridge ratio* the maximum of absolute value of the *bridge factor* is $\frac{1}{2}$ for $F = \pm j$. For $F = -1$ the *bridge factor* can be infinitely high.

Example 4.3

1. In Fig. 4.15a a bridge with imaginary bridge ratio is shown. The bridge ratio is:

$$F = jwRC\,. \tag{4.25}$$

It can be seen that the optimum value 1 of the bridge ratio can be reached at one frequency only.

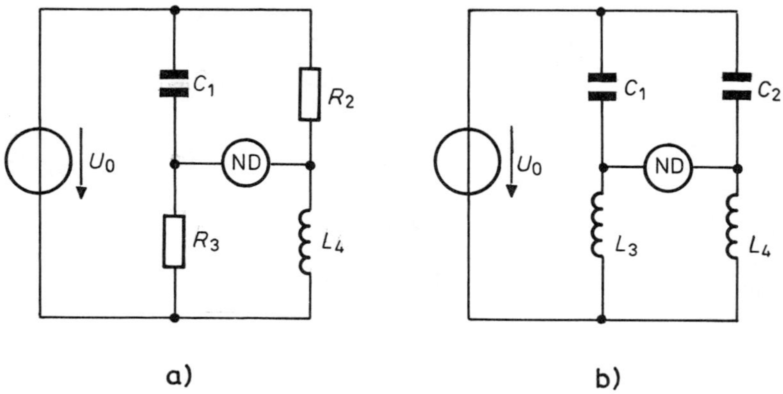

Fig. 4.15 Bridges with *a*) imaginary bridge ratio *b*) negative bridge ratio

2. The bridge shown in Fig. 4.15b is a so-called *resonance bridge.* The bridge ratio is:

$$F = -\omega^2 LC\,. \tag{4.26}$$

At a given frequency the value of F can be -1, therefore the bridge factor will be infinitely high. Actually, the L and C elements always have losses, therefore the *bridge ratio* can never be exactly -1. In practice very high sensitivity can be reached by a resonance bridge, at a given (resonance) frequency. It is very interesting to note that with exchanged source and detector the bridge ratio becomes real and frequency independent.

3. Actually the impedance to be measured has a complex structure. In the simplest case it has a frequency independent real part (R_x) and a frequency dependent imaginary part ($X_x = \omega L$):

$$Z_x = R_x + jX_x\,.$$

In (4.20) the absolute value of $\Delta Z_x / Z_x$ is used. It can be expressed as a function of the relative deviation of R_x and X_x:

$$\left|\frac{\Delta Z_x}{Z_x}\right|^2 = \left(\frac{\Delta R_x}{R_x}\right)^2 \frac{1}{1+Q^2} + \left(\frac{\Delta X_x}{X_x}\right)^2 \frac{Q^2}{1+Q^2}\,, \tag{4.27}$$

where $Q = X/R$.

If the impedance is dominantly reactive ($Q \gg 1$), the sensitivity to the resistive part will be Q times smaller than on the reactive part!

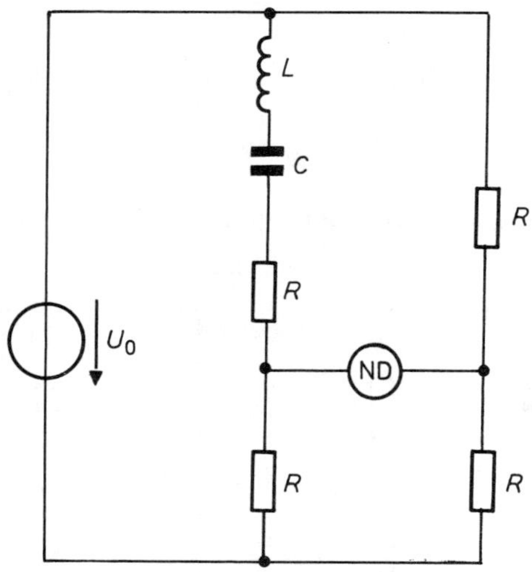

Fig. 4.16 Resonance bridge

Frequency dependence

A class of bridges, the so-called resonance bridges, are structurally frequency dependent.

The bridge shown in Fig. 4.16 can be balanced at the angular frequency ω_0 only:

$$j\omega_0 L + \frac{1}{j\omega_0 C} = 0 \qquad \Rightarrow \qquad \omega_0 = \frac{1}{\sqrt{LC}} .$$

Bridges of a quite larger class are frequency independent. The bridges shown in Fig. 4.17 have frequency independent balancing conditions:

$$a) \;\; R_{\mathrm{x}} = R; \qquad b) \;\; C_{\mathrm{x}} = C_2 \frac{R_3}{R_4}, \quad R_{\mathrm{x}} = \frac{C_4}{C_2} R_3 .$$

Frequency independence refers to the balancing conditions only. The trajectories of the Schering bridge (Fig. 4.17b) do depend on the frequency applied, since the impedances of the capacitances are frequency dependent.

The bridges in Fig. 4.17 are only *theoretically* frequency independent. The stray impedances of bridge elements cause errors at higher frequencies. This effect is described in Subsection 4.2.1. The connecting leads cause errors too, as described in the same subsection. If the connecting leads are short, the effects of the earth[1] impedances only must be taken into consideration.

In Fig. 4.18 the earth impedances Z_{A}, Z_{B}, Z_{C} and Z_{D} are conducting current from and to the nodes, therefore they disturb the balancing conditions.

If node D is connected to earth, the effects of Z_{A} and Z_{D} are eliminated, but Z_{B} and Z_{C} are connected parallel with the bridge impedances Z_3 and Z_4,

[1] For the definition of earth, see Subsection 2.2.1.

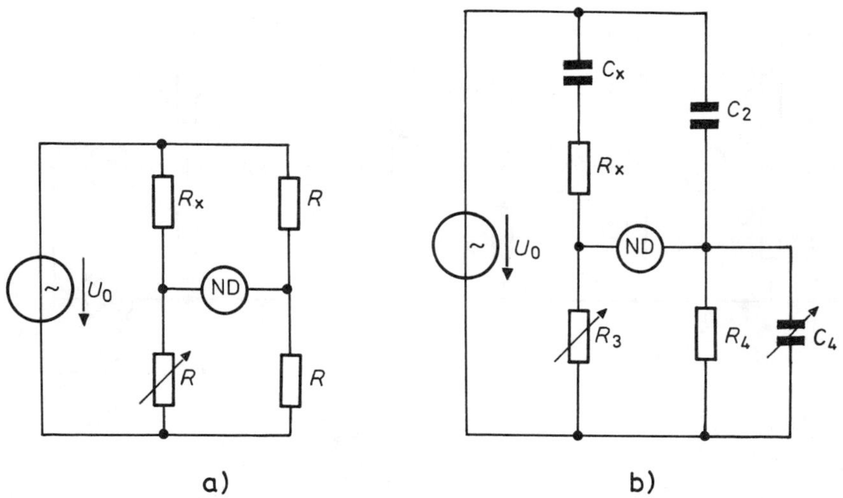

Fig. 4.17 Frequency independent bridges *a)* Wheatstone bridge *b)* Schering bridge

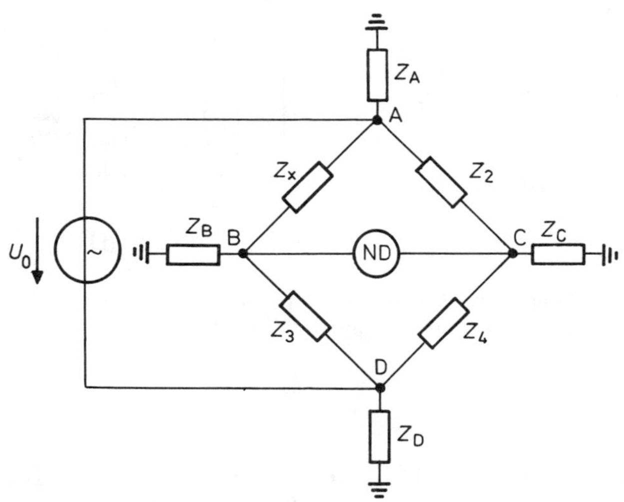

Fig. 4.18 Earth impedances of a bridge

respectively, and they cause calculable errors. If C or B are connected to the earth, the impedances Z_A and Z_D cause errors.

For a compensation of the effect of earth impedances, the potentials of bridge nodes must be controlled. In Fig. 4.19 four solutions are presented.

In Fig. 4.19a the *Wagner earths device* (Wagner, 1911) is shown. A half-bridge is connected in parallel to the supply, to adjust the potential of node C to the earth potential. If ND2 displays zero potential difference, there is no current from node C to earth, and if at the same time the indication of ND1 is zero, node B

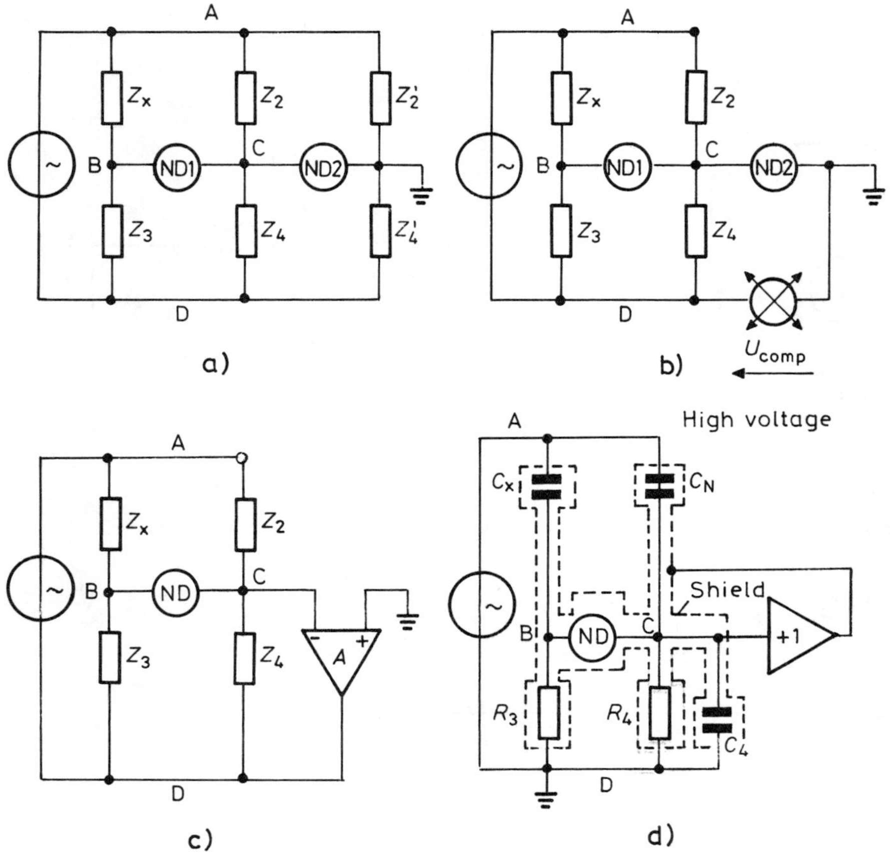

Fig. 4.19 Compensation of the effect of earth impedances *a)* Wagner earths device *b)* complex compensator *c)* automatic solution *d)* solution with voltage follower

will be also at earth potential. Therefore no current flows from B to earth. In this way the effects of earth impedances are eliminated. It is important to note that *neither node B nor C is connected to earth,* only their potential is adjusted to the earth potential.

A similar solution is shown in Fig. 4.19b. The potentials of B and C are brought to earth potential by a (usually electronic) complex voltage source, connected between node D and the earth. This is a more modern solution, and offers higher flexibility and better convergence.

The proposal of Brueckel (1962), shown in Fig. 4.19c, is the automation of the above solution. The potential difference between node C and earth is detected by an amplifier A. The output of the amplifier is connected to node D. If the gain of amplifier is infinitely high, its output forces the potential difference between C and earth to zero. The main problem is to guarantee the stability of the amplifier–bridge structure. The large range of the unknown impedance Z_x makes the feedback loop variable and it is difficult to stabilize the loop with a fixed compensation.

The solution of Fig. 4.19d is used in high-voltage technology, notably in high-voltage Schering bridges. This bridge must be connected to earth because the high-voltage source is also connected to earth. The low-voltage part of the bridge is shielded and the shield potential is adjusted to the potential of the bridge arm B–C. The zero potential difference between the bridge nodes B, C, and the shield do not allow an error current to flow from B to earth, or from C to earth.

Connections

The Wheatstone bridge is a tool used typically for two-wire measurement. The three-wire measurement is possible if earth compensation is used. Four- and five-wire measurements are possible e. g. by the method of Kelvin (see Section 4.4).

The speed of measurement

A bridge circuit is balanced by successive approximation. If an absolute value indicating null detector is used, the operator has to reach the balanced state by successive search of minimal deflections. The detector voltage of a four arm bridge can be expressed as a function of the two balancing parameters (u and v):

$$U_d = U_0 \frac{Z_x Z_4 - Z_2 Z_3}{(Z_x + Z_2)(Z_3 + Z_4)} = \frac{a + bu + cv + euv}{f + gu + hv + kuv} . \tag{4.28}$$

where a, b, c are complex constants.

If the parameter v is fixed, the vector U_d, as function of u, follows a circle in the complex plane:

$$U_d = \frac{\alpha + \beta u}{\gamma + \sigma u} . \tag{4.29}$$

If the parameter u is fixed, U_d traces another circle:

$$U_d = \frac{\alpha' + \beta' v}{\gamma' + \sigma' v} . \tag{4.30}$$

As an example, in Fig. 4.20 the trajectories of the complex vector of the output voltage of a Schering bridge (Subsection 4.4.3) are shown.

The horizontal axis represents the real part of the detector voltage, the vertical axis the imaginary one. The parameters of the trajectories are the resistor R_3 and the capacitor C_4. The operator starts for example from the point A and, changing R_3, reaches a minimal deflection at point B. Next C_4 is changed to a second minimum, then R_3 again, and so on. The fewer the steps are, necessary to reach a certain domain around the zero point, the higher the speed is.

The linearized trajectories are often used for the characterization of convergence. The simplest means of linearization is to use the nominator of (4.28) only,

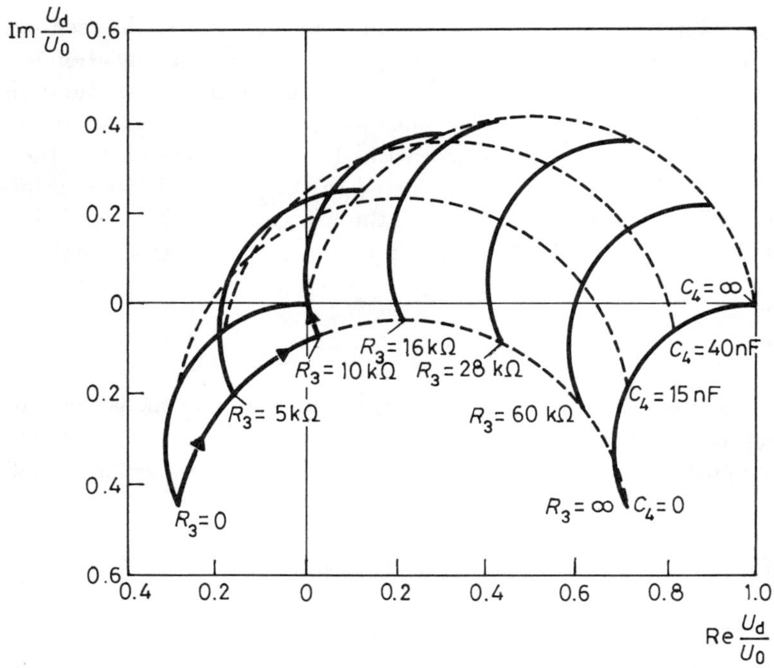

Fig. 4.20 Convergence trajectories of a Schering bridge (see Fig. 4.17b), for $R_x = 3\,\text{k}\Omega$, $C_x = 10\,\text{nF}$, $R_4 = 10\text{k}\Omega$, $f = 1\,\text{kHz}$. The bridge is balanced at $R_3 = 10\,\text{k}\Omega$ and $C_4 = 3\text{nF}$.

since near to the optimum the denominator is approximately constant. Near the zero value of U_d, this means a rotation of all the trajectories:

$$\frac{U_d}{U_0} K = Z_x Z_4 - Z_2 Z_3 = \alpha + \beta u = \alpha' + \beta' v\,, \tag{4.31}$$

where K is a complex constant.

In Fig. 4.21 the linearized trajectories near to the balanced state are shown.

The balancing procedure is illustrated in Fig. 4.21a for the case when the angle between the two trajectory directions is $\varphi \neq 90°$. To reach the point E from A, four steps are necessary. In Fig. 4.21b the angle is $\varphi = 90°$, therefore (theoretically) only 2 steps are necessary to balance the bridge. The *convergence angle* φ is a measure of the speed of convergence. If it is equal to 90°, the speed is maximal.

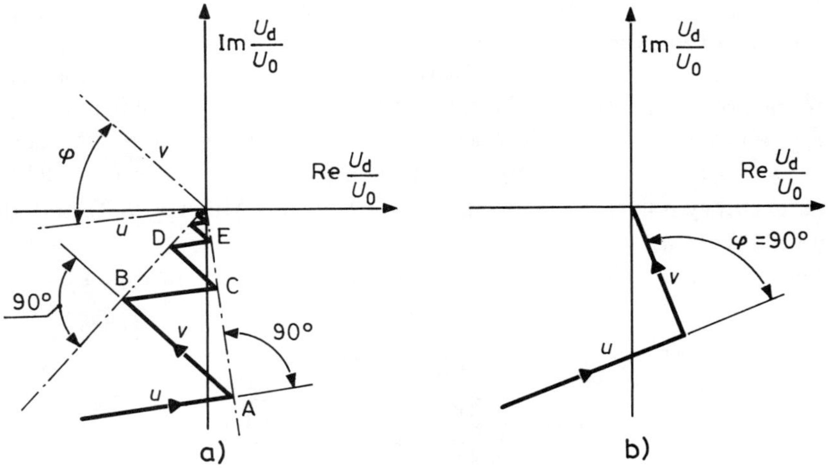

Fig. 4.21 Convergence vectors *a)* $\varphi \neq 90°$, *b)* $\varphi = 90°$

4.3.2 Bridges with Ratio Transformer and Inductive Divider

The ratio transformer

The inductive ratio transformer is essentially a voltage transformer with two or more windings. These transformers are wound on a toroidal core made of a high-permeability material. Such a transformer is shown in Fig. 4.22.

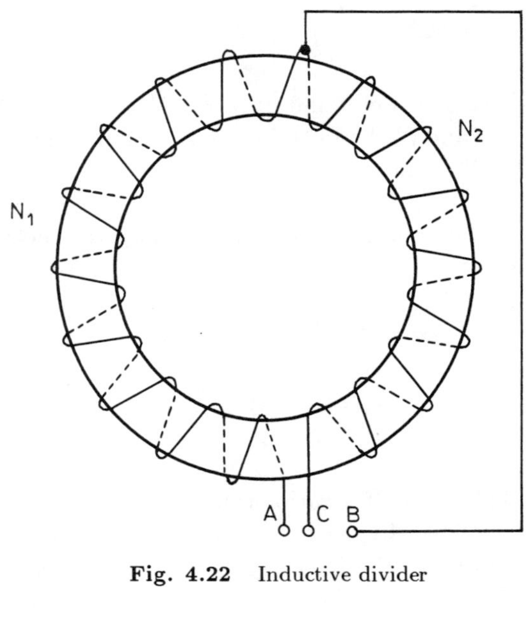

Fig. 4.22 Inductive divider

The ideal (not loaded) ratio transformer produces two voltages with a ratio proportional to the ratio of the two numbers of turns. This is the main advantage of the ratio transformer over the resistive divider; the resistance ratio is dependent on temperature, aging etc, but the ratio of the numbers of turns is not.

The method was first proposed by Walsh (1930), and was introduced into practice by Watton and Pemberton (1949) and Clark and Vanderlyn (1949), while the high-accuracy solution was proposed by Hill and Miller (1962) and Hill and Deacon (1968).

The principle of the ratio transformer bridge is shown in Fig. 4.23.

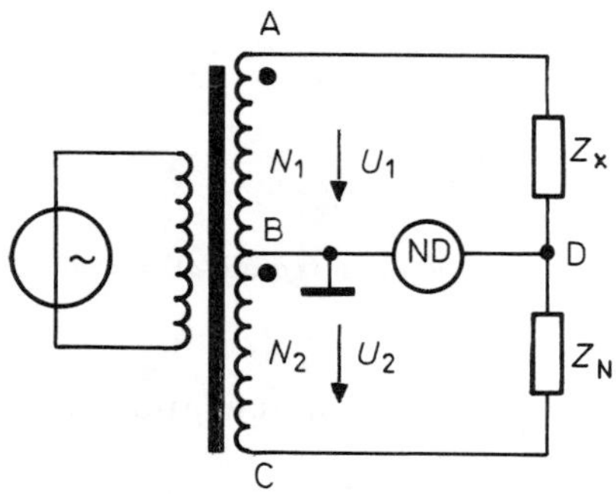

Fig. 4.23 The ratio transformer bridge

The bridge is balanced by adjusting the two numbers of turns N_1 and N_2 and the standard impedance Z_N. In the balanced state the null detector has zero deflection,therefore the potential of node D is equal to the potential of node B:

$$\frac{Z_\mathrm{X}}{Z_\mathrm{N}} = \frac{N_1}{N_2}\,; \qquad Z_\mathrm{X} = Z_\mathrm{N}\frac{N_1}{N_2}\,.$$

A more flexible practical solution is shown in Fig. 4.24. The standard side has two balancing windings (N_C and N_G) and the standard admittances are the decades C_N and G_N. The winding N_x is normally the *range winding* with numbers of turns of 1, 10, 100 etc.

In the balanced state the following expression gives the value of the admittance to be measured:

$$Y_\mathrm{X} = \frac{1}{N_\mathrm{X}}(N_\mathrm{G}G_\mathrm{N} + j\omega N_\mathrm{C}C_\mathrm{N})\,. \tag{4.32}$$

Structure of bridges

The coarse and fine balancing of a ratio transformer bridge can be solved in two ways. In the first one, a fixed divider is used and a few decades of standard admittances and capacitances make the balancing possible. This is not a very practical method, because the high variety of standard impedances makes it very expensive.

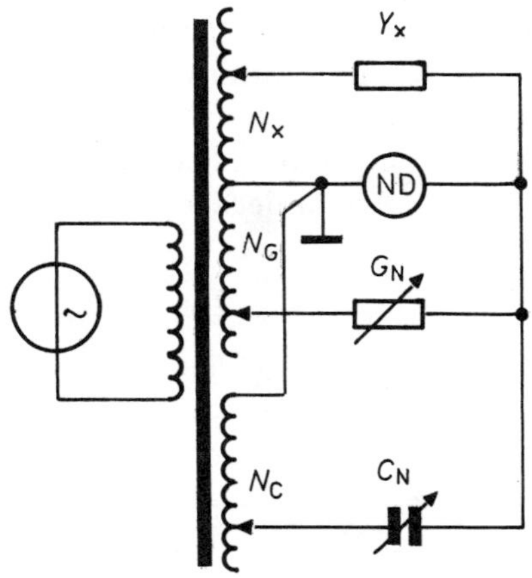

Fig. 4.24 Practical ratio transformer bridge

In a more practical solution a fixed standard admittance and a fixed standard capacitor are used. In Fig. 4.24 the windings N_C and N_G can be made of serially connected decade windings. The advantages of this solution are the flexibility and the low cost. On the other hand, using this method it is difficult to accomplish fine balancing, and the complex structure of the windings on the core increases the errors (see later).

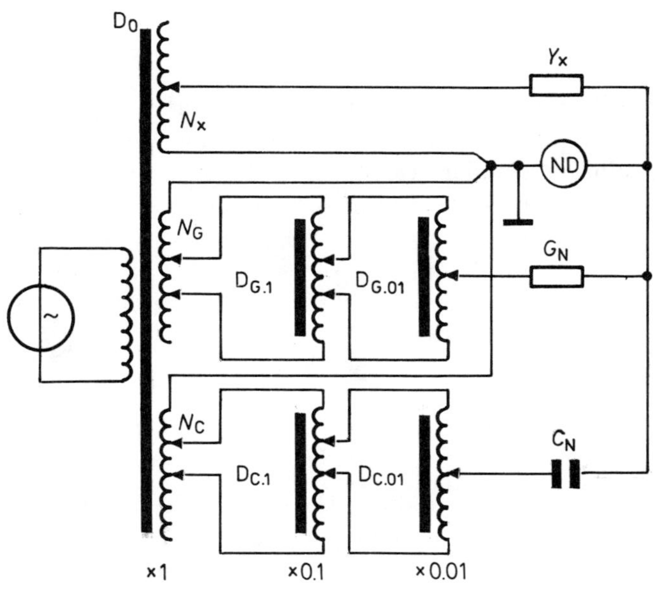

Fig. 4.25 High-accuracy ratio transformer bridge

A high-accuracy solution patented by Cutler and Rorden (1958) is shown in Fig. 4.25. The main ratio transformer contains the most significant windings (N_X, N_C and N_G) only. Fine balancing is realized by inductive dividers, connected in cascade (see later). In Fig. 4.25 only two additive decades are shown ($D_{G.1}$, $D_{G.01}$, $D_{C.1}$ and $D_{C.01}$), but in practice several decades may be applied. This solution is accurate but expensive.

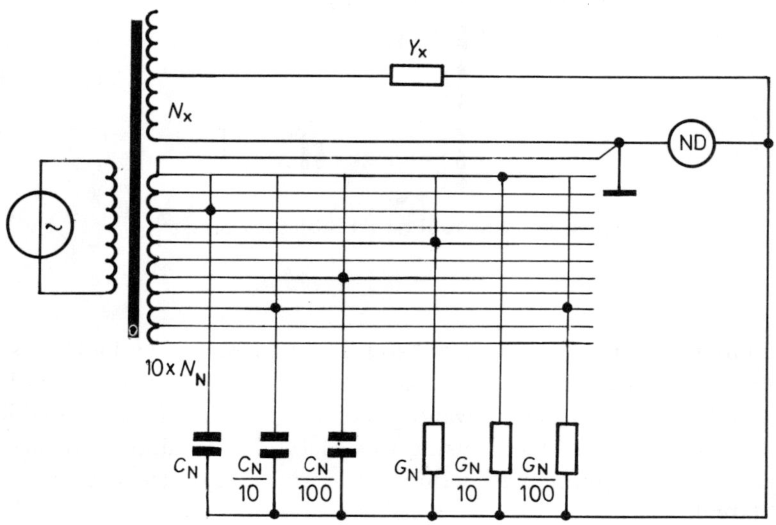

Fig. 4.26 Combined ratio transformer bridge

In Fig. 4.26 (Genrad, 1971) the combination of the two principles is shown. The ratio transformer has only ten taps on side N, and the decades are realized using decreasing values of standard capacitors and admittances. This solution has higher accuracy than the one shown in Fig. 4.24, and is less expensive than the one in Fig. 4.25.

The errors of ratio transformer bridges

a) Basic errors and sensitivity

The unloaded ratio transformer produces the ratio with an error, dependent on the inhomogeneity of the core permeability, on the imperfect coupling between the windings, and on the interwinding capacitances. These sources of error are analyzed by Binnie and Foord (1967). For $f = 1\,\text{kHz}$ the errors of a well constructed and not loaded ratio transformer are in the range of $10^{-7} - 10^{-8}$, well below the practical requirements.

b) Further errors and frequency dependence

The situation is different if the ratio transformer is loaded. All measuring circuits act as loads and thus cause errors. In the followings three practical setups will be analyzed.

The simple ratio transformer

The ratio transformer model shown in Fig. 4.27 has three windings. The first one is the excitation winding, the two others are the ratio windings. Both ratio windings have serial stray impedances: Z_{s1} and Z_{s2} are composed of the copper resistances and stray inductances of coils. The impedances Z_1 and Z_2 represent the stray (earth) capacitances of the windings. Z_X is the impedance to be measured, and Z_N is the standard impedance. The common node of Z_X and Z_N is not connected to earth, because it is the balanced node, i. e. in the balanced state there is no potential difference between this point and the earth, thus no current flows to the earth. The currents I_1, I_2 and I cause voltage drops on the impedances Z_{s1} and Z_{s2}, therefore the ratio of the voltages on nodes A and B will not be equal to the ratio N_1/N_2. The quantity of interest is the difference between the ratio of the two voltages on the terminals A and B (U_1/U_2) and the ratio of the numbers of turns (N_1/N_2).

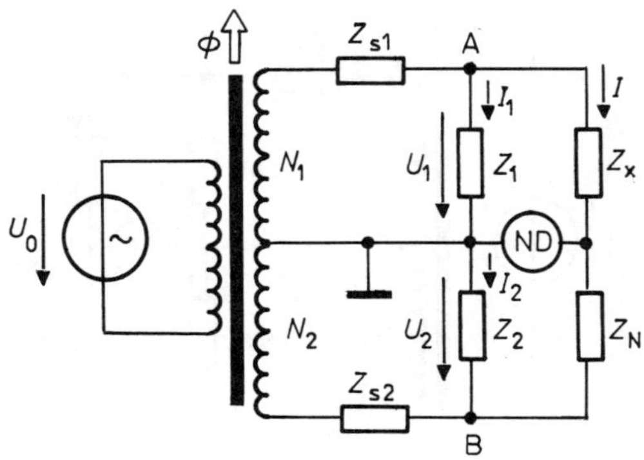

Fig. 4.27 Ratio transformer

The circuit equations are:

$$I_1(Z_1 + Z_{s1}) + IZ_{s1} = j\omega N_1 \Phi\,, \tag{4.33}$$

$$I_2(Z_2 + Z_{s2}) + IZ_{s2} = j\omega N_2 \Phi\,, \tag{4.34}$$

$$I(Z_{s1} + Z_{s2} + Z_X + Z_N) + I_1 Z_{s1} + I_2 Z_{s2} = j\omega (N_1 + N_2)\Phi\,. \tag{4.35}$$

If the stray impedances Z_{si} are significantly smaller than the other impedances, the error term (the difference of N_1/N_2 and U_1/U_2) will be the following:

$$\epsilon_r \approx \frac{Z_{s2}}{Z_2} - \frac{Z_{s1}}{Z_1} + \frac{N_1 + N_2}{Z_x + Z_N}\left(\frac{Z_{s2}}{N_2} - \frac{Z_{s1}}{N_1}\right). \tag{4.36}$$

The inductive voltage divider

The circuit shown in Fig. 4.28 is very similar to the one above. The only difference is the connection of the supply. The transformer is driven directly from an ac source, therefore it is working as an *inductive divider.* The circuit equations are:

$$\Phi = \Lambda(N_1 I_{s1} + N_2 I_{s2}), \tag{4.37}$$

$$U_0 = U_1 + U_2, \tag{4.38}$$

$$U_1 = j\omega\Phi N_1 + I_{s1} Z_{s1}, \tag{4.39}$$

$$U_2 = j\omega\Phi N_2 + I_{s2} Z_{s2}, \tag{4.40}$$

$$0 = I_{s2} + U_2/Z_2 - U_1/Z_1. \tag{4.41}$$

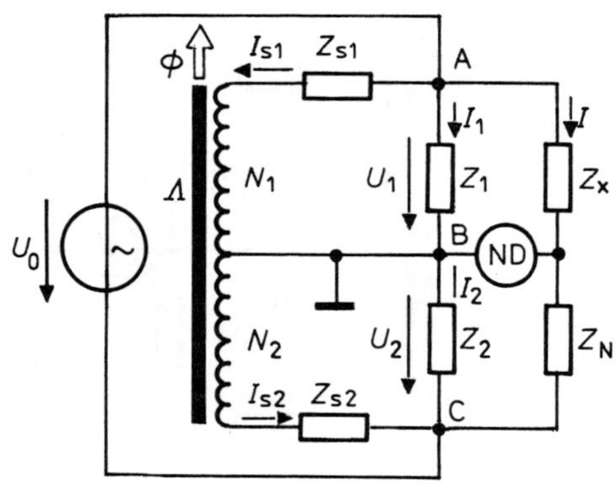

Fig. 4.28 Loading error of inductive voltage divider

The ratio error is:

$$\epsilon_r = \left(\frac{Z_{s1} Z_{s2}}{j\omega(L_{12} + L_1) Z_1} + \frac{N_2}{N_2 + N_1} \cdot \frac{Z_{s1}}{Z_1}\right)\left(\frac{N_2 Z_1}{N_1 Z_2} - 1\right)$$

$$+ \frac{N_1}{N_1 + N_2} \cdot \frac{Z_{s2}}{Z_2}\left(1 - \frac{N_1}{N_2} \cdot \frac{Z_2}{Z_1}\right) + \frac{Z_{s2}}{j\omega(L_{12} + L_2)}\left(\frac{N_2}{N_1} \cdot \frac{Z_{s1}}{Z_{s2}} - 1\right). \tag{4.42}$$

Comparison of the two types of ratio transformer

Comparing (4.36) and (4.42) it is obvious that the inductive divider is not sensitive to the values of impedances Z_X and Z_N, i. e. it is not load sensitive.

Example 4.4
To demonstrate the practical limits of ratio transformers, a two-layer toroidal construction will be studied. In Fig. 4.29 one layer is shown.

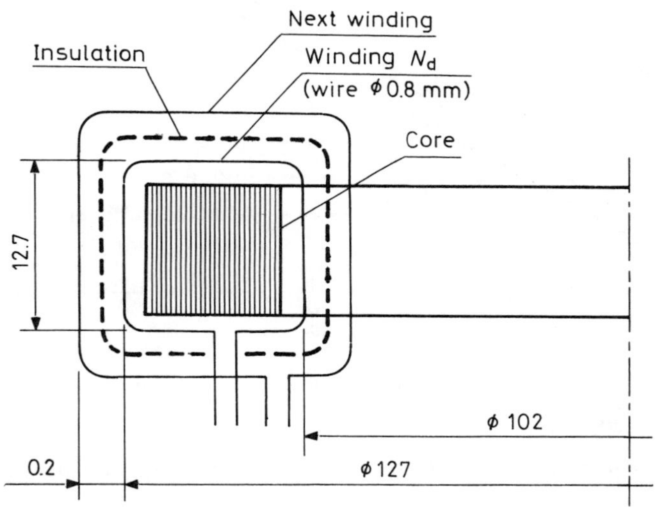

Fig. 4.29 A typical toroid ratio transformer

The wire has a cross section of $0.5\,\mathrm{mm}^2$ i. e. a diameter of 0.8 mm. With filling factor 0.8 the number of turns on the first layer is $N_0 = 240$. The inductance of each winding is: $\Lambda = 6.3 \cdot 10^{-5}$ H, if the relative permeability is 100 000. The 0.2 mm thick insulation between the layers gives a stray capacitance of $C_s = 360$ pF ($C_{\text{layer}/2}$) and a stray inductance of one turn $\lambda = 4 \cdot 10^{-11}$ H. The resistance of one turn is $r = 1.78 \cdot 10^{-3}\,\Omega$. The expressions of the stray impedances are:

$$Z_{s1} = rN_d + j\omega\lambda N_d^2; \qquad Z_1 = Z_2 = \frac{1}{j\omega C_s}. \tag{4.43}$$

If the sum of the turns of the windings is constant in each layer: $N_1 + N_2 = N_0$, and the ratio is $x = N_1/N_2$,

$$N_1 = \frac{x}{1+x}; \qquad N_2 = \frac{1}{1+x}. \tag{4.44}$$

With these values, the error equation (4.36) of the *ratio transformer* is:

$$\epsilon = \frac{1-x}{1+x}\left(j\omega rC_s N_0\left(1 + \frac{j\omega\lambda N_0}{r}\right) + \frac{j\omega\lambda N_0}{Z_X + Z_N}\right). \tag{4.45}$$

In case of $x = 1$ the error is zero. For a small error the number of turns should be kept low.

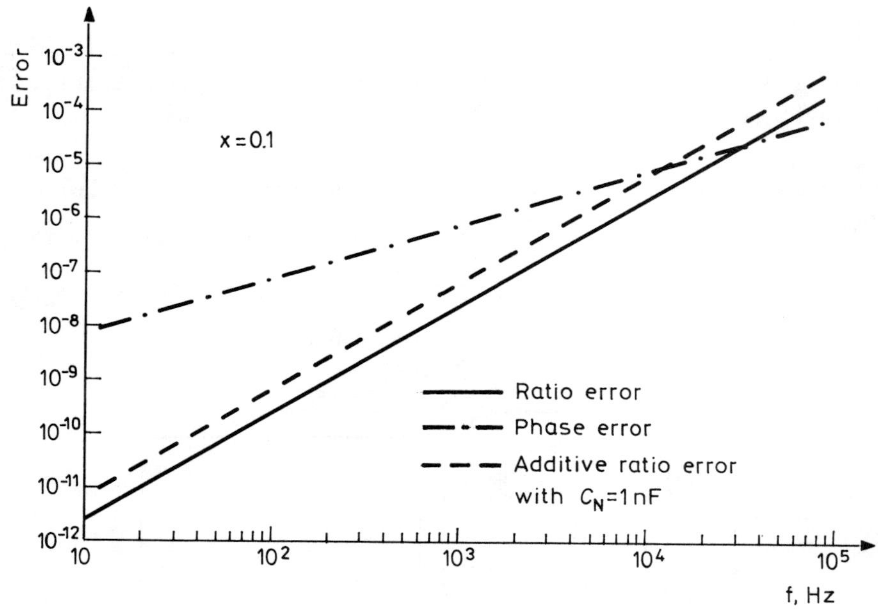

Fig. 4.30 Ratio and phase error of an inductive divider

The error equation of the *inductive divider* with the practical parameters is:

$$\epsilon = \frac{1-x}{1+x}\left(j\omega r C_s N_0 \left(1 + \frac{j\omega\lambda N_0}{r}\frac{1}{1+x}\right)\right). \tag{4.46}$$

The second term of (4.45) is missing. The frequency dependence of the real part (ratio error) and imaginary part (phase error) of (4.46) is shown in Fig. 4.30.

The error caused by the second term of (4.45) for a C_N value of 1 nF is shown on the same diagram. In this case $Z_N = 1/j\omega C_N$ and $Z_x = xZ_N$.

In practice, the inductive divider is used if high accuracy is required. The simple ratio transformer offers higher flexibility because it is also suitable for more than two winding structures. A special but very useful variant of the inductive divider is shown in Fig. 4.31.

The divider works as a true divider; the winding N_1 is loaded by an external impedance Z_L (i. e. by the impedance to be measured). The error-free ratio is:

$$y = \frac{N_1}{N_1 + N_2} \qquad \text{with} \qquad N_1 + N_2 = N_0\,.$$

With these substitutions the ratio error is:

$$\epsilon = \frac{\lambda}{\Lambda}(1-y) - \frac{rN_0}{Z_L}y(1-y)\left(1 + 2\frac{j\omega\lambda N_0}{r}y(1-y)\right).$$

The ratio error has its maximum at $y = 0.5$. With $Z_L = 10\,\mathrm{k\Omega}$, using the parameters of transformer of the example, the error is 10 ppm.

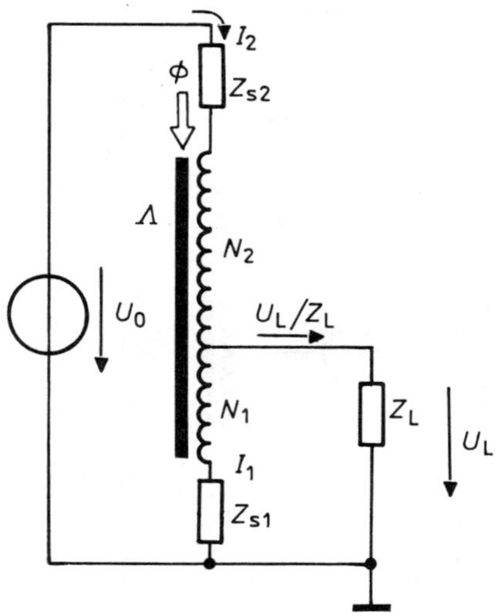

Fig. 4.31 Autotransformer type inductive voltage divider

As a conclusion, it can be stated that the inductive ratio transformer is a high-accuracy ratio device with no temperature and aging errors. It gives higher accuracy and higher flexibility than the resistive divider, and the solution shown in Fig. 4.31 is not sensitive to the load.

The disadvantages of the ratio transformer are its high cost and the reduced accuracy over 100 kHz. The high-frequency problem can be partially solved using high-frequency ferrite cores and a reduced number of turns, but despite these improvements, at frequencies over 100 MHz, a large inductive core is not comparable to a small, low-capacitance bulk metal[2] resistor.

Connections

The ratio arm bridges are ideal for three-terminal measurements. As it is shown in Fig. 4.23, terminal D is not connected to earth though it is at earth potential. Therefore, if a shield around the impedance Z_x is connected to the node B, in connection with the earth, the current flowing to the shield goes directly to earth, and therefore does not disturb the potential distribution in Z_x.

It is more difficult to solve the problem of four- or five-terminal measurements. A solution, proposed by Hill and Deacon (1968), is shown in Fig. 4.32.

The main inductive divider D_1 is connected between the terminals P_{xH} and P_{NL}; therefore it divides the potential between the two remote potential terminals. The secondary inductive divider D_2 is connected to the terminals P_{xL} and P_{NH}, therefore it divides the potential difference on the connection leads between the

[2]Bulk metal is a trademark of Vishay Intertechnology, Inc. It means a special technology by which high-quality resistors are produced.

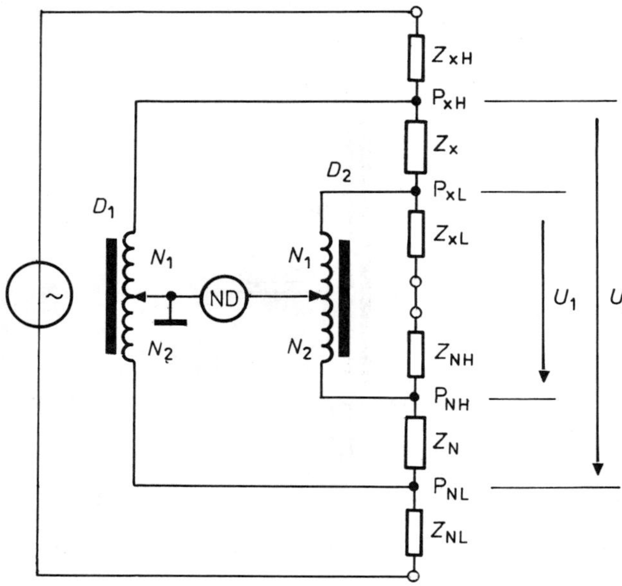

Fig. 4.32 Four-wire measurement with two inductive dividers

impedances. If the settled ratios of the two dividers are the same, and the bridge is balanced, the effect of the connection leads is suppressed:

$$\frac{Z_{\mathrm{X}}}{Z_{\mathrm{N}}} = \frac{N_1}{N_2} \,. \tag{4.47}$$

The solution works only if the reactance of the second divider is significantly higher than the impedances of the leads.

Another solution, proposed by Hill and Miller (1963), is shown in Fig. 4.33.

The potential drops on the connection leads are subtracted from the potential difference between the nodes $P_{xH} - P_{NL}$. If the two transformers T_X and T_N have high impedances, the measurement errors caused by them are negligible.

The bridge ratio of these circuits is always real (because of the ratio transformer), therefore the optimum bridge factor is unity.

Further solutions were proposed by Hall (1970).

Speed of measurement

These bridges have a perpendicular convergence angle, therefore they have a high balancing speed.

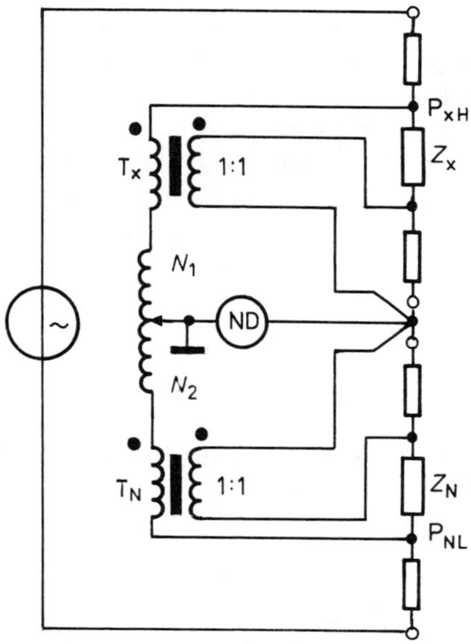

Fig. 4.33 Four-wire measurement with one inductive divider

4.3.3 Bridges with Inductive Current Comparator

The comparison unit

The inductive current comparator consists of three or more windings on a ferromagnetic core. The electromagnetic (and not the mechanical) structure of a inductive current comparator is shown in Fig. 4.34.

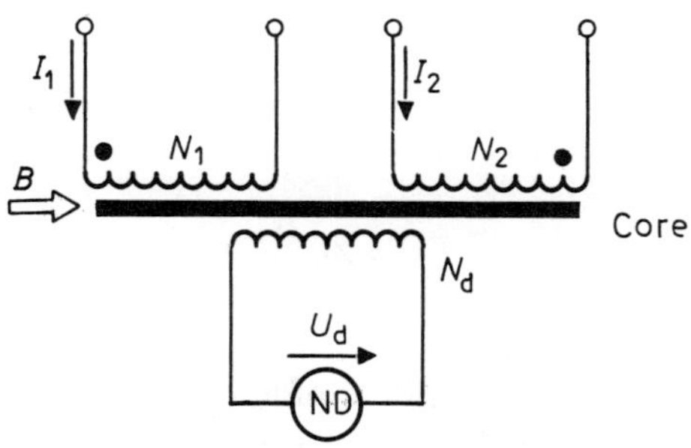

Fig. 4.34 The inductive current comparator

The currents I_1 and I_2 in windings N_1 and N_2 produce opposite magnetic fields in the core. If the two ampere-turns are equal, the resulted induction will be zero: if $I_1N_1 = I_2N_2$, $B = 0$.

The zero induction is detected by a third winding on the core. If the induction is zero, the induced voltage in the third winding ($N_{\rm d}$) will also be zero. In this case the null detector ND has a zero deflection. The so-conceived inductive current comparator is a current comparison unit with a very high inherent accuracy. The principle was introduced by Baker (1917) but the realization only became possible with the appearance of very high-permeability ferromagnetic materials (Miljanic et al, 1962; Kusters and Moore 1964). N_1 and N_2) are the *working windings*, and $N_{\rm d}$ is the *detection winding.*

Structure of bridges

The practical form of an impedance measuring bridge, based on a current comparator, is shown in Fig. 4.35.

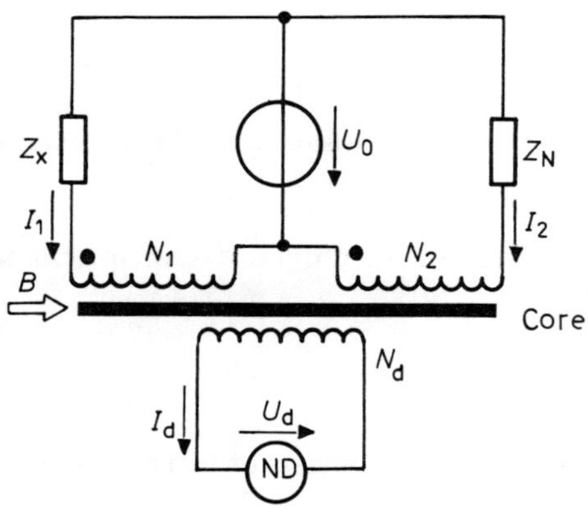

Fig. 4.35 Bridge with inductive current comparator

In the case of balance ($U_{\rm d}=0$):

$$Z_{\rm x} = Z_{\rm N}\frac{N_1}{N_2}\,.$$

The result of measurement is independent of temperature, aging and other physical effects because the balance is realized by a ratio of number of turns, a well defined quantity.

Another usual form of this bridge type is a bridge with combined inductive ratio transformer and inductive current comparator, as shown in Fig. 4.36. This is not the highest-accuracy solution, because the two arms of the voltage divider are individually loaded by the impedances, but the flexibility is high, and the measurement range of the bridge is very large.

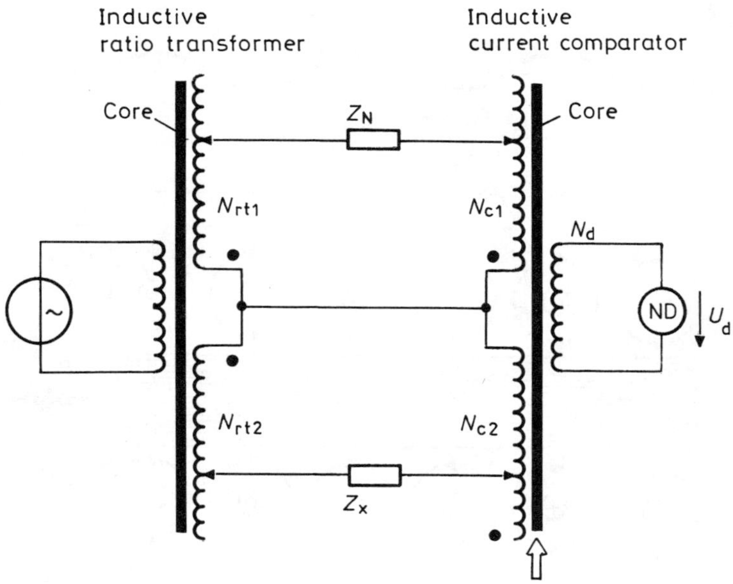

Fig. 4.36 Combined ratio-transformer current-comparator bridge

The errors of current comparator measurements

a) Basic errors and sensitivity

The sources of error in an inductive current comparator are the magnetic inhomogeneity of the core–winding structure and the stray capacitances.

b) Magnetic errors

The first study of a toroidal core was published by Kennelly and Alger (1917). Although the ampere-turns are zero in the balanced state, this does not mean that the flux is everywhere zero in the core. A simplified diagram is shown in Fig. 4.37.

The magnetic induction B_x is shown in an infinitely long linear core. The curve-integral of the field is zero, therefore in the detector winding theoretically zero voltage is induced. In spite of the zero integral, the magnetic field is oscillating with a high amplitude. If the spatial distribution of the null detector winding is perfectly uniform, the alternating field does not induce voltage in it. But if the winding N_d has inhomogeneity, the integrally zero ampere-turn can induce a voltage in it. In practice the special construction of current comparators guarantees the accurate null detection: the core has a toroidal form and it is made from a high-permeability material (*ultraperm*). The relative permeability of ultraperm is very high at zero induction, normally over 100 000. The high-permeability has a *smoothing* effect,

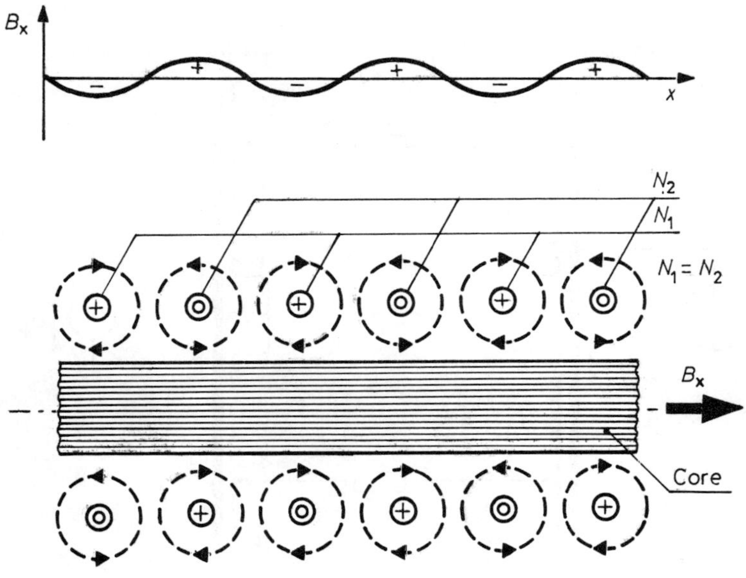

Fig. 4.37 Field distribution in an inductive current comparator

i. e. it reduces the amplitude of field oscillation in the core. The *detection winding* is always the first one on the core. If the distance of working windings from the detector winding is larger, the field oscillation in the core will be smaller. By a special construction this distance can even be virtually increased. In Fig. 4.38 a so-called shielded core structure is shown.

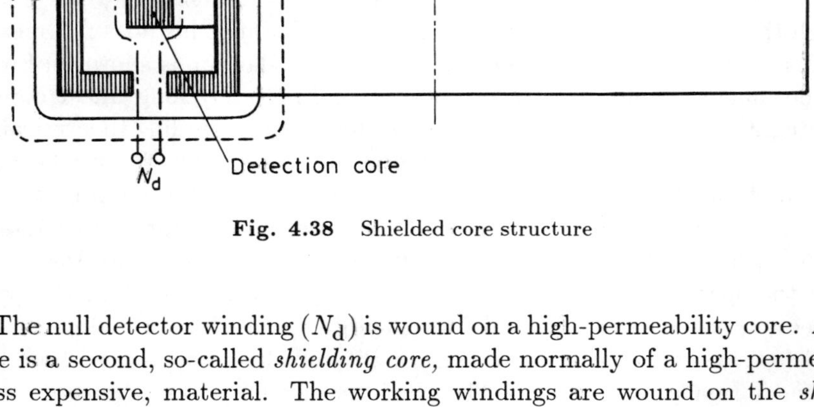

Fig. 4.38 Shielded core structure

The null detector winding (N_d) is wound on a high-permeability core. Around it there is a second, so-called *shielding core,* made normally of a high-permeability, but less expensive, material. The working windings are wound on the *shielding core*. The field fluctuation appears mainly in the shielding core, and in the inner core (called *detection core*) the field is smoother. The smoothing effect is inversely proportional to the permeability of the shielding core, the cross section of it, and to the distance between the shielding and detection core.

The errors caused by magnetic effects can be smaller than 10^{-8} in laboratory applications. The industrial solutions have an uncertainty of $1-5\cdot10^{-6}$.

c) Capacitive errors

In the windings of a current comparator, in a balanced state, there is no induced voltage. However, if the coupling is not perfect between the windings, a stray inductance will be present. The windings have ohmic resistance, too. The currents flowing in the windings result in voltage drops on the stray impedances, and the potential difference on the stray capacitances induces error currents. A detailed, lumped parameter equivalent of a two winding group of the two working windings of Fig. 4.34 is shown in Fig. 4.39.

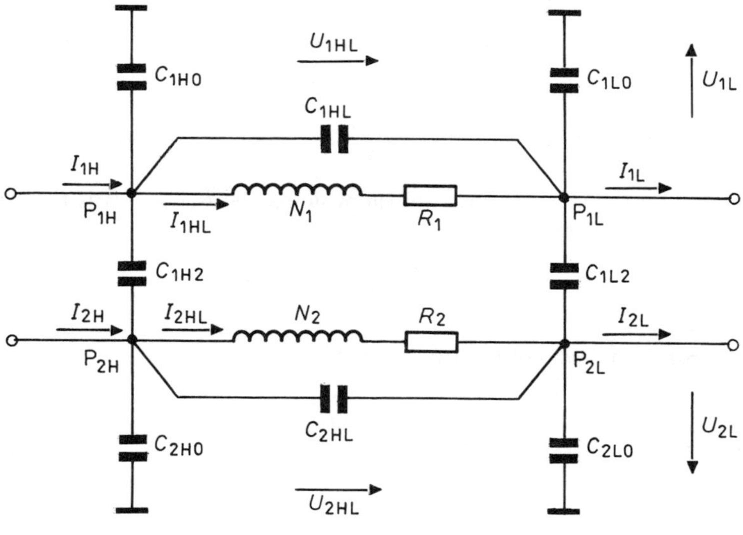

Fig. 4.39 Lumped parameter equivalent of a winding group

The low-voltage node of a winding $P_{i\mathrm{L}}$ has normally zero potential. If the connection leads to earth have finite resistance or a balancing circuit has any potential on this terminal, the potential $U_{i\mathrm{L}}$ on capacitance $C_{i\mathrm{L0}}$ results in an earth current from the node. Therefore the current $I_{i\mathrm{HL}}$, flowing to the winding N_d is not the same as the current $I_{i\mathrm{L}}$ flowing from the terminal. The current in the winding results in a potential drop on the stray inductances and copper resistance. The potential at the node $P_{i\mathrm{H}}$ is composed from the potential $U_{i\mathrm{L}}$ of the low terminal and the potential drop on the winding. This potential forces a current through the stray capacitance C_{iH0}. Therefore, the current, which flows in the winding, is not equal to the input current $I_{i\mathrm{H}}$ of terminal $\mathrm{P}_{i\mathrm{H}}$. If there is a potential difference between the nodes $\mathrm{P}_{1\mathrm{H}}$ and $\mathrm{P}_{2\mathrm{H}}$, a second error current will be forced through the capacitor $C_{1\mathrm{H}2}$.

The error currents from the high terminals have the most dangerous effect, because the input currents of these terminals come directly from the impedances, thus any deviation from them causes errors.

The voltage drop on the winding can be reduced by suitable construction. A low number of turns makes it possible to reduce the winding resistance, and at the same time, to reduce the stray inductances. Reducing the thickness of insulation between the layers is not a good idea because it increases the stray capacitances. Another possibility is to use potential regulated shields. A shield, connected to the low-voltage terminal, and covering the winding and the measuring cable, eliminates the effect of capacitances C_{iH0} and $C_{1\mathrm{H}2}$. In a well designed current comparator the main danger comes from the potential difference between the low-voltage terminal and earth.

Example 4.5
Let us investigate a high-voltage capacitance measuring circuit, which measures a capacitance $C_{\mathrm{x}}=1\,\mathrm{nF}$, with a measurement voltage of $U_0=1\,\mathrm{kV}$ (Fig. 4.35). The current in the winding N_{x} is:

$$I_{\mathrm{x}} = 2\pi f C_{\mathrm{x}} U_0 = 3.14 \cdot 10^{-4}\,\mathrm{A} \qquad \text{if} \qquad f = 50\,\mathrm{Hz}\,.$$

Let the potential on the common point of the windings be $U_{1\mathrm{L}}=1\,\mathrm{V}$. If the stray capacitance is $C_{\mathrm{s}}=100\,\mathrm{pF}$, the error current I_{s} is:

$$I_{\mathrm{s}} = 2\pi f U_{\mathrm{s}} C_{\mathrm{s}} = 3.14 \cdot 10^{-8}\,\mathrm{A}\,.$$

The relative error is the ratio of the two currents:

$$\epsilon_{\mathrm{r}} = \frac{I_{\mathrm{s}}}{I_{\mathrm{x}}} = 100\,\mathrm{ppm}\,,$$

a very high value!

The sensitivity of the current comparator bridge

The output voltage of a simple inductive comparator is computed on the basis of Fig. 4.40. The ampere-turn on the core is:

$$\Theta = I_1 N_1 - I_2 N_2 - I_{\mathrm{d}} N_{\mathrm{d}}\,. \tag{4.48}$$

The induced voltage on the detector winding N_{d} is:

$$U_{\mathrm{d}} = j\omega \Lambda N_{\mathrm{d}} \Theta\,, \tag{4.49}$$

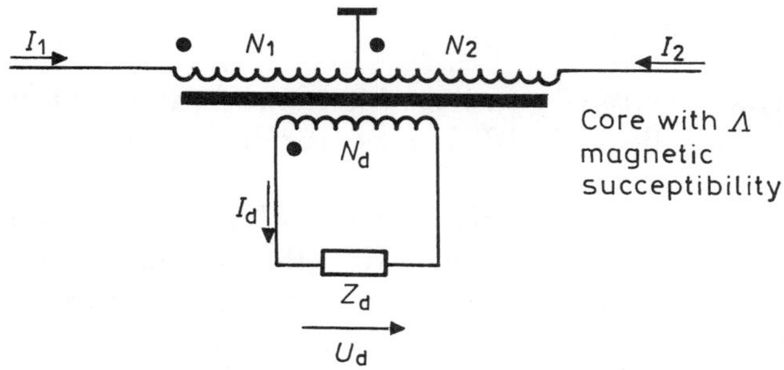

Fig. 4.40 Current comparator

where Λ is the magnetic succeptibility of the core. With

$$I_{\mathrm{d}} = U_{\mathrm{d}}/Z_{\mathrm{d}}\,, \tag{4.50}$$

$$U_{\mathrm{d}} = \frac{j\omega\Lambda N_{\mathrm{d}}(I_1 N_1 - I_2 N_2)}{1 + \dfrac{j\omega\Lambda N_{\mathrm{d}}^2}{Z_{\mathrm{d}}}}\,. \tag{4.51}$$

If the current comparator is used in an impedance measuring circuit, shown in Fig. 4.35, the bridge factor can be computed:

$$H = \frac{\Delta U_{\mathrm{d}}/U_0}{\Delta Z_{\mathrm{x}}/Z_{\mathrm{x}}} = \frac{j\omega\Lambda N_1 N_{\mathrm{d}}}{1 + \dfrac{j\omega\Lambda N_{\mathrm{d}}^2}{Z_{\mathrm{i}}}} \cdot \frac{1}{Z_{\mathrm{x}}}\,. \tag{4.52}$$

If a high input impedance null detector is used $\left(Z_{\mathrm{i}} \gg j\omega\Lambda N_{\mathrm{d}}^2\right)$:

$$H = \frac{j\omega\Lambda N_1 N_{\mathrm{d}}}{Z_{\mathrm{x}}}\,. \tag{4.53}$$

The bridge factor of the current comparator can be much higher than 1/4. The sensitivity is increasing with the frequency and with the number of turns. If the null detector is current sensitive, i. e. it has a low input resistance: $Z_{\mathrm{d}} \gg \omega\Lambda N_{\mathrm{d}}^2$, the circuit sensitivity (bridge factor) is:

$$H = \frac{N_2}{N_{\mathrm{d}}} \cdot \frac{Z_{\mathrm{d}}}{Z_{\mathrm{x}}}\,. \tag{4.54}$$

The voltage-sensitive detector ($Z_{\mathrm{d}} \to \infty$) gives a high sensitivity, but the output voltage depends on the iron permeability. If the circuit is far away from the balancing point, the saturation of the core can reduce the sensitivity and disturb the convergence. The linearity of the bridge with current sensitive detector ($Z_{\mathrm{d}} \to 0$) is better, but the sensitivity is lower. A solution to the sensitivity problem is shown later in the subsection concerning the speed of measurement.

Frequency dependence and connections

At low frequencies the errors are mainly due to the magnetic effects and the sensitivity of the bridge. Below 16 2/3 Hz it is very difficult to construct a sensitive bridge.

A very important field of application of current-comparator bridges is in the range of industrial frequencies (50–60 Hz). Here the magnetic properties have theoretically higher influence on the errors than the capacitive effects. The reduced sensitivity in the case of capacitance measurement needs a higher number of turns, therefore the influence of stray capacitances will be increased. An industrial-frequency current comparator bridge has usually an error of 1 − 10 ppm.

At higher frequencies (over 10 kHz) the capacitive error dominates. At 100 kHz the magnetic quality degrades, the capacitive effect is high, therefore an error over 100 ppm can be expected.

The optimum working frequency of an inductive current comparator is 1 kHz. Here the sensitivity is high, and the effect of capacitances is low. A well constructed comparator can have an error of 10^{-8} at 1 kHz.

Structurally the current comparator allows two- and three-wire measurements. In special cases the five-terminal measurement is also possible. In Fig. 4.41 such a capacitance measuring bridge is shown.

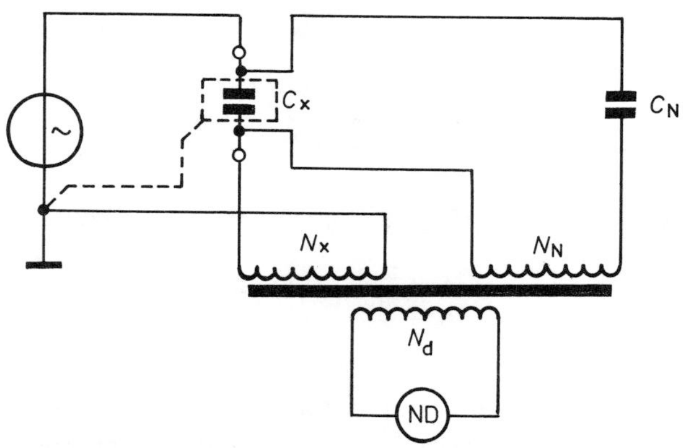

Fig. 4.41 Five-wire measurement with current comparator

The capacitance C_X has a much higher value than the capacitor C_N. The high value of C_X necessitates a four- or five-terminal connection because the resistance of the connecting wire and the winding N_X can cause a high phase error.

The circuit C_N is connected on the potential terminals of C_X, thus the bridge compares the voltage on the unknown terminal to the current of C_X. In this way the effects of the leads are eliminated. The fifth wire connects the shield of C_X to the earth.

An important error source is the potential on the winding N_N. If C_X is carrying a large current, the potential on the low terminal of N_N can be large, and therefore a large error current can flow from the high terminal of the winding N_N to the earth.

Further errors

The high-permeability core of the current comparator can produce further noises. The mechanical deformation of the core (e. g. vibration) produces so-called magnetostrictive voltage. If the frequency of this voltage is the same as the measurement frequency, it can cause very large errors.

The core is also very sensitive to external magnetic fields. A comparator of a sensitive bridge must be magnetically shielded. These interferences are the most dangerous in the industrial frequency range (50 – 60 Hz), which coincides with the vibration frequency of motors and transformers. The stray fields of transformers also include industrial frequency components.

The speed of measurement

A typical balancing circuit is shown in Fig. 4.42.

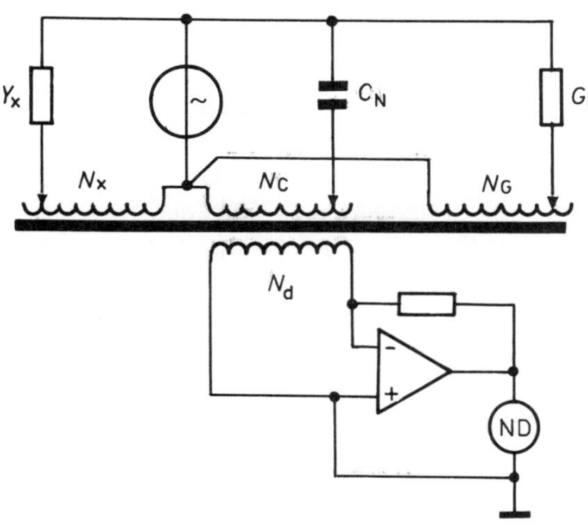

Fig. 4.42 Balancing with perpendicular convergence angle

Two separate windings serve for the in-phase (N_G) and quadrature (N_C) balancing. The convergence is perpendicular, therefore the balancing speed is maximal. Perpendicular convergence is only realized if the detection circuit is electronically enhanced as shown in Fig. 4.42. The comparator is loaded with low impedance, therefore the core cannot be saturated. The sensitivity can be high if the feedback resistor in the current-voltage converter is high.

4.3.4 Electronic Impedance Measuring Circuits

Electronic circuits for ratio generation and signal comparison

Electronic voltage ratio generation

The circuit shown in Fig. 4.43 produces an output voltage U_{out}, proportional to the input voltage and the ratio of the two feedback impedances:

$$U_{\text{out}} = -U_{\text{in}}\frac{Z_2}{Z_1}\,, \tag{4.55}$$

if the gain of the operational amplifier is infinitely high: $A=\infty$.

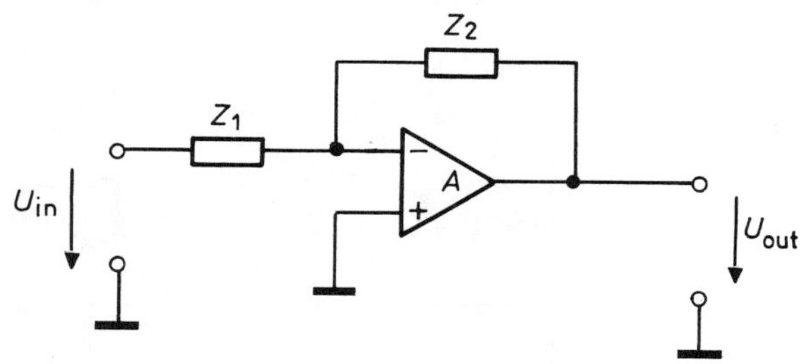

Fig. 4.43 Electronic voltage ratio generator

The produced voltage U_{out} has reversed polarity compared with the input voltage U_{in}, therefore the circuit is useful for methods of current comparison.

Electronic units of current comparison

The circuit shown in Fig. 4.44 gives an output signal U_{i} proportional to the sum of two (or more) input currents:

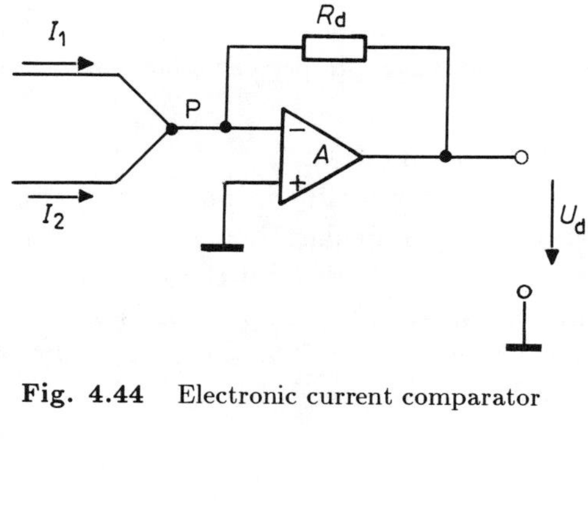

Fig. 4.44 Electronic current comparator

$$U_i = R_i(I_1 + I_2)\,. \tag{4.56}$$

If $I_1 = -I_2$, the output voltage is zero, therefore the unit works as a null current detector. Obviously, the circuit of Fig. 4.45 has the same effect as the one in Fig. 4.44, but the latter has a nonzero input voltage in the unbalanced case, therefore it has a feedback to the measurement circuit.

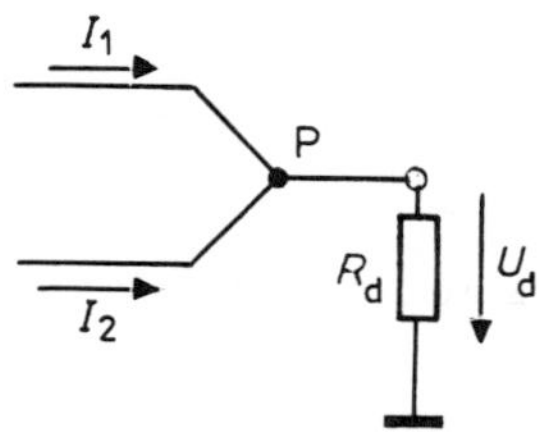

Fig. 4.45 Passive current comparator

Electronic ac ratio detectors

The circuit shown in Fig. 4.46a was introduced by Hall (1973) and Hashimoto and Tamamura (1976). It is actually an ac dual-slope voltmeter, working as follows. During the time t_1, switch S_1 is in position X, and switch S_2 is in position 0°. The *phase sensitive detector* (PSD) is a controlled switch which passes the signal U_a during the positive period of U_{ref} and passes $-U_a$ during the negative period of U_{ref}. The mean value of the output signal of the PSD is proportional to the real component of the voltage U_x. This mean value is transformed to a dc voltage by the so-called *sectional average integrator.* This unit integrates the voltage U_b during a whole number of periods. The output voltage U_c is integrated by a dual-slope voltmeter. After the time t_1, the integrated value is

$$U_r = \frac{t_1}{T_d}\mathrm{Re}\{U_x\}\,, \tag{4.57}$$

where T_d is the integration time constant of the dual-slope voltmeter.

Now switch S_1 is switched to position N, switch S_2 remains in position 0°. After a whole number of periods, the sectional average integrator will produce a dc voltage, proportional to the absolute mean value of the voltage U_N. The dual-slope voltmeter 'integrates back' this voltage from U_r to zero in the time t_2:

$$U_r = \frac{t_2}{T_d}U_N\,. \tag{4.58}$$

The ratio of the real component of U_x to U_N is:

$$\frac{\mathrm{Re}\{U_x\}}{U_N} = \frac{t_2}{t_1}\,. \tag{4.59}$$

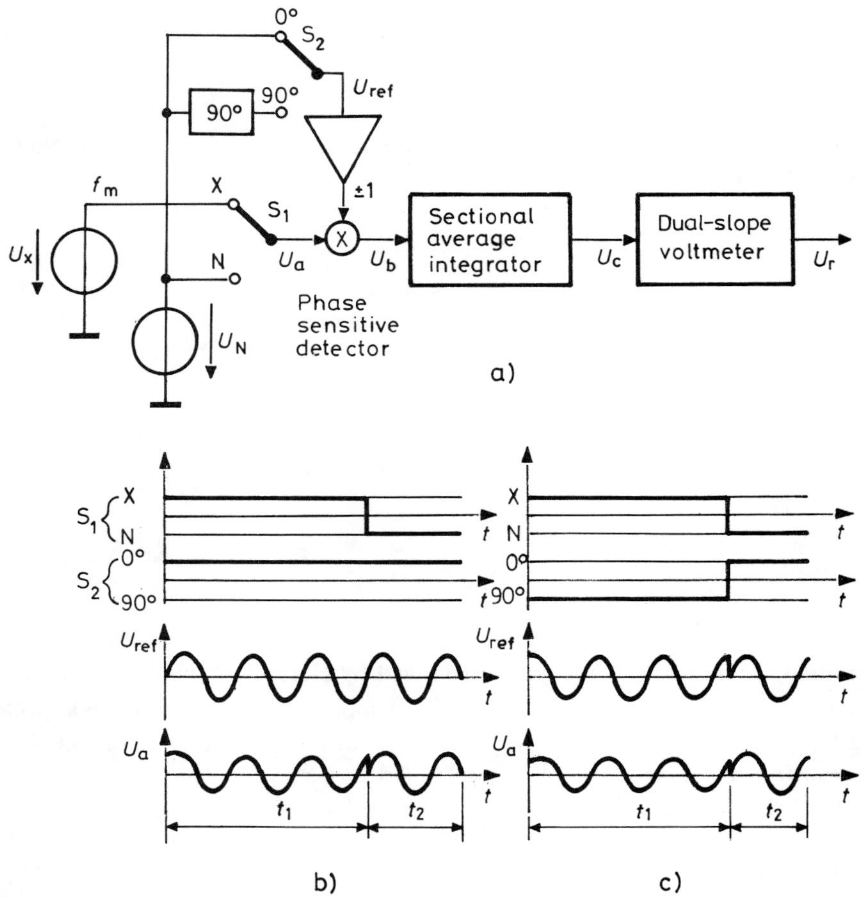

Fig. 4.46 Electronic ac ratio detector *a)* Block diagram *b)* Time diagrams of the measurement of the real component *c)* Time diagrams of the measurement of the imaginary component

In this way the measurement of the ratio of the real component of the voltage U_X to the absolute mean value of the voltage U_N is reduced to a time ratio measurement.

The imaginary component of U_X is measured in a similar way. As shown in Fig. 4.47c, during the phase t_3 switch S_1 is in position X and switch S_2 is in position 90°. The output of the phase sensitive detector is proportional to the imaginary component of U_X. The integral of this signal after time t_3 is:

$$U_\mathrm{i} = \frac{t_3}{RC}\mathrm{Im}\{U_\mathrm{X}\}\,. \tag{4.60}$$

Now switch S_1 is switched to position N and S_2 to 0°. The *integrator* integrates again until zero, and the following equality holds:

$$U_i = \frac{t_4}{RC} U_N \,. \tag{4.61}$$

The ratio of the imaginary component to the voltage U_N corresponds to the ratio of time intervals:

$$\frac{\mathrm{Im}\{U_x\}}{U_N} = \frac{t_4}{t_3} \,. \tag{4.62}$$

Now the value t_4 will be displayed.

A simplified solution is shown in Fig. 4.47.

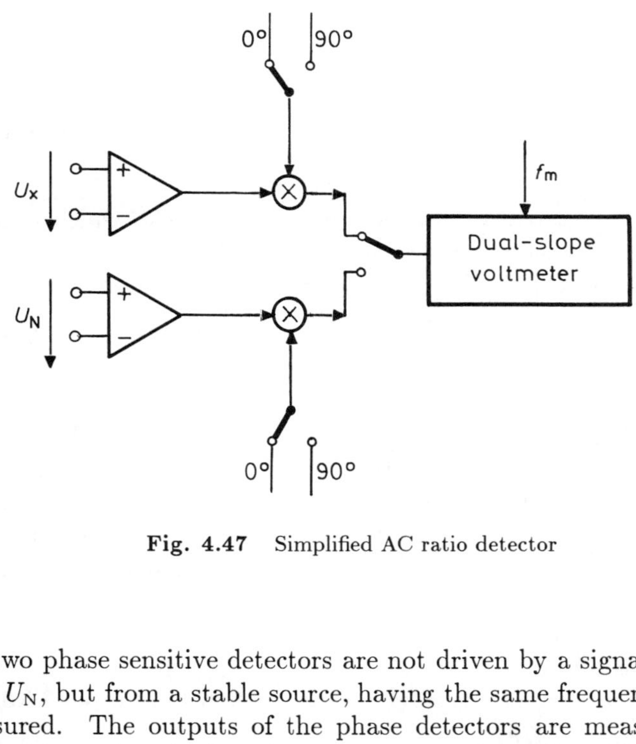

Fig. 4.47 Simplified AC ratio detector

The two phase sensitive detectors are not driven by a signal proportional to the voltage U_N, but from a stable source, having the same frequency as the signals to be measured. The outputs of the phase detectors are measured by a dual-slope voltmeter, synchronised to the measurement frequency. The digital outputs of the voltmeter are processed by a computer. By successive sampling, the real and imaginary components of the signals X and N are measured. The reference vector is always the same. The computer evaluates the outputs of the voltmeter, and computes the complex ratio X/N from the measured complex vectors of voltages. The advantage of this method is the simple control structure and the high accuracy of the 90° phase shift. The reference signal is generated by a PLL (Phase Locked Loop) circuit in the supply voltage generation part of the instrument.

Structure of electronic impedance measurement circuits

The *measurement circuit with ratio generator and current comparator* (Osada and Suehiro, 1975; Maeda, 1974), shown in Fig. 4.48, is the electronic equivalent of the bridges with inductive ratio transformer and current null detector.

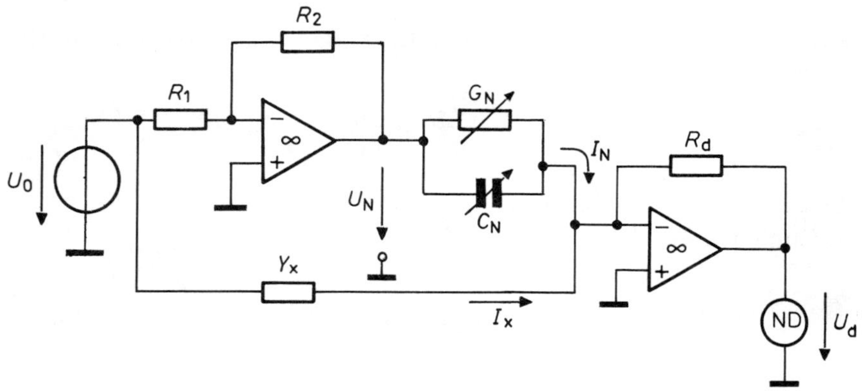

Fig. 4.48 Measurement circuit with electronic ratio generator and current comparator

The ratio generator produces the output voltage U_N, which is proportional to the input U_0 and to the resistance ratio R_2/R_1. The circuit is balanced by the adjustment of G_N and C_N. In the balanced state the measured values are:

$$\mathrm{Re}\{Y_x\} = \frac{R_2}{R_1} G_N ; \qquad \mathrm{Im}\{Y_x\} = \frac{R_2}{R_1} \omega C_N .$$

The advantage of the method is the simple earth structure. In the case of an inductive current comparator either the voltage source or the common node of impedances can be connected to earth, but not both of them at the same time. In Fig. 4.48 the source is connected to earth, and one node of Y_x is on earth potential.

A disadvantage of the given solution is the clumsy balancing unit. A better one uses electronically controlled multiplying D/A converters before the balancing impedances, and makes possible the automation of the method. The ratio generator is generally used as a range selector.

A *measurement circuit with ratio detector* is shown in Fig. 4.49 (Maeda, 1974; Maeda and Naramitsa, 1979).

The signal X, proportional to the voltage $U_x = I_0 Z_x$, is divided by the signal proportional to $U_N = I_0 R_N$. From the measured ratios:

$$\mathrm{Re}\{Z_x\} = R_N \frac{A_2}{A_1} \mathrm{Re}\left\{\frac{X}{Y}\right\}; \qquad \mathrm{Im}\{Z_x\} = R_N \frac{A_2}{A_1} \mathrm{Im}\left\{\frac{X}{Y}\right\}. \tag{4.63}$$

The method is theoretically clear, because the two amplifiers detect directly the voltage drops on the impedances, and only one standard impedance is required.

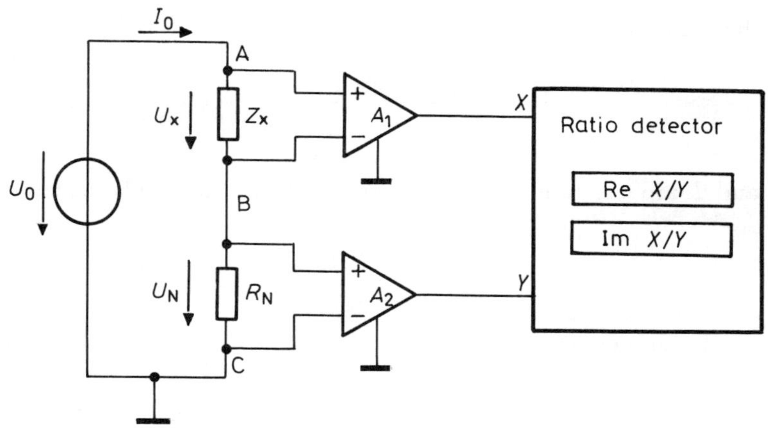

Fig. 4.49 Measurement circuit with electronic ratio generator and current comparator

Otherwise the ratio measurement procedure and the measuring circuits are clearly separated, and have no reciprocal effects. However, any stray impedance between the low-potential terminal of Z_x and the guard causes an error in the current measurement, and thus in the impedance measurement as well.

The errors of electronic impedance measurement circuits

a) Basic errors and sensitivity

An electronic solution cannot override the limits of uncertainties of standard impedances because it is also built of impedances and other elements. The amplifiers have finite gain and a finite output impedance. In Fig. 4.49 the common mode voltage on amplifier A_1 causes an error. Despite this, electronic circuits provide the highest accuracy, because only one standard resistor is used. The cost of only one high-accuracy, wide frequency band resistor is essentially lower than the costs of multiple resistor and capacitor blocks.

The phase detection system must be sufficiently accurate. Dual-slope voltage detection can guarantee the required accuracy, and the phase-sensitive detectors are the main sources of inaccuracy.

b) Frequency dependence and connections

In the case of an *electronic ratio generator,* shown in Fig. 4.43, the frequency dependent transfer function of the operation amplifier yields a further systematic error.

The transfer function of the unit in the case of $Z_1 = R_1 = Z_2 = R_2$ is:

$$R(f) = \frac{U_{\text{out}}}{U_{\text{in}}} = \frac{-1}{1 + \frac{2(1+j\omega T)}{A}}, \tag{4.64}$$

where A and T are the gain and the time constant of the amplifier, respectively:

$$H(\omega) = \frac{A}{1 + j\omega T}. \tag{4.65}$$

The ratio and phase errors of the unit are:

$$\epsilon_{\text{r}} = -2\left(\frac{1}{A} + \left(\frac{f}{f_{\text{t}}}\right)^2\right); \qquad \delta = -2\frac{f}{f_{\text{t}}}, \tag{4.66}$$

where

$$f_{\text{t}} = \frac{A}{2\pi T}$$

is the transit frequency of the amplifier.

Example 4.6
An amplifier has a transit frequency of 5 MHz and a gain of 10^5. At 50 Hz the ratio error will be 20 ppm, and the phase error 20 μrad. These are large values compared with the phase error of an inductive voltage divider. To reach the 1 ppm accuracy an amplifier with 100 MHz transit frequency would be required. Presently it is not a problem to get a commercial amplifier of this specification, but the technological problems are immense: for 50 Hz use it is required to implement an electronic circuit working in the dc–100 MHz frequency range.

A second problem of the electronic ratio generation and current ratio detector lies in the stability of circuits. The impedance to be measured is included in high-frequency electronic feedback loops. The load of the ratio generator on the amplifier can result in errors and instability at higher frequencies. Only a 1 metre piece of coaxial cable on the output of an operation amplifier may make it unstable. The impedance to be tested is included in the feedback loop of the current ratio detector. The structure of the test object is unpredictable, in extreme cases it can be the inductance of a large power transformer or the output impedance of an amplifier to be tested. In practice it is impossible to find a feedback compensation working correctly in all cases. For example, measuring an inductive object with an inexpensive ohmmeter, the autoranging will fail because of the oscillating current generator.

The current comparison method is appropriate first of all for *three-terminal measurements,* but with some compromises it is possible to accomplish *five-terminal* measurements, too. In Fig. 4.50 a five-terminal connection is shown.

The ratio unit gets the input signal from the potential terminal P_{H} of the test object, and the current balance is detected on the low-potential terminal P_{L}. The current detector does not load the terminal, because in the balanced state the

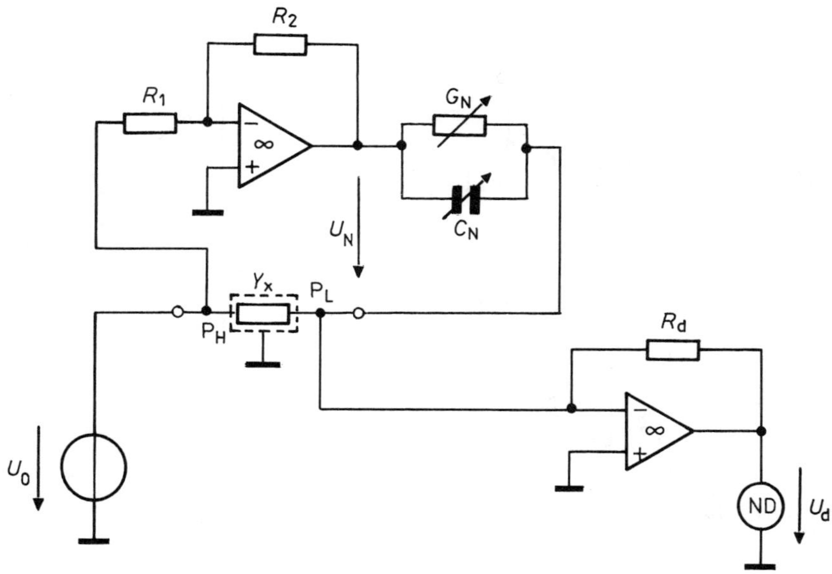

Fig. 4.50 Five-terminal measurement with electronic current comparator

detector current is zero. The point P_H is loaded by the resistance R_1, but in most cases this causes no problem, because this is the reference point. Only the current distribution can cause errors in the case of measurement of very low resistances with high currents. The shield of Y_x, connected to earth, does not produce an error, because the terminal P_L of the test object is only *virtually* brought to earth potential. The standard admittance(s) must work in two-terminal mode.

In the case of *electronic ratio detection* (Fig. 4.51) the situation is theoretically better. In the circuit of the test object no amplifier is fed back, therefore there is no stability problem. The amplifiers A_1 and A_2 are connected in noninverting mode, therefore their feedback is out of the current loop of the test object. These amplifiers have of course frequency dependent errors coming from the frequency dependence of their gain and common-mode rejection. The gain error can be compensated by calibration procedures, but the common-mode signal and rejection is strongly dependent on the structure of the test object and the attempts of compensation are subject to failure. A solution of the common-mode problem is shown in Fig. 4.51.

The operation amplifier marked with gain '∞' sets the potential of the terminal P_L to the potential of the earth. The two matching amplifiers A_1 and A_2 have no more excessive common-mode voltage, therefore a *five-terminal measurement* is realized.

The price of this common-mode voltage rejection is the stability problem of the amplifier, fed-back by the standard resistor. A common solution is the use of a so-called transposition amplifier. The circuit shown in Fig. 4.52 can stabilize the measuring circuit even in the case of an impedance of extremely complex structure.

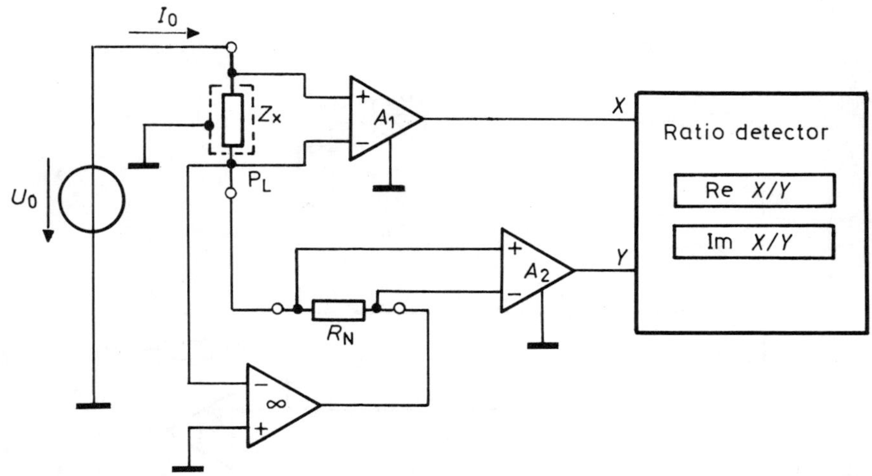

Fig. 4.51 Rejection of common-mode voltage in a voltage ratio measurement circuit

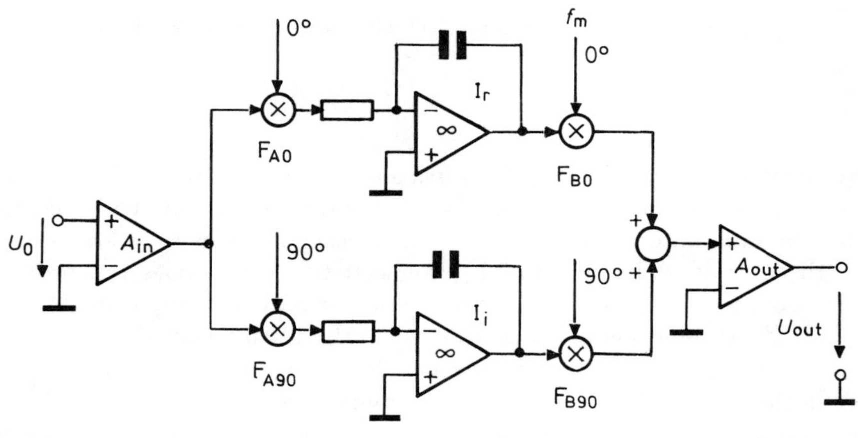

Fig. 4.52 The transposition amplifier

The amplified input signal is transformed by the phase-sensitive detectors F_{A0} and F_{A90} to dc voltages, proportional to the real and imaginary components. The outputs of the phase sensitive detectors are integrated by the integrators I_r and I_i. The modulators F_{B0} and F_{B90} produce ac signals, proportional to the DC outputs of the integrators. The two outputs of the modulators are added, and after amplification the output is generated.

Obviously, at carrier frequency f_m, the gain of the transposition amplifier is infinitely high, therefore high accuracy is guaranteed. At other frequencies the amplifier can be very 'slow', if the time constants of the integrations are high, therefore the critical poles and zeros of the feedback loop can be neutralized. The disadvantage of this solution is the requirement for a modulating signal. In the case of an instrument, where the measuring voltage is generated locally, this is not a problem, but in heavy current measurements, where the supply voltage is generated externally, the production of a modulating signal can be difficult.

The offset and drift currents of the integrators cause an error in the impedance measurement.

4.3.5 T- and Double T-Circuits

Structure of circuits

The *T-circuits* and *double T-circuits* are special variants of current comparison circuits. Fig. 4.53 presents a *T-circuit.*

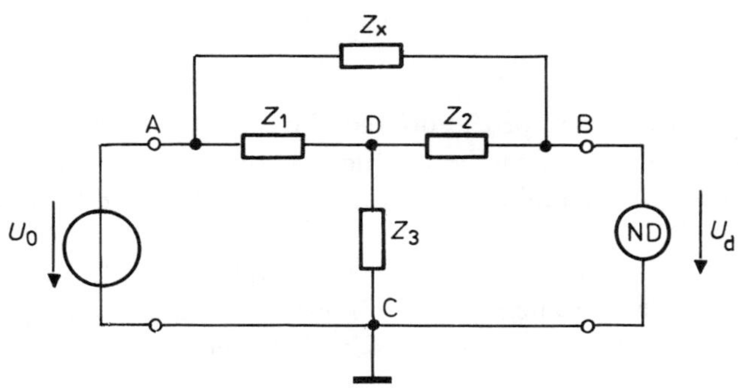

Fig. 4.53 The T-circuit

The Δ-equivalent of the Y part, composed from the impedances Z_1, Z_2 and Z_3, can show a negative admittance in a given frequency domain, between the nodes A and B. The null detector displays zero current if

$$Y_x + \frac{Y_1 Y_2}{Y_1 + Y_2 + Y_3} = 0\,. \tag{4.67}$$

The *double T-circuit*, shown in Fig. 4.54, is composed of two T-circuits.

The balancing equation is:

$$\frac{Y_1 Y_2}{Y_1 + Y_2 + Y_3} + \frac{Y_4 Y_5}{Y_4 + Y_5 + Y_6} = 0\,. \tag{4.68}$$

The errors of T-circuit measurements

T-circuits are not the most accurate measuring circuits. The balancing is frequency dependent; in the case of the simple T-circuit two- and three-wire measurements

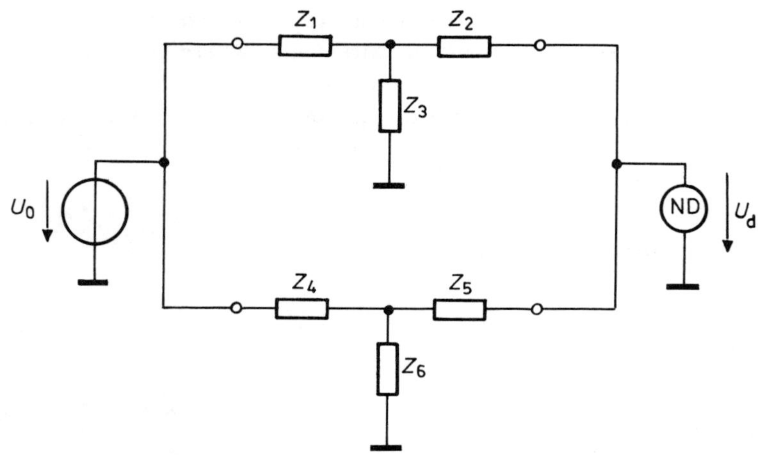

Fig. 4.54 The double T-circuit

are only possible; in the case of the double T-circuit two-wire measurements can only be realized. However, the method is the only one suitable for higher-frequency measurements, because the null detector and the signal source have common earth terminals.

Over a few 100 kHz all other methods give dubious results, even the balancing can be impossible. The balancing of the T-circuit is almost always possible and therefore it makes possible the use of substitution methods.

4.3.6 Resonance Methods

The series resonance method

The current, flowing in the circuit in Fig. 4.55, has a maximum at the angular frequency

$$\omega_0 = \frac{1}{\sqrt{LC}}\,; \qquad I_{\max} = \frac{U_0}{R}\,. \tag{4.69}$$

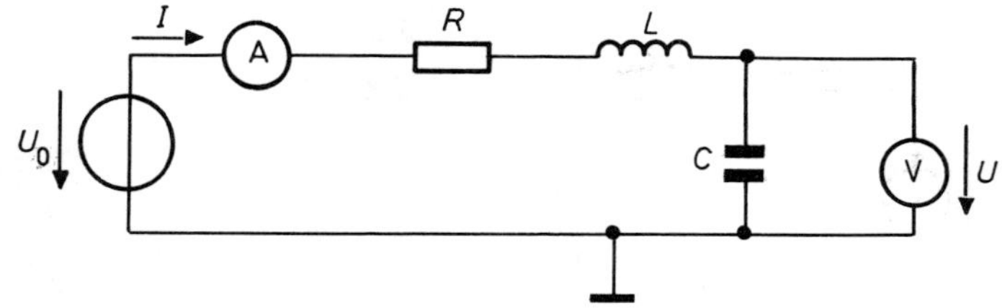

Fig. 4.55 The series resonance method

If the impedance represented by R and L is to be measured, their values are calculable from the known values of C and ω_0. The situation is similar if the values C and R are to be measured; in this case the value of L must be known.

The parallel resonance method

In Fig. 4.56 the *parallel resonance method* is shown. At the angular frequency ω_0 the output voltage U has a maximum:

$$\omega_0 = \frac{1}{\sqrt{LC}}; \qquad U_{\max} = I_0 R .$$

Either R and L from C and ω_0, or R and C from L and ω_0 are calculable.

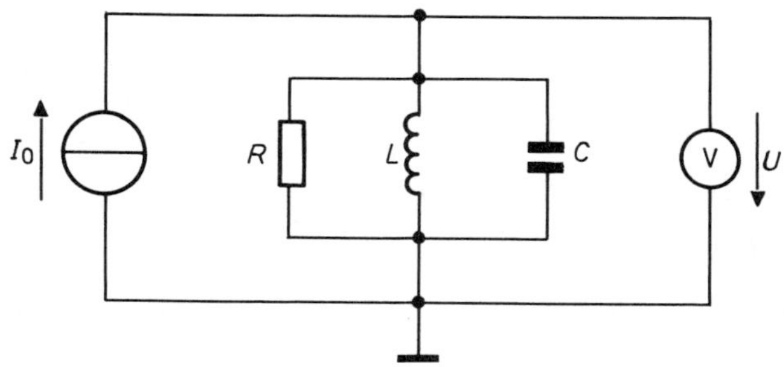

Fig. 4.56 The parallel resonance method

The errors

Resonance methods are not very accurate, but they are the most commonly used ones for high-frequency measurements. The earth nodes of the signal source and the voltage detector are always common, and in the case of the parallel resonance method, one terminal of each impedance is connected to the common earth. The 'balancing', i. e. the search for a maximum output signal of the circuit, is much easier than the balancing of a T-circuit. Only two-wire measurement is possible. Substitution procedures are used in order to reduce the errors.

The current measuring device has no common node with the earth in either case, therefore it can decrease the frequency band of measurement. To override this problem, the current is not used in the evaluation. Usually some typical points of the resonance curve are measured (e. g. the frequency and amplitude of the maximal output voltage and the 3 dB points) and from these values the impedances are calculated.

4.4 MEASUREMENT OF BASIC IMPEDANCES

4.4.1 Measurement of Resistances

The simplest method is the *measurement with voltmeter and ampermeter,* shown in Fig. 4.57.

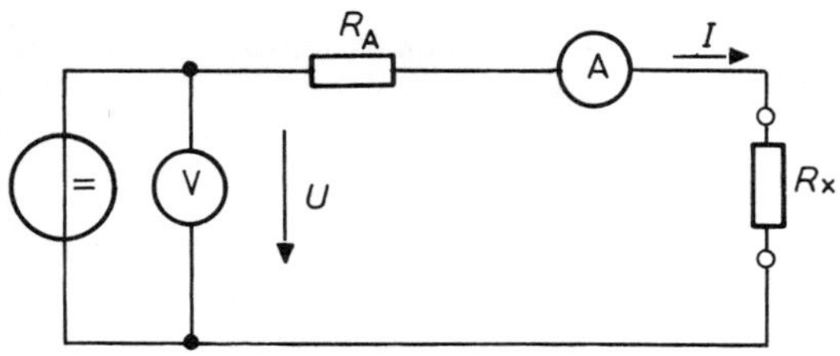

Fig. 4.57 Resistance measurement by voltmeter and ampermeter

In the given setup the internal resistance of the ammeter causes a systematic error, therefore a correction is necessary:

$$R_X = \frac{U}{I} - R_A . \tag{4.70}$$

The *ohmmeter in series,* shown in Fig. 4.58, uses a voltage source with stable dc voltage U_0.

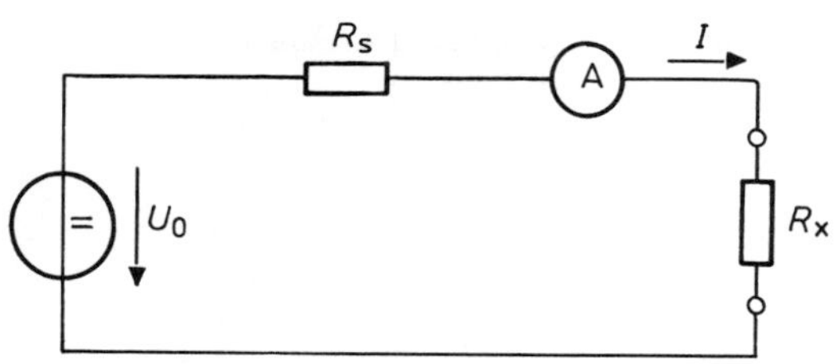

Fig. 4.58 Ohmmeter in series

With an ohmmeter of resistance R_s in series with R_X, the meter is scaled in Ohms, according to

$$I = \frac{U_0}{R_s + R_X} . \tag{4.71}$$

The *parallel ohmmeter,* shown in Fig. 4.59, also uses a stable voltage source. The voltmeter, measuring the voltage drop on the resistance R_X, is scaled in ohms according to the nonlinear expression of the voltage:

$$U = U_0 \frac{R_X}{R_s + R_X} . \tag{4.72}$$

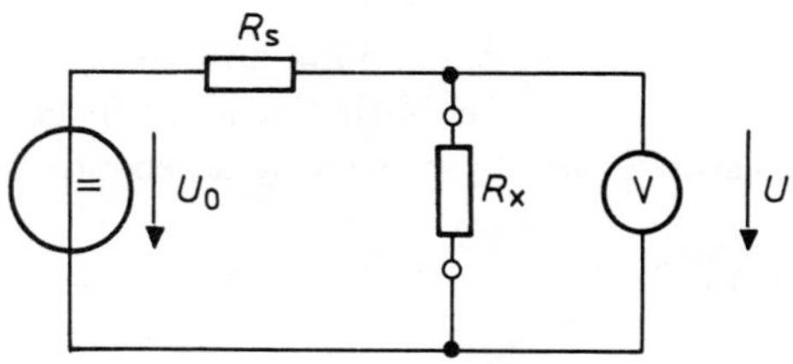

Fig. 4.59 Parallel ohmmeter

The *Wheatstone Bridge* (Wheatstone, 1843), shown in Fig. 4.60, is the simplest bridge circuit. In the balanced state

$$R_x = \frac{R_2 R_3}{R_4}. \tag{4.73}$$

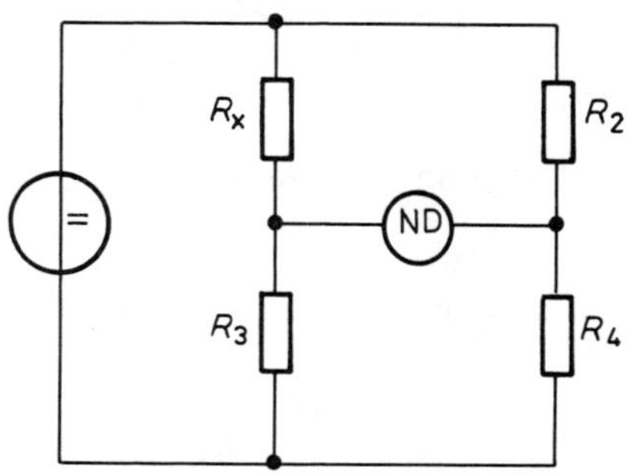

Fig. 4.60 The Wheatstone bridge

The smallest uncertainty of the bridge is 10^{-5}, the range of measurement is $10\,\Omega$–$10^{10}\,\Omega$.

The *Thomson-Kelvin bridge* (Hague, 1962), shown in Fig. 4.61, is designed for four-wire measurements of small resistances.

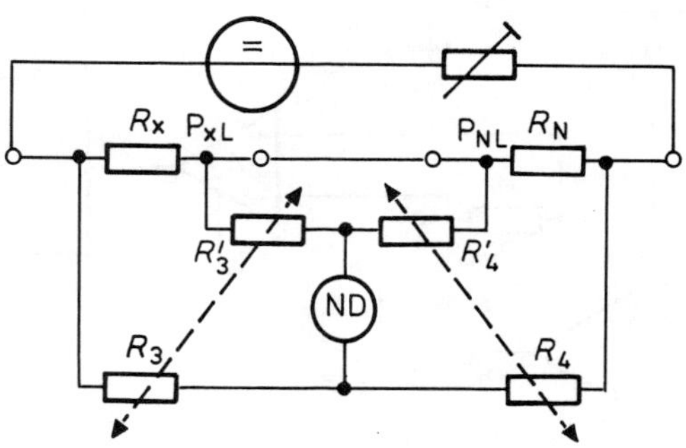

Fig. 4.61 The Thomson-Kelvin bridge

The resistances arise from connections and internal leads between the two potential terminals P_{xL} and P_{NL}, for both the resistance to be measured and the standard one. The resistances are by-passed by a secondary bridge arm. If

$$\frac{R_3'}{R_4'} = \frac{R_3}{R_4}, \qquad \text{then} \qquad R_x = \frac{R_N R_3}{R_4}, \tag{4.74}$$

the resistances of the connections are not measured.

The best resolution is $0.1\,\mu\Omega$, the maximum measurable value is $100\,\Omega$. The smallest uncertainty may reach the value of 10^{-5}.

The *electronic ohmmeter,* shown in Fig. 4.62, is the modernized version of the parallel ohmmeter.

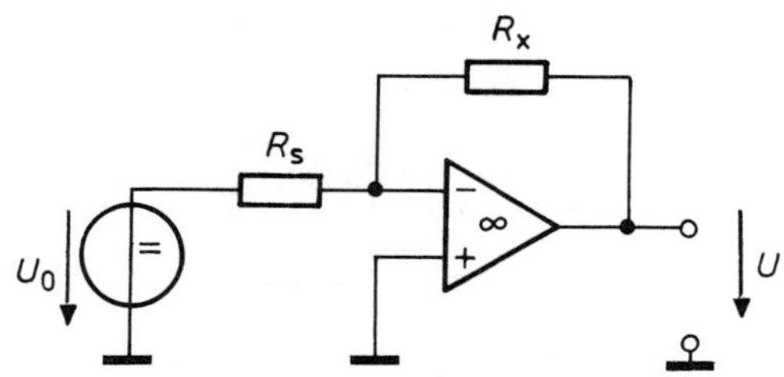

Fig. 4.62 Electronic ohmmeter

The output voltage is usually measured by a digital voltmeter, and it is proportional to the value of R_x:

$$U_{out} = \frac{U_0}{R_s} \cdot R_x \,. \tag{4.75}$$

Using a dual-slope voltmeter, the range of this method is $1\Omega - 10\,M\Omega$. The uncertainty is smaller than 10^{-4}. The method is sensitive to the structure of R_x: an inductive test object can cause oscillations of the high-gain amplifier.

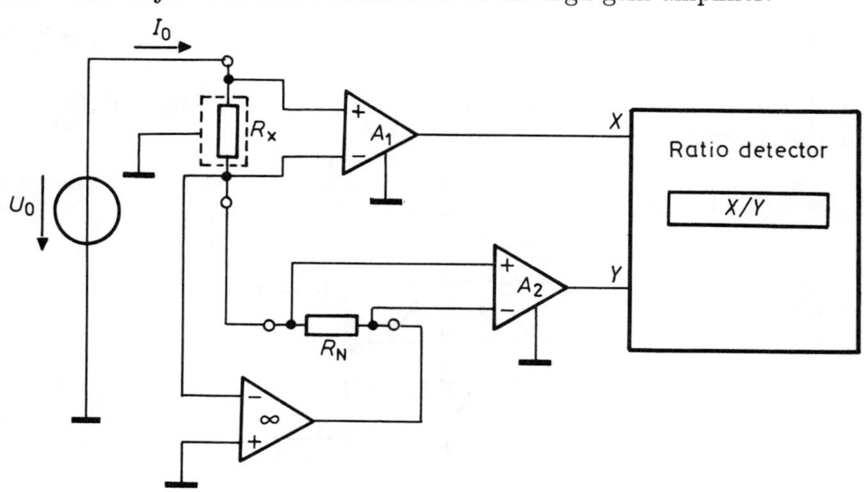

Fig. 4.63 Resistance measurement by voltage comparison

The method of *voltage comparison,* shown in Fig. 4.63, is the most flexible method of modern resistance measurement.

The four-wire connections of the resistors R_X and R_N make the measurement of low values of resistance possible. If the leakage currents of the amplifiers are low, high-resistance measurements are also possible. The range of this method is $0.1\,\mu\Omega - 10^{14}\,\Omega$. The uncertainty is smaller than 10^{-4}.

4.4.2 Measurement of Inductances

The *Maxwell-Wien Bridge* (Maxwell, 1873; Wien, 1891; Hague, 1962), shown in Fig. 4.64, is the most popular classical bridge for inductance measurement.

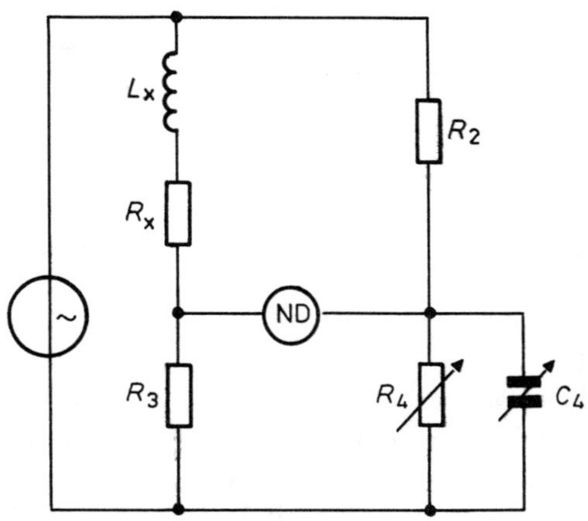

Fig. 4.64 The Maxwell-Wien bridge

In the balanced state:

$$L_X = R_2 R_3 C_4\,; \qquad R_X = \frac{R_2 R_3}{R_4}\,, \tag{4.76}$$

$$Q = \frac{\omega L_X}{R_X} = \omega R_4 C_4\,. \tag{4.77}$$

The best resolution is $0.1\,\mu$H, the maximum measurable value is 100 H, the uncertainty is smaller than 10^{-4}.

The optimum range of the measurement frequency is between 50 Hz and 10 kHz. Commercial bridges are specified up to 300 kHz, but with reduced accuracy. At very high Q values R_4 must be very large, therefore the loss of C_4 causes an error: a low-loss capacitance decade is required. The method detects the serial model of the inductance, and only the two-wire connection is possible.

The *Hay bridge* (Hay, 1910; Hague, 1962), shown in Fig. 4.65, measures the parallel equivalent, and for high quality factors a small resistance R_4 is required:

$$L_X = R_2 R_3 C_4\,; \qquad G_X = \frac{R_4}{R_2 R_3}\,, \tag{4.78}$$

$$Q = \frac{1}{\omega G_x L_x} = \frac{1}{\omega R_4 C_4}\,. \tag{4.79}$$

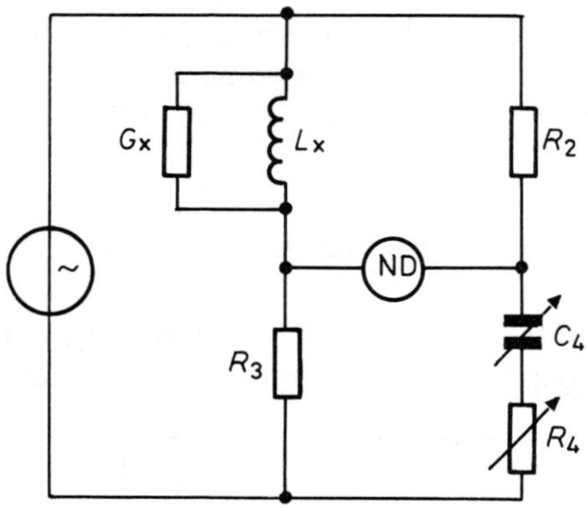

Fig. 4.65 The Hay bridge

The frequency domain and the range are the same as in the case of the Maxwell-Wien bridge.

An *inductance measuring circuit with electronic ac ratio meter* (Maeda, 1974) is shown in Fig. 4.66.

In the first phase of measurement switch S_1 is in position X, switch S_2 in position G. The signal U_d is proportional to the voltage on the impedance to be measured. The phase-sensitive detector PSD1 gets the 90° shifted version of the voltage U_0 as reference. Switch S_3 is switched on and the integrator I_a integrates the signal U_d. After the time T_1 the voltage at the output of the integrator is equal to:

$$U_a = G_s \omega L_x \frac{U_0}{R_N} T_1\,. \tag{4.80}$$

Next S_1 is switched to position N, S_2 to C, and the inverted voltage U_d (the inverter is not shown in the Figure) is integrated back to zero:

$$U_a = U_0 \omega C_s T_2\,. \tag{4.81}$$

From the equivalence of the two output voltage differences the inductance to be measured is calculable:

$$L_x = \frac{R_N C_s}{G_s} \cdot \frac{T_2}{T_1}\,. \tag{4.82}$$

In the second phase of the measurement the series resistance is measured. Switch S_1 is in position X, S_2 is in G. The phase sensitive detector PSD2 gets the voltage U_0.

After a time of integration T_1 (phase III) the output voltage of the second integrator is:

$$U_b = G_s R_x \frac{U_0}{R_N} T_1\,. \tag{4.83}$$

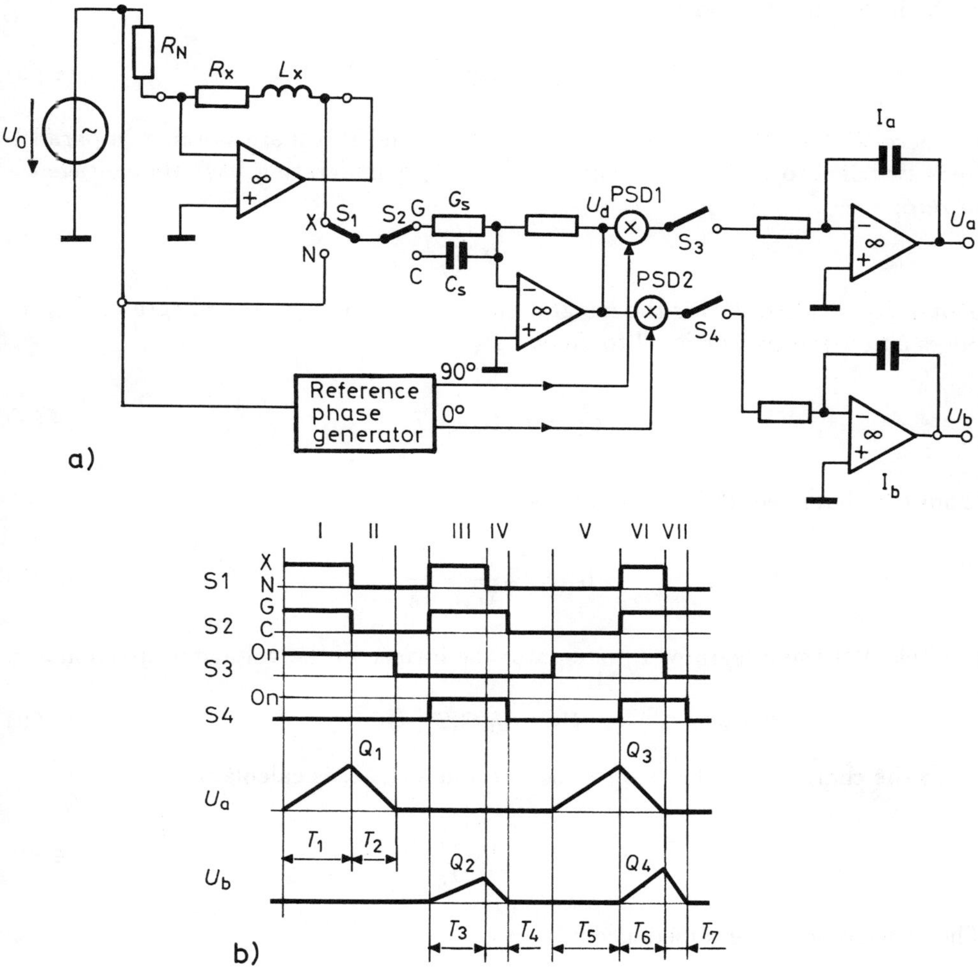

Fig. 4.66 Inductance measuring circuit with electronic ac ratio meter *a)* block diagram *b)* timing

Then in phase IV, S_1 is switched to N, and S_2 stays in position G. The voltage U_d is inverted (not indicated in the figure), and the voltage U_b is integrated back to zero:

$$U_b = U_0 G_s T_3 \,. \tag{4.84}$$

The resistance R_x is calculable from the known values of the components and from the measured time T_3:

$$R_x = R_N \frac{T_3}{T_1} \,. \tag{4.85}$$

In the last phase the loss tangent of L_x is measured. In phase V the voltage U_d is proportional to the voltage U_0, but its phase is shifted by 90°. The output voltage of the integrator I_a after the time T_5 is:

$$U_a = U_0 \omega C_s T_5 \,. \tag{4.86}$$

In phase VI the phase-sensitive detectors PSD1 and PSD2 are working in parallel. S_1 is switched to X, S_2 to G. The integrator I_a is integrating back the voltage U_d to zero:

$$U_a = \omega L_x \frac{U_0}{R_N} T_6 \,. \tag{4.87}$$

Switch S_4 is closed during the integration time T_6, therefore the integrator I_b produces an output proportional to T_6 and R_x:

$$U_b = R_x G_s \frac{U_0}{R_N} T_6 \,. \tag{4.88}$$

From the charge equivalence of phases V and VI:

$$L_x = C_s \frac{R_N}{G_s} \cdot \frac{T_5}{T_6} \,. \tag{4.89}$$

In phase VII the integrator I_b integrates the inverse of the voltage proportional to U_0:

$$U_b = U_0 G_s T_7 \,. \tag{4.90}$$

From the charge equivalence of phases VI and VII, R_x is calculable:

$$R_x = R_N \frac{T_7}{T_6} \,. \tag{4.91}$$

The loss factor of the impedance is:

$$\tan\delta = \frac{R_x}{\omega L_x} = \frac{G_s}{C_s} \cdot \frac{T_7}{T_5} \cdot \frac{1}{\omega} \,. \tag{4.92}$$

The voltage U_0 and the timing T_5 have the same clock source, therefore ωT_5 is a known number. Similarly, the position of the counter, counting T_7, is directly proportional to $\tan\delta$.

The decoded counter position can be displayed directly, as a measured value.

The method can be used in four-wire connection. The best resolution is $0.1\,\mu$H, the maximum measurable value is 1 000 H. The main error source is the uncertainty of the phase-sensitive detectors. In realized instruments the absolute phase error of measurement is equal to or higher than 10^{-3}. The frequency range is between 100 Hz and 1 MHz, but the highest accuracy is reached at 1 kHz.

The *inductance measuring circuit with a dual-slope ratio meter* (Hashimoto and Tamamura, 1976), shown in Fig. 4.67, has a simpler structure than the one above.

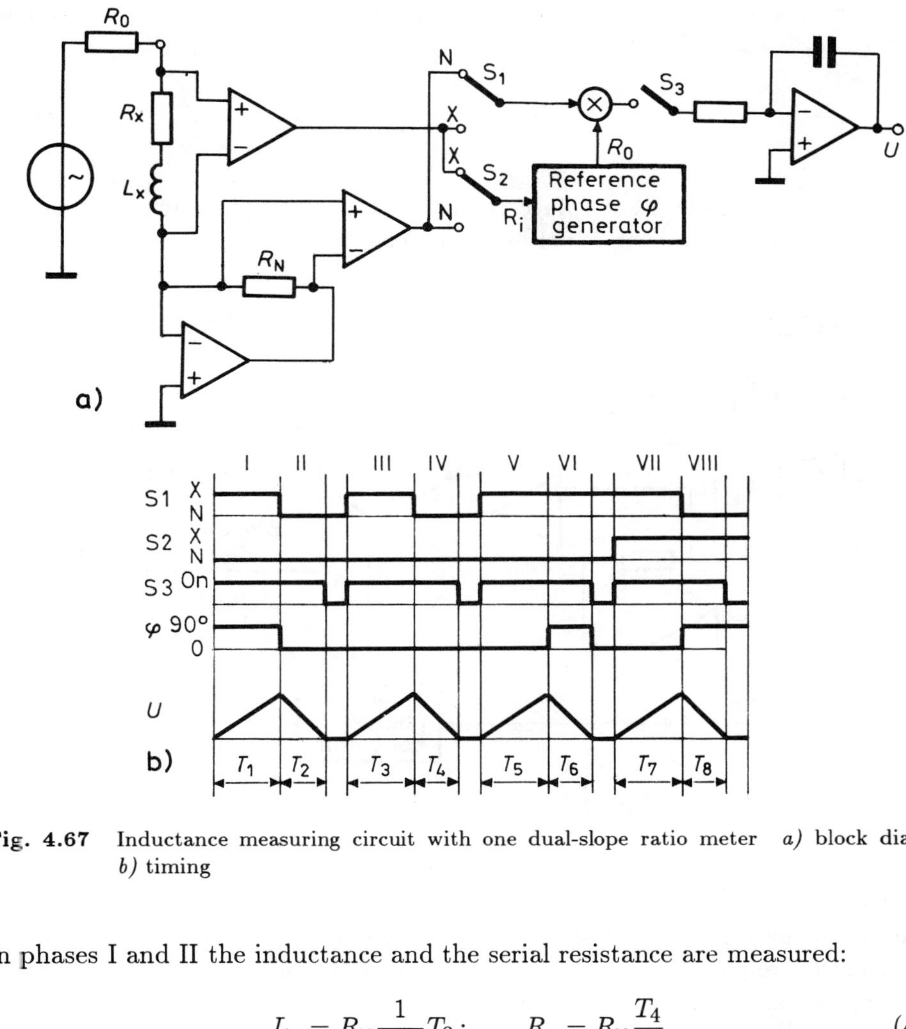

Fig. 4.67 Inductance measuring circuit with one dual-slope ratio meter *a)* block diagram *b)* timing

In phases I and II the inductance and the serial resistance are measured:

$$L_X = R_N \frac{1}{\omega T_1} T_2 ; \qquad R_X = R_N \frac{T_4}{T_3} , \tag{4.93}$$

where ωT_1 is an integer number. In phases V and VI the $\tan\delta$ of the impedance is measured:

$$\tan\delta = \frac{T_6}{T_5} . \tag{4.94}$$

In phases VII and VIII the parallel equivalent of the inductance is measured. In phase VII the voltage U_X on the impedance $R_X - L_X$ is detected with its own phase direction: the output of the integrator is proportional to the amplitude of the signal U_X. In phase VIII the component of the voltage on R_N, in-phase with the quadrature component of U_X, is integrated. The parallel inductance (L_p) is:

$$L_p = R_N \frac{T_8}{\omega T_7} . \tag{4.95}$$

If the voltage U_N is measured in phase VIII with reference to U_X, the parallel resistance R_p is calculable from the measured time intervals:

$$R_p = R_N \frac{T_8}{T_7}. \tag{4.96}$$

The solution shown in Fig. 4.68 (Maeda and Naramitsa, 1979) is a simplified version of the two former methods of ac voltage ratio measurements.

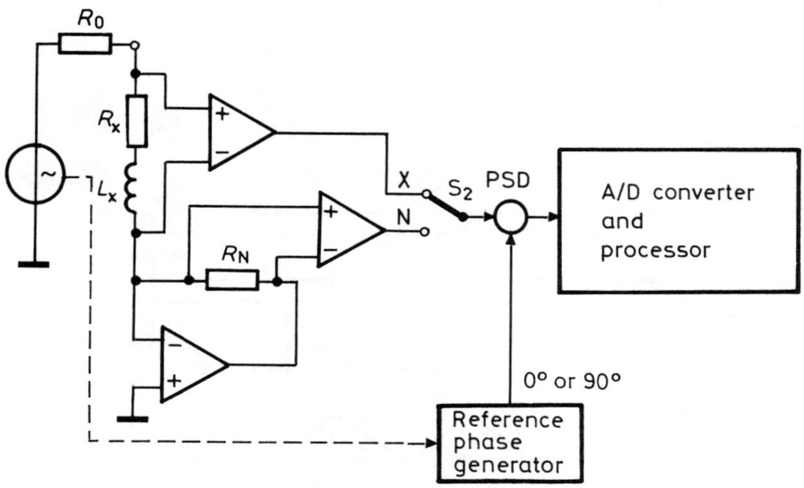

Fig. 4.68 Simplified ratio meter

The tricky methods of direct counting are useful for the simple display of the measured parameters. Presently computer technology makes possible having simpler hardware parts in the measuring circuit at the price of more complicated software processing. The voltages, decoupled from Z_X and R_N, are phase detected. The reference direction of the phase-sensitive detector is optional, only the frequency of this signal must be the same as the frequency of the supply of the impedance measurement circuit. The in-phase and quadrature components of both voltages are measured by a dual-slope or a successive approximation A/D converter. From these four results the values of L_X and R_X are computed by different methods of vector algebra.

The above three methods are similar because of the use of a phase-sensitive detector. The last method is the most modern and the cheapest one. In the future an enhanced method can be anticipated, where the phase sensitive detector is omitted, and fast A/D converters sample the signal forms. The detected record can be processed by a digital signal processor using all the sophisticated methods of signal analysis.

The T-circuit, shown in Fig. 4.69, measures the series equivalent of an inductance.

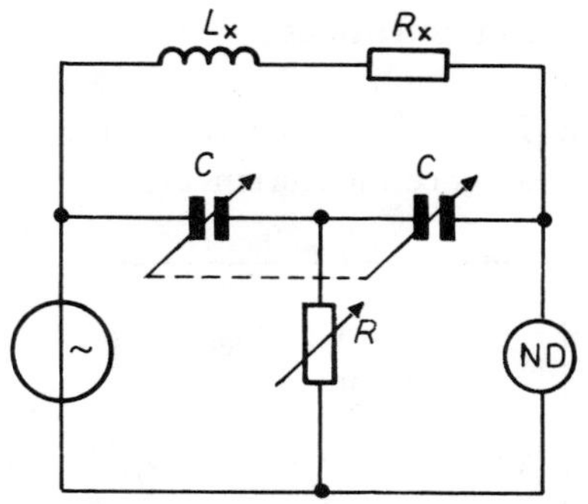

Fig. 4.69 T-circuit for inductance measurement

The balancing equations are:

$$L_\mathrm{x} = \frac{2}{\omega^2 C}\,; \qquad R_\mathrm{x} = \frac{1}{\omega^2 R C^2}\,.$$

The error of the method is in the order of 1%, the double capacitance C makes the design complicated, the convergence is bad. It was used mainly in classical measurement technology for the testing of nongrounded radio-frequency inductors.

The *double T-circuit*, shown in Fig. 4.70, has mainly historical importance. The balancing equations are:

$$L_\mathrm{x} = \frac{C_1'}{\omega^2(C^2 + 2CC_1' + C_1'C_3)}\,; \tag{4.98}$$

$$R_\mathrm{x} = \frac{C_1'}{\omega^2 C^2 R_2'(C_1' + C_3')}\,. \tag{4.99}$$

This method was used to test grounded radio frequency inductances.

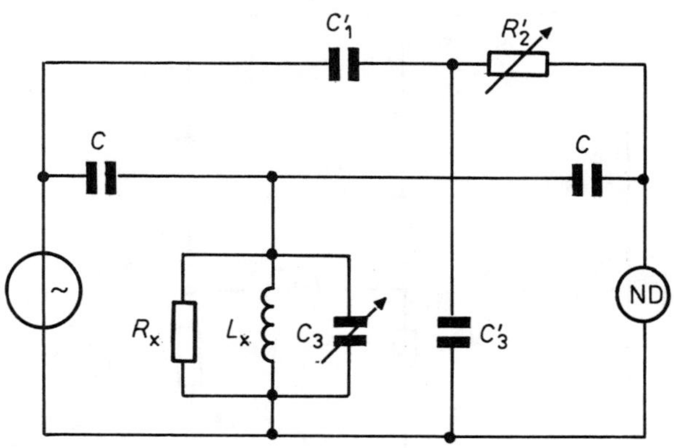

Fig. 4.70 Double T-circuit for inductance measurement

4.4.3 Measurement of Capacitances

The *direct method,* shown in Fig. 4.71, uses a voltmeter and an ammeter, similarly to the simplest method of resistance measurement.

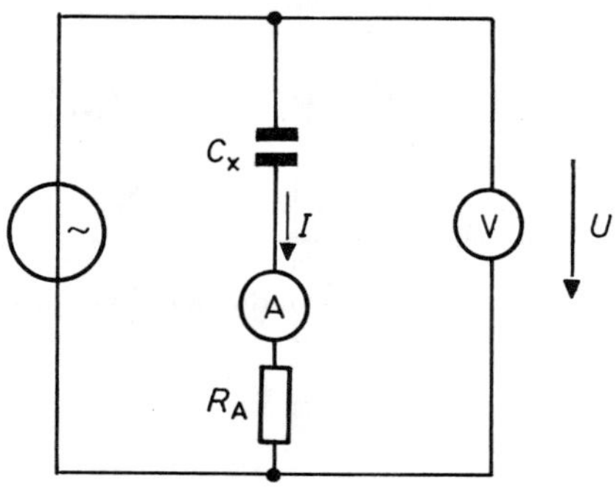

Fig. 4.71 Direct method of capacitance measurement

The measured value of capacitance:

$$C_\mathrm{x} = \frac{I}{U} \cdot \frac{1}{\omega} \cdot \frac{1}{\sqrt{1 - \frac{R_\mathrm{A}^2 I^2}{U^2}}} \approx \frac{I}{U} \cdot \frac{1}{\omega}\left(1 + \frac{R_\mathrm{A}^2 I^2}{2U^2}\right). \tag{4.100}$$

The necessary correction is usually small because of the perpendicular direction of two impedance phasors.

The method has moderate accuracy; it is used for rough estimations only in heavy current technology.

The *high-voltage Schering bridge,* patented by Thomas (1915), which was introduced into high-voltage measurements by Schering (1931; see also PTB, 1920), is shown in Fig. 4.72.

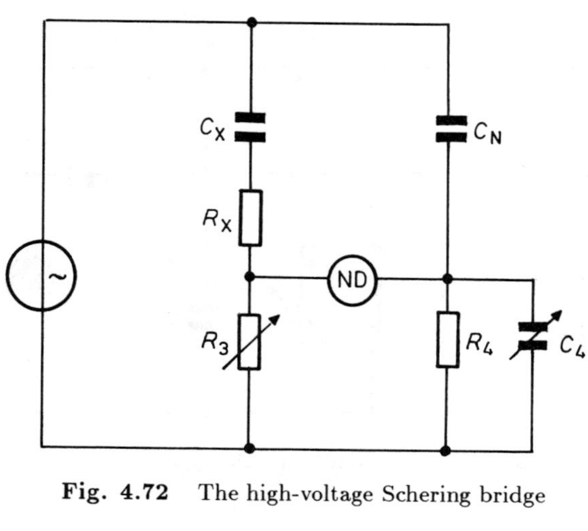

Fig. 4.72 The high-voltage Schering bridge

This is currently the most often used high-accuracy capacitance measuring circuit. At balance

$$\left(\frac{1}{j\omega C_X}\right) = \frac{R_3}{j\omega C_N} \cdot \frac{1 + j\omega R_4 C_4}{R_4},$$

and from the separation of the real and imaginary terms,

$$C_X = C_N \frac{R_4}{R_3}; \qquad \tan\delta_X = \omega R_4 C_4 . \tag{4.101}$$

This form of the Schering bridge is the so-called *high-voltage Schering bridge,* because the main voltage drop is on the capacitors C_X and C_N. On the bridge balancing elements there is no more than 100 V. In practice the resistance R_4 is realized as a multiple of $1000/\pi\,\Omega$, making possible the direct reading of $\tan\delta$. This is a typical two-wire measurement, three- and four-wire connections are realizable with compromises only. The range of measurement is in the interval of $1\,\text{pF} - 100\,\mu\text{F}$. The minimal error is 10^{-5}, if the bridge ratio R_3/R_4 is 1. With other ratios the uncertainty is higher, typically 10^{-4}.

The frequency range is the range of industrial frequencies (50 – 60 Hz). The maximal test voltage depends on the nominal voltage of C_N only, it can be over 1 MV.

The *high-frequency Schering bridge*, shown in Fig. 4.73, is a high-voltage Schering bridge, rotated by 90°.

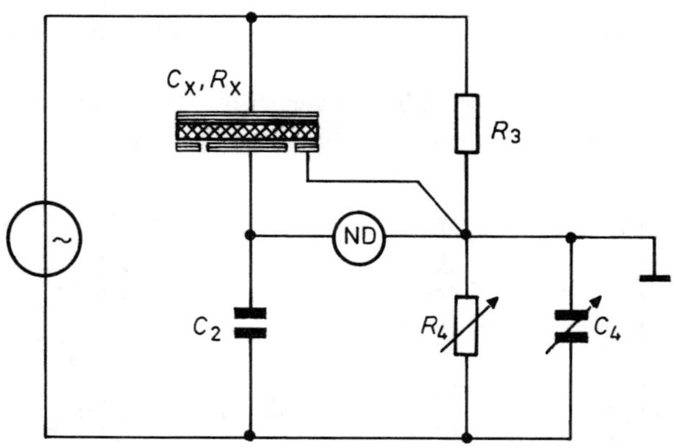

Fig. 4.73 The high-frequency Schering bridge

The bridge ratio is better at high frequencies, and the grounded detector node makes possible the simple use of guard-ring capacitors. The range of measurement is the same as in the case of the high-voltage bridge. The error is not smaller than 10^{-4}, and the frequency range is 100 Hz – 300 kHz. This is a typical three-wire method.

The *ratio transformer capacitance measuring bridge* shown in Fig. 4.74 is the most accurate capacitance measuring circuit in the frequency range of 1 kHz – 10 kHz.

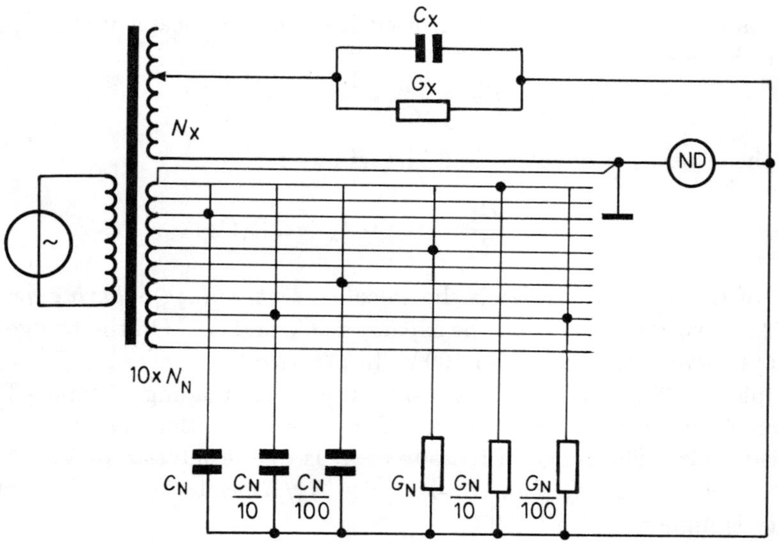

Fig. 4.74 Ratio transformer capacitance measuring bridge

Near 1 kHz the minimal error is 10^{-8}. Outside of this frequency range the error increases fast, at 100 kHz it is higher than 10^{-3}, at low frequencies the sensitivity is low. This is also a typical three-wire method, working in the range of 10 pF – 100 nF.

The *Glynne bridge* (Glynne, 1952; Baker, 1962; 1965), shown in Fig. 4.75, is also a high-voltage capacitance measuring bridge.

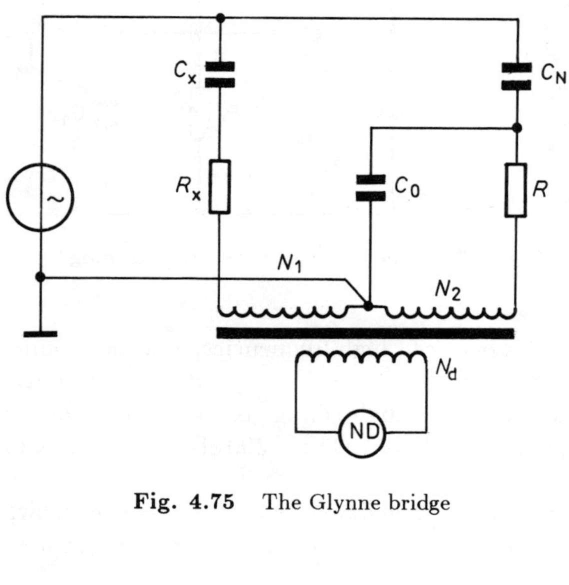

Fig. 4.75 The Glynne bridge

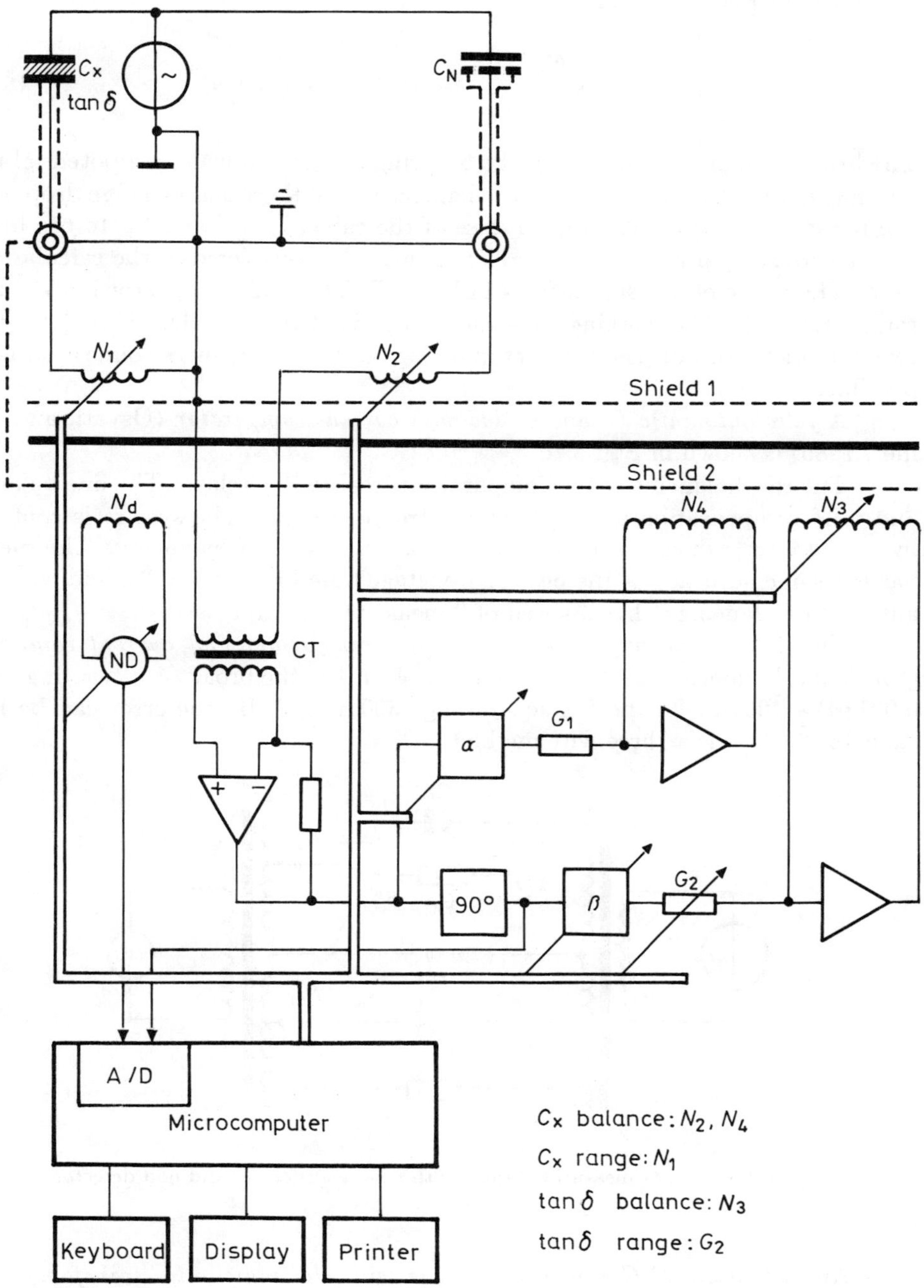

Fig. 4.76 Automatic high-voltage current comparator bridge

The balancing equations are:

$$C_\mathrm{x} = C_\mathrm{N}\frac{N_2}{N_1}\,; \qquad \tan\delta_\mathrm{x} = \omega R(C_0 + C_\mathrm{N})\,. \tag{4.102}$$

The bridge is more accurate than the Schering bridge and usually a potential regulation is not necessary, because in the balanced state there is no voltage drop on the comparator windings. The capacitance of the cable, connecting C_N to the bridge, is connected in parallel to C_0, therefore it must be considered in the calculation of $\tan\delta$. The range of measurement is $1\,\mathrm{pF} - 1\,\mu\mathrm{F}$. The achievable error is 10^{-5} in all ranges (ratios). The working frequency range is $50\,\mathrm{Hz}\ -1\,\mathrm{kHz}$. This is a typical two-wire method, but using the setup of Fig. 4.41, the four-wire connection is also possible.

A *fully automatic C-tanδ bridge with current comparator* (Osváth and Widmer, 1986) is shown in Fig. 4.76.

The raw balancing of the comparator is realized by relays. The fine balancing and $\tan\delta$ compensation are made by electronic circuits. The system is controlled by a computer, which simulates the manual procedure of balancing. The method has the same accuracy as the best high-voltage $\tan\delta$ bridges (10^{-5}), and with fully automatic balancing it has a speed of 2 measurements in a second.

The *C-G measuring bridge with ratio transformer and current comparator* (Clark and Vanderlyn, 1949), shown in Fig. 4.77, has the broadest measuring range: $0.001\,\mathrm{pF} - 100\,\mu\mathrm{F}$. In the frequency range $300\,\mathrm{Hz} - 3\,\mathrm{kHz}$ the error can be lower than 10^{-5}. This is a three-wire method.

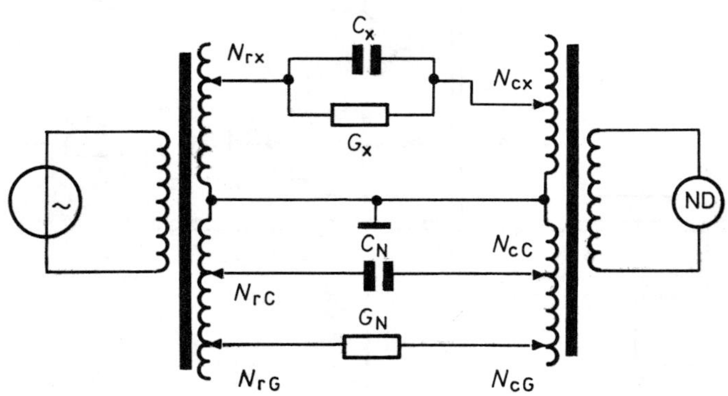

Fig. 4.77 C-G measuring bridge with ratio transformer and null detector

An *automatic C-G measuring bridge with electronic ratio transformer and null detector* (Osada and Suehiro, 1975) is shown in Fig. 4.78.

If switch S is in position G, the computer controlled D/A converters supply voltages to the standard impedances C_N and G_N. In the balanced state:

$$C_\mathrm{x} = \alpha C_\mathrm{N} \qquad \text{and} \qquad G_\mathrm{x} = \beta G_\mathrm{N}\,. \tag{4.103}$$

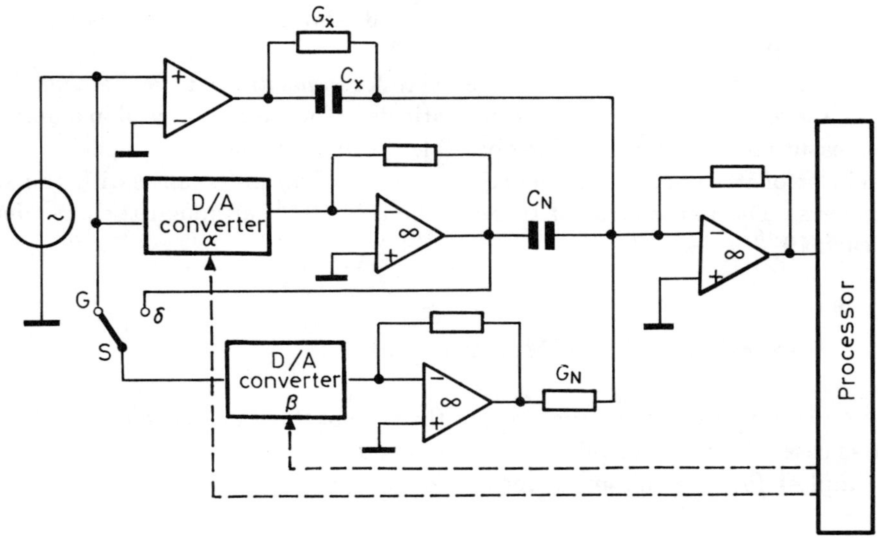

Fig. 4.78 Automatic C-G measuring bridge with electronic ratio transformer and null detector

If the switch is in position 'δ', the value in the D/A converter β is proportional to tanδ:

$$\tan\delta_{\mathrm{X}} = \beta\frac{G_{\mathrm{N}}}{\omega C_{\mathrm{N}}}\,.$$

The smallest achievable error of the method is 10^{-3}, the range of measurement is 10 pF–100 nF in the frequency range of 100 Hz – 10 kHz. The amplifier, feeding C_{X} and the operational amplifier of the null detector must be well compensated to avoid the instability caused by an eventual complex structure of the impedance to be measured.

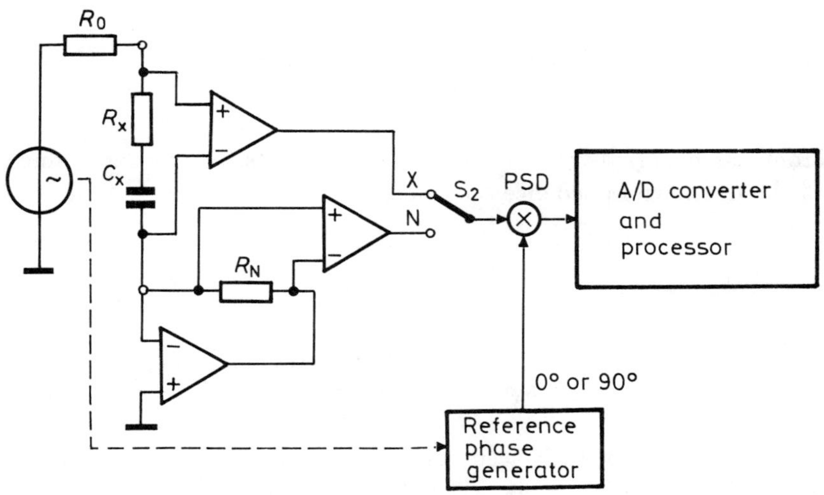

Fig. 4.79 Capacitance measurement with the electronic voltage comparison method

The *electronic voltage comparison method*, shown in Fig. 4.79, is essentially the same as the one shown in Fig. 4.68.

The accuracy of this method, used here for capacitance measurement, is the same as above. In practice the electronic ratio detector method is used in a universal RLC measuring instrument, making possible the measurement of a capacitance and its $\tan\delta$, or of its serial or parallel equivalent, or of an inductance with the same parameters. The phase error of these methods is 10^{-3}, although their resolution can reach 10^{-5}.

4.4.4 General Impedance Measurement

In some practical cases the structure and the type of the impedance are not known. In these cases the absolute value and the phase of the impedance can be measured. The simplest form of this measurement is the *three voltmeter method*, shown in Fig. 4.80.

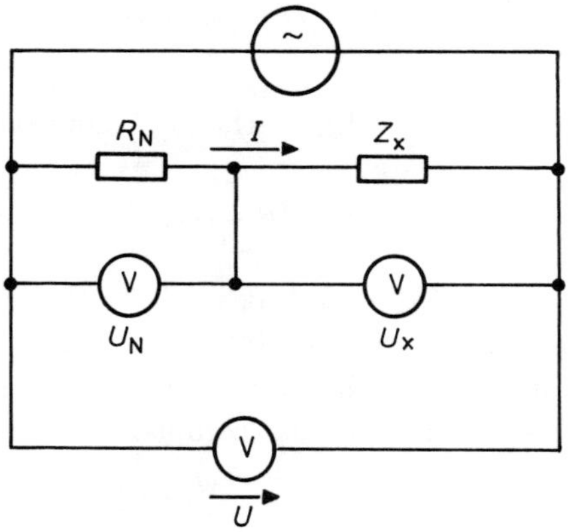

Fig. 4.80 Three voltmeter method

From the displayed values of the three voltmeters the absolute value and the phase of the unknown impedance are calculable:

$$|Z_x| = \frac{U_x}{U_N} R_N\,; \tag{4.104}$$

$$\cos\varphi = \frac{U^2 - (U_x^2 + U_N^2)}{2U_x U_N}\,. \tag{4.105}$$

With an ac voltmeter of high accuracy the method can have a surprisingly low uncertainty (10^{-3}). It can work without restrictions with internal and external supplies and in a wide frequency range (10 Hz – 10 MHz).

4.5 IMPEDANCE ANALYZERS

Throughout the previous discussions of this chapter, the impedance under test was modelled by a simple structure: two pure elements, which may be resistance, capacitance or inductance, were connected in series or in parallel. For example, a capacitor was modelled by a pure capacitance C_s and a pure resistance R_s, connected in series, or a pure capacitance C_p and an admittance G_p, connected in parallel. The frequency dependence of the complex impedance of such models differs significantly (see Fig. 4.81). In the series case the phase angle is –90° at low frequencies (capacitive impedance), and it asymptotically tends to 0° in the high-frequency range (resistive impedance):

$$\tan \varphi = -\frac{1}{\omega R_s C_s} .$$

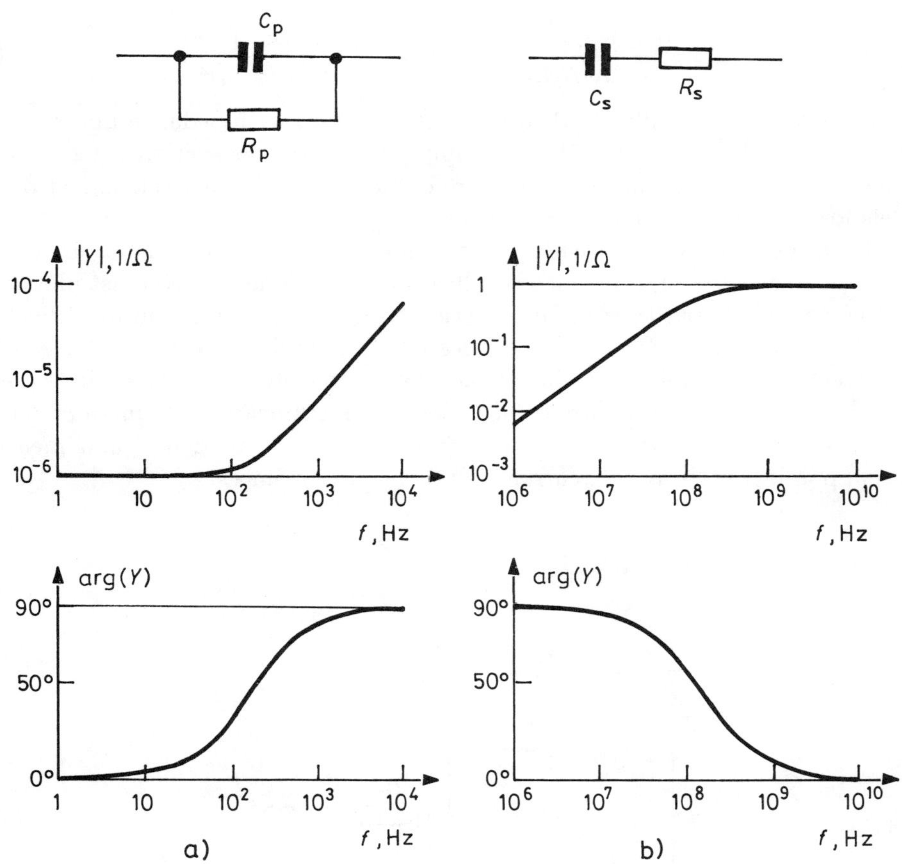

Fig. 4.81 Frequency dependence of two-element capacitance models *a)* parallel model, $C_p = 1\,\text{nF}$, $R_p = 1\,\text{M}\Omega$ *b)* Serial model, $C_s = 1\,\text{nF}$, $R_s = 1\,\Omega$

In the parallel model the effect is the opposite: the phase is resistive at low frequencies, and it is capacitive at high frequencies:

$$\tan\varphi = -\frac{\omega C_p}{G_p}.$$

It is very important to note that the behaviour of the true impedance is determined by the physical nature of the capacitor, and usually one of the above models will better approximate this behaviour than the other.

The elements of an inappropriate model will strongly depend on the frequency. For example, if the series model is a good one, a parallel model measuring method will yield the following results:

$$Y_p = G_p + j\omega C_p = \frac{1}{R_s + \frac{1}{j\omega C_s}}, \tag{4.106}$$

$$C_p = \frac{C_s}{1+(\omega R_s C_s)^2}, \qquad G_p = \frac{R_s(\omega C_s)^2}{1+(\omega R_s C_s)^2}. \tag{4.107}$$

The problem is more difficult when the behaviour of the impedance under test is more complex. The models of the basic impedances (see Subsection 2.3.2) usually consist of not only two, but three or more elements, and further refinement of the models leads to very complex structures.

In the general case the frequency domain analysis of the impedance under test is to be performed. The result may be illustrated in a Bode or a Nyquist plot. The appropriate model structure can be determined by transfer function identification methods (see Section 6.8). The impedance is interpreted as a two-port, e. g. with the current as the input and the voltage as the output. Advanced impedance analyzers may offer the possibility of wide-band identification of different model structures. In such instruments it is also often possible to detect and measure eventual nonlinearities and dc current or voltage dependences.

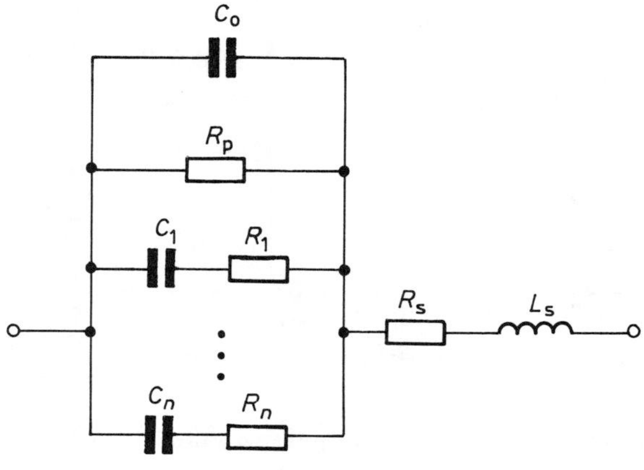

Fig. 4.82 Wide-band model of a capacitor

A typical example of complex models is the wide-band model of capacitors (Fig. 4.82). Polarisation in the dielectric material increases the ability of the capacitor to absorb electrical charge in the electrodes at the given voltage. There are different physical processes that represent polarisation: ions, dipoles, electrons etc. can be rearranged inside or at the surface of the dielectric. These rearrangements have different speed, and can be well modelled by means of serial R–C circuits as in Fig. 4.82. The time constants vary from 10^{-16} seconds to a few hundred seconds.

The wide-band model can be measured by means of special signal analyzers. Charge/discharge with adequate time lengths can be used to measure the different time constants of the R–C branches.

References

Baker, H. S. (1917): Current Transformer Ratio and Phase Error by Test Ring Method. *AIEE Proceedings,* Vol. XXXVI, No. 12, pp. 1173–1183.

Baker, W. P. (1962): Recent Developments in 50 c/s Bridge Networks with Inductively Coupled Ratio Arms for Capacitance and Loss-Tangent Measurement. *Proc. IEE, Part A,* Vol. 109, pp. 243–247. June 1962.

Baker, W. P. (1965): Modern 50 c/s Bridge Networks for Dielectric Loss Measurements. *Associated Electrical Industries (AEI) Engineering,* Vol. 5, No. 2, pp. 98–102, March-April 1965.

Binnie, A. J. and Foord, T. R. (1967): Leakage Inductance and Interwinding Capacitance in Toroidal Ratio Transformers. *IEEE Trans. on Instrumentation and Measurement,* Vol. 16, No. 4, pp. 307–314. Dec. 1967.

Brueckel, W. (1962): The Commutable Schering Bridge of Tettex AG. Zurich, Tettex Information.

Clark, H. A. M. and Vanderlyn, P. B. (1949): Double-Ratio AC Bridges with Inductively Coupled Ratio Arms. *Proc. IEE, Part II,* Vol. 96, No. 51, pp. 365–378. Jan. 1949.

Cutler, L. S. and Rorden, R. J. (1958): Precision Selective-Ratio Transformers. US Patent 2 832 036, Ser. No. 524 544.

Genrad (1971): Capacitance Bridge Type 1615. Manual. General Radio Company. US patent No. 2 548 457.

Glynne, A. (1952): Bridges with Inductively Coupled Ratio Arms. *Bulletin of Electrical Engineering Education,* No. 8, pp. 69–73. Manchester, College of Technology. May 1952.

Hague, B. (1962): Alternating Current Bridge Methods. 5th, revised edition. London, Sir Isaac Pitman & Sons.

Hall, H. P. (1970): Four-Terminal Equal-Power Transformer-Ratio-Arm Bridge. *IEEE Trans. on Instrumentation and Measurement,* Vol. 19, No. 4, pp. 308–311. Nov. 1970.

Hall, H. P. (1973): An AC-DC Ratiometer an Its Use in a CRL Meter. *IEEE Trans. on Instrumentation and Measurement,* Vol. 22, No. 4, pp. 387–390, Dec. 1973.

Harris, F. K. (1952): Electrical Measurements. New York, John Wiley & Sons Inc.

Hashimoto, S. and Tamamura, T. (1976): An Automatic Wide Range Digital LCR Meter. *Hewlett-Packard Journal,* Vol. 28, No. 1, pp. 9–15, Sept. 1976.

Hay, C. E. (1910): Measurement of Self-Inductance and Effective Resistance at High Frequencies. *The Electrical Review,* Vol. 67, No. 1, pp. 965–966.

Hill, J. J. and Deacon, T. A. (1968): Voltage-Ratio Measurement with a Precision of Parts in 10^{-9} and Performance of Inductive Voltage Divider. *IEEE Trans. on Instrumentation and Measurement,* Vol. 17, No. 4, pp. 269–278, Dec. 1968.

Hill, J. J. and Miller, A. P. C. (1962): A Seven Decade Adjustable-Ratio Inductively Coupled Voltage Divider with 0.1 part per million Accuracy. *Proc. IEE,* Vol. 109B, pp. 157–163.

Hill, J. J. and Miller, A. P. C. (1963): An AC Double Bridge with Inductively Coupled Ratio Arms for Precision Platinum Resistance Thermometry. *Proc. IEE,* Vol. 110, No. 2, pp. 453–458.

Kennelly, A. E. and Alger, P. L. (1917): Magnetic Flux Distribution Annular Steel Laminae. *Conference of the American Institute of Electrical Engineers,* pp. 999–1017. New York, A.I.E.E.

Kusters, N. L. and Moore, W. J. M. (1964): The Compensated Current Comparator: a New Reference Standard for Current-Transformer Calibrations in Industry. *IEEE Trans. on Instrumentation and Measurement,* Vol. 13, No. 2-3, pp. 107–114, June-Sept. 1964.

Maeda, K. (1974): An Automatic, Precision 1-MHz Digital LCR Meter. *Hewlett-Packard Journal,* Vol. 25, No. 7, pp. 2–9, March 1974.

Maeda, K. and Naramitsa, Y. (1979): Multi-Frequency LCR Meters Test Components under Realistic Conditions. *Hewlett-Packard Journal,* Vol. 30, No. 2, pp. 24–31, Feb. 1979.

Maxwell, J. C. (1873): Treatise of Electricity and Magnetism. Oxford, Clarendon Press.

Miljanic, P. N., Kusters, N. L. and Moore, W. J. M. (1962): The Development of the Current Comparator, a High-Accuracy AC Ratio Measuring Device. *Communication and Electronics,* No. 62, pp. 359–368, Nov. 1962.

Oliver, B. M. and Cage, I. M. (1971): Electronic Measurements and Instrumentation. New York, McGraw-Hill.

Osada, K. and Suehiro, J. I. (1975): Digital High-Capacitance Measurements to One Farad. *Hewlett-Packard Journal,* Vol. 26, No. 6, pp. 10–16.

Osváth, P. and Widmer, S. (1986): A High-Voltage High-Precision Self-Balancing Capacitance and Dissipation Factor Measuring Bridge. *IEEE Trans. on Instrumentation and Measurement,* Vol. 35, No. 1, pp. 19–23, March 1986.

PTB (1920): Die Tätigkeit der Physikalisch-Technischen Reichsanstalt im Jahre 1919 (The Activity of the Imperial Institute of Physics and Technology in 1919). *Zeitschrift für Instrumentenkunde,* Vol. XXXX, p. 113, June 1920. (In German)

Schering, H. (1931): Die Empfindlichkeit einer Wechselstrombrücke (The Sensitivity of an AC Bridge). *Elektrotechnische Zeitschrift,* Vol. 52, No. 36, pp. 1133–1134. (In German)

Shubis, T. and Met, A. (1989): A Mathematical Model of the Admittance between Tapped Windings and their Shield for Ratio Instruments. *IEEE Trans. on Instrumentation and Measurement,* Vol. 38, No. 3, pp. 764–768.

Siemens, W. (1892): Scientific and Technical Papers. Vol. I. John Murray, London.

Stout, M. B. (1960): Basic Electrical Measurements. 2. ed. Englewood Cliffs, Prentice-Hall, Inc.

Thomas, P. (1915): Method of Measuring Capacity and Power Factor. US Patent 1 166 159, Ser. No. 804 755.

Tóth, E. (1966): Új módszer váltakozóáramú mérőhidak földpotenciál-kiegyenlítésére (New Method for Earth Potential Balancing of AC Measuring Bridges). Candidate's Thesis, Hungarian Academy of Sciences, Budapest, Hungary. (In Hungarian)

Wagner, B. W. (1911): Zur Messung dielektrischer Verluste mit der Wechselstrombrücke (On the Measurement of Dielectrical Losses Using the AC Bridge). *Elektrotechnische Zeitschrift,* Vol. 32, No. 40, pp. 1001–1002. (In German)

Walsh, R. (1930): Inductive Ratio Arms in Alternating Current Bridge Circuits. *Philosophical Magazine,* Vol. 10, No. 62, pp. 49–71, July 1930.

Watton, B. Sc. and Pemberton, M. E. (1949): A Direct-Capacitance Aircraft Altimeter. *Proc. IEE,* Vol. 96, No. 2, pp. 379–389.

Wheatstone, C. (1843): An Account of Several New Instruments and Processes or Determining the Contents of a Voltaic Circuit. *Philosophical Trans. of the Royal Society of London,* Vol. 133, pp. 303–327.

Wien, M. (1891): Messung der Induktionskonstanten mit dem optischen Telephon (Measurement of the Induction Constant Using the Optical Telephone). *Annals der Physik,* Vol. 44, pp. 689–712.

Zoltán, I. (1977): Numerikusan vezérelt impedanciamérő hálózatok pontossága és frekvenciafüggése (Accuracy and Frequency Dependence of Numerically Controlled Impedance Measuring Networks). Doctoral dissertation, Technical University of Budapest. (In Hungarian)

5

Time and Frequency Measurement

P. Pataki

Editorial introduction

Time and frequency are two quantities closely related to each other. Frequency is by definition the number of occurrences of a given periodic phenomenon *in a given time*, and can also be measured as the reciprocal of the time length between successive occurrences. Therefore, time and frequency measurements are usually made by the same instrument, and frequency standards are also time standards at the same time.

'Time' has two meanings in practice: it can mean on the one hand the position of an event on a conventional time axis, on the other hand the time elapsed between two events (time-length or time-interval).

The first meaning is important in astronomy, space research, navigation, telecommunications etc. In order to determine the actual time, it is usually sufficient in practice to receive radio signals emitted from a very accurate frequency standard (an 'Atomic Clock'), synchronized to a well defined starting point (1st January 1958, $0^h0'0''$), or to have an independent frequency standard, somehow synchronized to the international ones.

It is the second meaning of time which has outstanding importance in measurements (and this is the concept closely related to frequency). In this chapter time-interval and frequency measurement methods are discussed: the most important measurement principles and devices are treated, and the achievable accuracy is analyzed. The discussion will be focused on instruments based on counters, since these are the most accurate tools; moreover, because of their relatively low price, they can be found in practically every electrical laboratory.

Beside time-interval and frequency measurements, phase measurement is the third topic to be discussed in this chapter, because it can be defined in terms of time shift and frequency, and can be measured via measurement of a time-interval or the ratio of time-intervals.

Technology of Electrical Measurements, Edited by L. Schnell
© 1993 John Wiley & Sons Ltd.

5.1 ACCURACY OF FREQUENCY STANDARDS AND THE CHARACTERIZATION OF THEM

Fig. 5.1 illustrates the evolution of the accuracy of frequency standards in the last 60 years. Frequency accuracy of 10^{-13} can be taken as a theoretical limit in the Earth-fixed coordinate system, because the relativistic effects and the change of the gravitational potential cause a time shift in this range. These effects influence even more the clock of a spaceship, where the change of the gravitational potential results in 10^{-10} frequency change with respect to a similar Earth-based clock (Reinhard, 1974; Glaze et al, 1974; Ramsey, 1987; Lewis, 1991; Winkler, 1991).

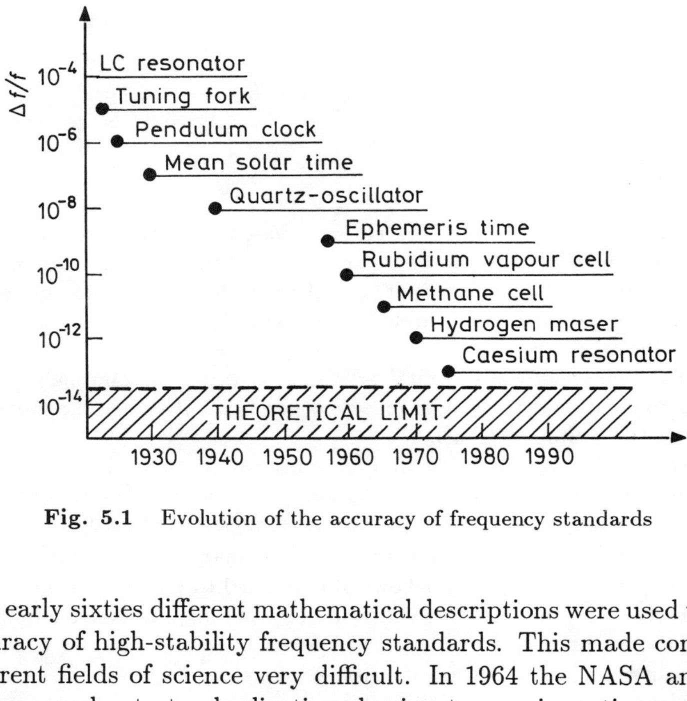

Fig. 5.1 Evolution of the accuracy of frequency standards

In the early sixties different mathematical descriptions were used to characterize the accuracy of high-stability frequency standards. This made communication among different fields of science very difficult. In 1964 the NASA and the IEEE had a conference about standardization, having two main notions: *a) fractional frequency offset, b) phase-time*, which can be defined as follows.

The instantaneous output voltage of a sinusoidal generator:

$$u(t) = U_{\mathrm{p}} \sin\bigl(2\pi f_1 t + \varphi(t)\bigr) \tag{5.1}$$

where U_{p} is the nominal value of the amplitude, f_1 the nominal value of the frequency, both independent of time, and $\varphi(t)$ is the phase fluctuation, depending on time. The amplitude fluctuation is not taken into account because in linear systems it does not cause phase fluctuation.

The instantaneous frequency of the sine wave is equal to

$$f(t) = f_1 + \frac{1}{2\pi} \cdot \frac{\mathrm{d}\varphi(t)}{\mathrm{d}t} = f_1 + \Delta f(t)\,, \tag{5.2}$$

where $\Delta f(t)$ is the time-dependent frequency fluctuation, the so-called frequency offset:

$$\Delta f(t) = \frac{1}{2\pi} \cdot \frac{\mathrm{d}\varphi(t)}{\mathrm{d}t}. \tag{5.3}$$

For a wide range of nominal frequencies, normalization is a very useful technique. The normalized instantaneous frequency fluctuation is as follows:

$$y(t) = \frac{\Delta f(t)}{f_1} = \frac{1}{2\pi f_1} \cdot \frac{\mathrm{d}\varphi(t)}{\mathrm{d}t} \tag{5.4}$$

where $y(t)$ is called *'fractional frequency offset'*.

Another useful quantity is the time integral of $y(t)$:

$$x(t) = \int_0^t y(t)\mathrm{d}t = \frac{1}{2\pi f_1}\varphi(t), \tag{5.5}$$

where $x(t)$ is the *'phase-time'*. $x(t)$ is proportional to the instantaneous phase, but its dimension is time. (The phase-time is the time difference between two clocks.)

$y(t)$ and $x(t)$ are unchanged by frequency multiplication and division if the devices are noiseless. They are useful for direct comparisons of the stability of sources with very different nominal frequencies (Barnes et al, 1971; Rutman and Walls, 1991).

5.2 FREQUENCY SOURCES

This section discusses the main features of high-accuracy time-interval and frequency generators and clocks.

5.2.1 Atomic Frequency Standard

The most precise way to produce a stable frequency generator is the *Atomic Clock*. Atomic frequency standards are based on the principles of quantum mechanics where the frequency is determined by an atomic resonator. The frequency of the microwave radiation is proportional to the energy difference between the discrete p and q energy levels:

$$f_1 = \frac{\Delta E}{h} = \frac{E_q - E_p}{h}, \tag{5.6}$$

where $h = 6.63 \cdot 10^{-34}$ Js is Planck's constant. In atomic clocks elements are used from the first column of Mendelejev's periodical system (alkali metals). These materials have only one valence electron on their outer energy level and after excitation the radiation energy, i. e. the frequency of radiation, depends on the energy difference between the two outer levels. Thus the radiation frequency is typical for these elements (Hyatt et al, 1973; Kartaschoff, 1978; Ramsey, 1991).

5.2.2 Quartz Crystal Oscillator

The most generally used frequency standard of practical measurement techniques is the *quartz crystal oscillator.* The piezoelectric effect of the quartz crystal makes it easy to excite mechanical vibrations by means of an oscillating electric field. The resonant frequency of the crystal oscillator is very stable (Kartaschoff, 1978).

5.2.3 Frequency Synthesizer

It is often necessary to produce a quartz-accurate but changeable frequency signal, in which case the frequency of the crystal oscillator is transferred into another frequency signal.

Direct frequency conversion can be done with a frequency multiplier and a frequency divider, or *indirect* frequency conversion can be done with a frequency synthesizer.

Frequency multipliers or harmonic generators are based on nonlinear components which distort the input signal. The distorted signal can be represented by means of a Fourier series with fundamental and harmonic frequencies. The desired harmonic term is extracted by a selective filter and an amplifier.

Frequency dividers are counters built of flip-flops.

Direct frequency conversion produces discrete frequencies only. In many applications signals of various frequencies are required with high accuracy and stability. For this purpose frequency synthesizers are used.

The *frequency synthesizer* performs all of the four basic arithmetic operations on the frequency generated by a crystal or atomic oscillator. These synthesizers are based on phase-locked loops (PLL). The basic elements of a phase-locked loop are a phase detector, a filter-amplifier and a voltage-controlled oscillator (VCO) as it can be seen in Fig. 5.3. The job of the phase-locked loop is to generate a clean waveform of appropriate shape that is locked-in to the average input frequency (Kroupa, 1973).

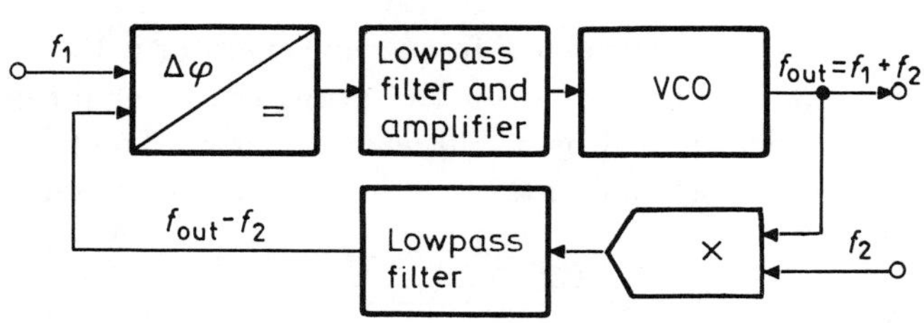

Fig. 5.2 Frequency summing phase-locked loop

Fig. 5.2 shows the scheme of the frequency summing loop. In steady-state the frequencies of the feedback signal is equal to the f_1 input frequency, hence:

$$f_{out} = f_1 + f_2 \,.$$

Fig. 5.3 shows the frequency multiplier loop. The feed-back arm divides the output frequency of the voltage controlled oscillator (VCO), thus in steady-state the divided frequency will be equal to the input frequency (f_{in}). The frequency of the output signal is

$$f_{\text{out}} = n f_{\text{in}} \, .$$

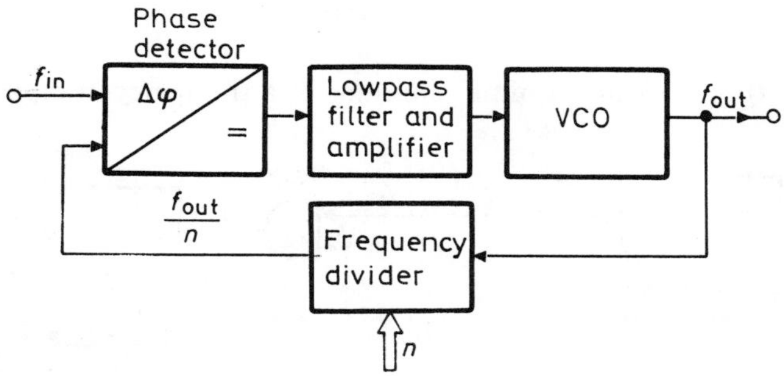

Fig. 5.3 Frequency multiplier phase-locked loop

Fig. 5.4 shows the block diagram of a frequency synthesizer. It is composed of decadic frequency dividers, filters, frequency multiplier and addition loops. The circuit can produce from the constant frequency signal of the quartz oscillator (e. g. 1 MHz) an arbitrary frequency signal with high accuracy and stability (Kroupa, 1973; McNamee and Platt, 1989).

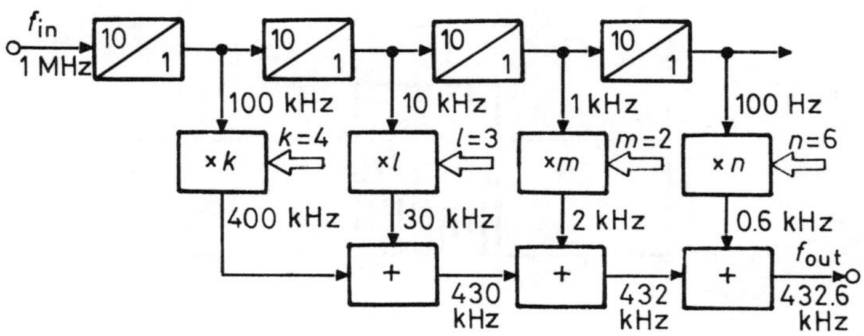

Fig. 5.4 Frequency synthesizer

Thanks to the recent VLSI technology there are new digital waveform synthesizers in which all waveform values are digitally calculated in real time up to 600 kHz (Ives, 1989).

5.3 FREQUENCY AND PERIOD TIME MEASUREMENTS

There are two main frequency measurement principles. One is the comparison method, in which the unknown frequency is compared to a known frequency standard. The other is the resonant method, where the unknown frequency will be determined by the help of any frequency dependent element.

Owing to the high-accuracy frequency standards, the comparison method is more accurate, so it will be discussed in detail.

5.3.1 Frequency and Period Comparison by Means of Counters

Fig. 5.5 shows the block and timing diagrams of a frequency comparing circuit.

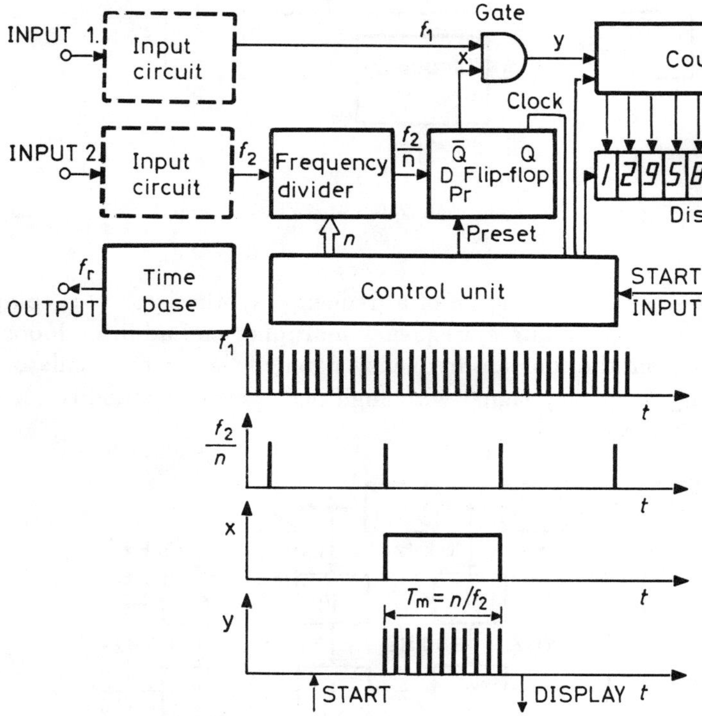

Fig. 5.5 Frequency comparison with counter

After the appearance of the starting signal (START) the control unit clears the counter and cancels the PRESET condition of the flip-flop. The next f_2/n frequency impulse opens the gate by setting the gate control flip-flop. The gate is opened during the $T_m = n/f_2$ measurement time-interval and the counter counts the periods of the f_1 frequency signal. After the T_m measurement time the gate is closed and the condition of the counter (N) is proportional to the ratio of the two frequencies. The arbitrary proportion factor is determined by the frequency divider (n). The displayed result is:

$$N = f_1 T_m = n \cdot \frac{f_1}{f_2} . \tag{5.7}$$

If one of the two frequencies (f_1 or f_2) is produced by a known frequency standard (e. g. quartz crystal oscillator or atomic clock) the other is calculable.

Frequency measuring instruments based on this principle have two different operating modes: direct frequency measurement mode and period measurement mode.

Direct frequency measurement mode

In this mode the signal of the unknown frequency is connected to the INPUT 1 through an input circuit which gives a waveform processing by means of a 'Schmitt trigger' circuit. The known frequency signal of the high stability frequency standard is connected to INPUT 2. Substituting $f_1 = f_x$ and $f_2 = f_r$ into (5.7), the unknown frequency is:

$$f_x = \frac{N}{n} f_r \,, \tag{5.8}$$

and the measurement time is:

$$T_m = n/f_r \,. \tag{5.9}$$

As soon as the measurement time has passed, the displayed number (N) is proportional to the unknown frequency. The multiplier factor is inversely proportional to the measurement time, thus by proper setting of the latter (1 s, 1 ms, 1 μs) the displayed number can be scaled in frequency (Hz, kHz, MHz).

The uncertainty of direct frequency measurement is basically limited by two factors: ± 1 count ambiguity and the accuracy of the frequency standard. Thus the relative error of the frequency measurement is:

$$\frac{\Delta f_x}{f_x} = \pm\left(\frac{1}{N} + \left|\frac{\Delta f_r}{f_r}\right|\right), \tag{5.10}$$

and by substitution of (5.8) and (5.9),

$$\frac{\Delta f_x}{f_x} = \pm\left(\frac{1}{T_m f_x} + \left|\frac{\Delta f_r}{f_r}\right|\right). \tag{5.11}$$

In Fig. 5.6 the frequency dependence of the relative error of direct frequency measurement is shown in log-log coordinates, with continuous line. It can be seen from Fig. 5.6 that the required measurement time is too long if the measured frequency is small.

The above mentioned input circuit processes the waveform. Fig. 5.7a shows the correct processing by means of a 'Schmitt trigger' bistable circuit. Fig. 5.7b shows the influence of the noises, resulting in measurement errors. To discover these, it is advisable to use always an oscilloscope. It will sometimes be necessary to insert a lowpass or a bandpass filter before the counter.

In direct frequency measurement mode the resolution is not sufficient at low frequencies. The resolution can be improved by measuring the period length and computing the frequency as its inverse.

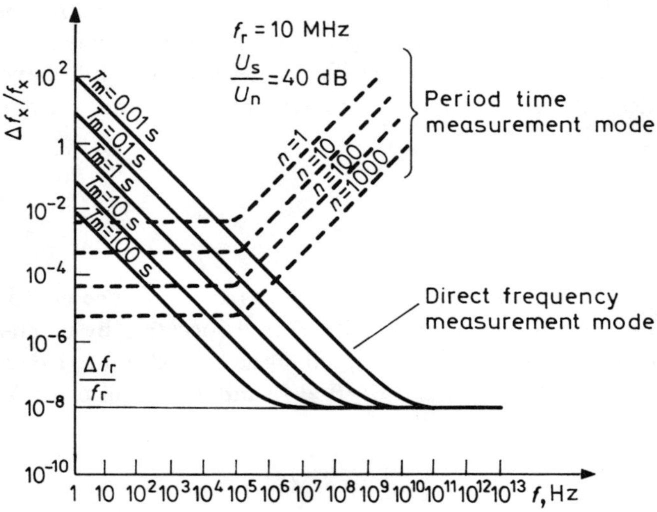

Fig. 5.6 Uncertainty of frequency and period length measurements

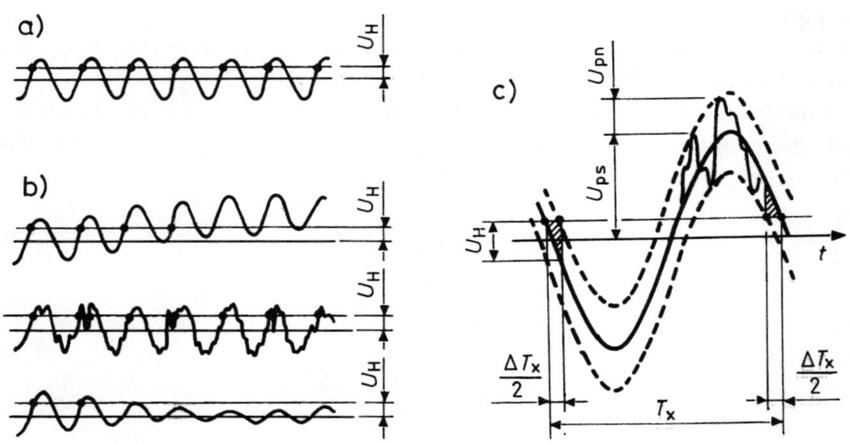

Fig. 5.7 The influence of noise on frequency measurement *a)* correct direct frequency measurement *b)* incorrect direct frequency measurement *c)* incorrect period length measurement

Period measurement mode

In this mode the signal of the unknown frequency is connected through the input circuit to INPUT 2, and the standard frequency is led to INPUT 1. This mode is the inverse of the direct frequency measurement mode. Substituting $f_1 = f_x$ into

(5.7), the unknown frequency is given by:

$$f_x = \frac{n}{N} f_r \,. \tag{5.12}$$

The period time is

$$T_x = \frac{N}{n} \cdot \frac{1}{f_r} \,, \tag{5.13}$$

and the measurement time is

$$T_m = nT_x \,. \tag{5.14}$$

This measurement mode with its high resolution provides high accuracy at low frequency.

The relative error of the unknown frequency is:

$$\frac{\Delta f_x}{f_x} = \left(\frac{\Delta T_x}{T_x}\right) = \pm\left(\frac{f_x}{n f_r} + \left|\frac{\Delta f_r}{f_r}\right|\right) . \tag{5.15}$$

This equation does not take into account the triggering error, caused by the noise (Fig. 5.7c). If the noise-to-signal ratio is low enough, this part of the error is as follows:

$$\frac{\Delta T_x}{T_x} \approx \pm \frac{1}{n\pi} \cdot \frac{U_{pn}}{U_{ps}} \,, \tag{5.16}$$

where n is the number of the measured periods, and U_p denotes the peak values of the signal and the noise. Measuring low frequencies in period-mode this error is predominant, e. g. if $n=1$ and the noise-to-signal ratio is -40 dB, the measurement error is $\pm 0.3\%$. If more periods (n) are measured and the average period is displayed, the measurement accuracy will be improved (in a modern instrument $n_{max}=10^5$). Consequently, the total relative error of the period measurement is:

$$\frac{\Delta f_x}{f_x} = \pm\left(\frac{1}{n f_r} f_x + \frac{1}{n\pi} \cdot \frac{U_{pn}}{U_{ps}} + \left|\frac{\Delta f_r}{f_r}\right|\right) . \tag{5.17}$$

Fig. 5.6 shows the relative error of the period measurement mode versus the frequency (dashed lines).

Extension of the frequency measurement range

Frequency can be measured with simple instruments up to 50–100 MHz only. To extend the measurement range, frequency dividers, frequency converters and transfer oscillators can be used.

a) Frequency divider

Fig. 5.8 illustrates the principle of a frequency divider. Over 100 MHz only binary dividers can be used, since because of the delay of the gates, no decimal divider is

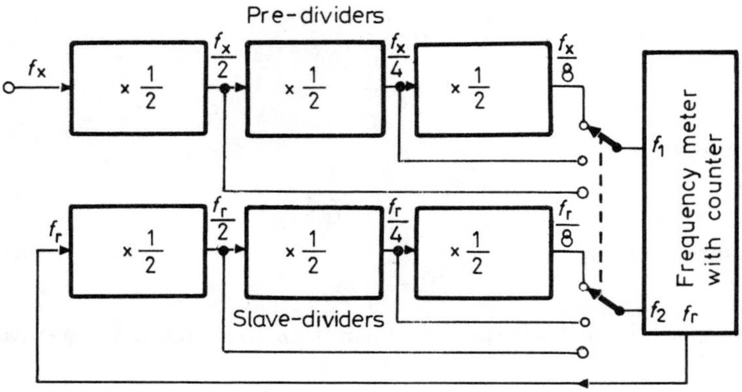

Fig. 5.8 Frequency division

realizable. The slave divider multiplies the measurement time with the same ratio by which the master divider divides the unknown frequency.

Consequently, the displayed value will be the unknown frequency and the ± 1 count ambiguity will be negligible. The number of digits of the counter and the display must be increased by the dividing ratio, but this is technically very simple. With frequency dividers the measurement range can be increased up to 1 ± 1.5 GHz.

b) Frequency converter

In this method a known frequency signal is mixed to the unknown frequency signal and the counter measures the difference of the two frequencies only. Fig. 5.9 shows the block diagram of the measurement principle.

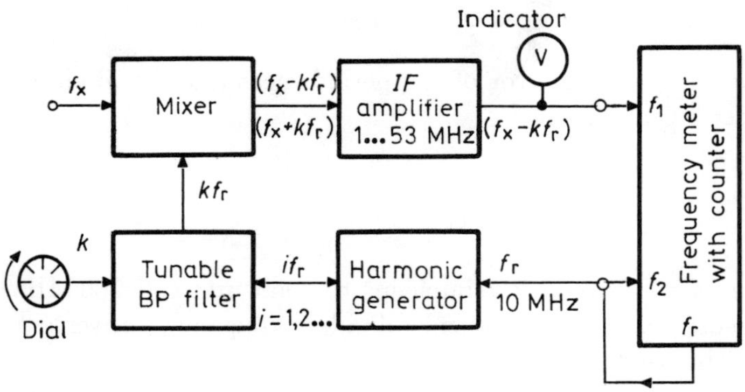

Fig. 5.9 Heterodyne frequency converter

With the tuneable bandpass filter the known frequency is increased from 50 MHz in 50 MHz steps, until the indicator shows that $f_m = f_x - k \cdot f_r < 50$ MHz.

In this case the frequency meter measures the f_m differential frequency and the number of the upper harmonics (k) can be seen on the tuner dial. The measured frequency is equal to

$$f_x = f_m + kf_r\,. \tag{5.18}$$

It can be proved that the error of f_x is in inverse proportion to the measurement frequency, i. e.:

$$\frac{\Delta f_x}{f_x} \approx \pm \frac{1}{T_m f_x}\,, \tag{5.19}$$

where T_m is the measurement time of the frequency counter ($T_m = n/f_r$). Thus, with this method a shorter measurement time and higher accuracy can be achieved than with frequency dividers.

Frequency converters can be used up to 10–15 GHz. To measure higher frequencies transfer oscillators can be used.

c) *Transfer oscillator*

With a transfer oscillator the measurement range can be extended up to 20–25 GHz. The oscillator is tuned until the frequency of one of the upper harmonics will be equal to the measurement frequency. The counter measures the fundamental harmonic of the oscillator signal. Fig. 5.10 shows the block diagram.

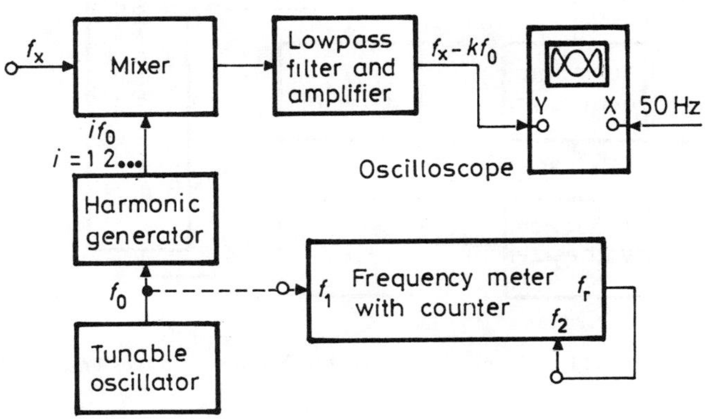

Fig. 5.10 Transfer oscillator

During the tuning process it can seen on the oscilloscope when one of the harmonic frequencies is equal to the measured frequency. In this case $f_0 = f_{01}$. Continuing the tuning of the oscillator until the oscilloscope shows that the frequencies are equivalent again, i. e. $f_0 = f_{02}$. From these two results it can be calculated which one of the harmonics of f_0 is equal to the unknown frequency:

$$k_1 = \frac{f_{02}}{f_{02} - f_{01}}\,. \tag{5.20}$$

Thus the unknown frequency is:

$$f_x = k_1 f_{01} \,. \tag{5.21}$$

Having multiplied the counting time of the counter by k_1, the display will show directly the value of the unknown frequency.

The error of the measurement process is determined by the ± 1 count ambiguity, the displaying error of the frequency equivalence (about $\pm 1\,\text{kHz}$), and the stability and the resolution of the transfer oscillator. With manually tuned transfer oscillator the measurement accuracy is about 10^{-7}. With PLL tuning the measurement accuracy is about 10^{-10}.

d) Harmonic heterodyne converter

This is an up-to-date piece of equipment. These counters combine the advantages of frequency converters and transfer oscillators. The operating principle is shown in Fig. 5.11.

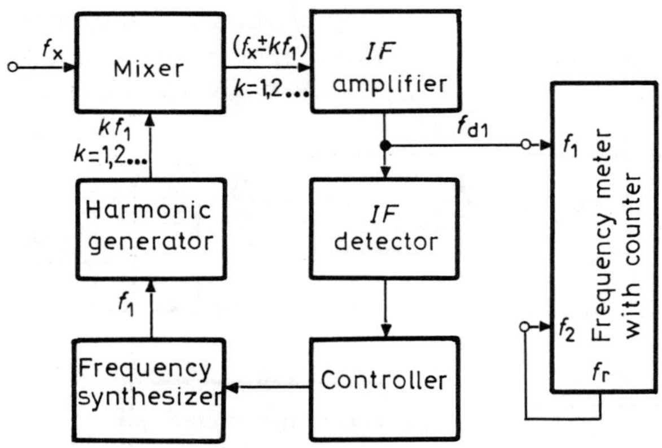

Fig. 5.11 Harmonic heterodyne frequency converter

The frequency synthesizer increases the frequency until on the output of the IF amplifier a signal with intermediate frequency f_{d1} appears. The IF detector detects it and the control unit stops the synthesizer. In this case the unknown frequency is:

$$f_{x1} = N f_1 \pm f_{d1} \,. \tag{5.22}$$

The counter measures the frequency difference f_{d1}, and f_1 is known from the state of the synthesizer. To determine the upper harmonic number (N) and the sign of the frequency difference, the system goes through a second measurement process, so that the synthesizer will be tuned by a frequency step Δf ($f_2 = f_1 - \Delta f$, where Δf is smaller than the bandwidth of the IF amplifier). In this case the counter measures the second frequency difference f_{d2}. From the two results the unknown frequency is calculable:

$$a) \qquad \text{if} \qquad f_{d1} > f_{d2}\,, \qquad f_x = N f_1 - f_{d1} \tag{5.23}$$

where

$$N = \frac{f_{d1} - f_{d2}}{f_1 - f_2}\,,$$

$$b) \qquad \text{if} \qquad f_{d1} < f_{d2}\,, \qquad f_x = N f_1 + f_{d1} \tag{5.24}$$

where

$$N = \frac{f_{d2} - f_{d1}}{f_1 - f_2}\,.$$

The harmonic heterodyne converter can be used up to 40 GHz.

So far frequency measurement with counters was discussed because of their simplicity and accuracy. These instruments develop continuously with the application of new advanced-technology solutions to advanced-technology problems (Wechsler, 1989).

Further on other frequency measurement methods will be briefly treated.

5.3.2 Frequency Measurement by Means of Resonators

The common feature of these frequency meters is that the unknown frequency is determined by the resonance of a resonator. Having resonance the unknown frequency is calculable from the known parameters of the resonator. Table 5.1. shows the properties of frequency meters of this type.

Table 5.1 Properties of resonator-based frequency meters

Type	Measurement range	Accuracy	Frequency span
RC bridges and RC selective circuits	10 Hz – 50 kHz	(0.1) – 1%	1 : 10
LC bridges and LC tuning circuits	50 kHz – 1.2 GHz	0.1 – 1%	1 : 3
Cavity resonators	1.1 GHz – 140 GHz	0.01 – 0.1%	1 : 2

5.3.3 Frequency Discriminators (Ratemeters)

These are frequency-to-voltage converters. In each period of the unknown frequency signal the ratemeter produces an impulse with constant amplitude and pulse width. The mean value of the impulse train is proportional to the unknown frequency. Fig. 5.12 shows the principle of this measurement mode. (A frequency discriminator can be used as a demodulator circuit for FM signal.)

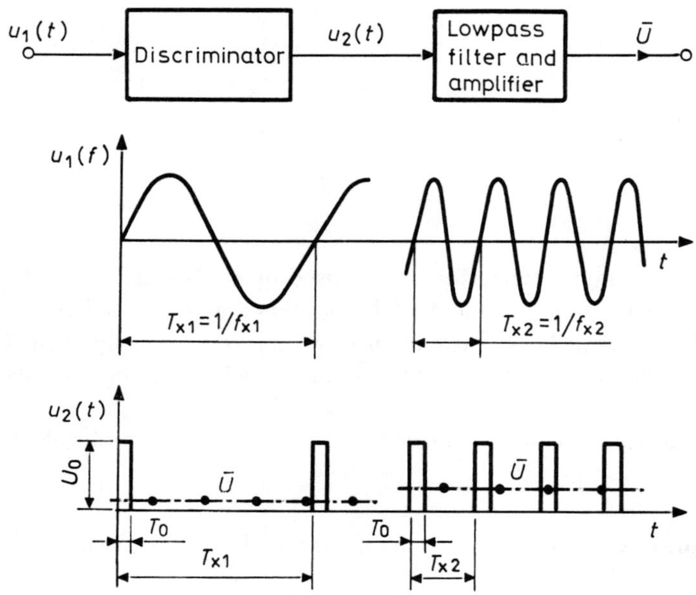

$$\bar{U} = \frac{1}{T_x} \int_{T_0}^{T_x} u_2(t)\, dt = U_0 T_0 f_x$$

Fig. 5.12 Ratemeter

5.3.4 Frequency Comparison by Means of an Oscilloscope

There are two methods:

a) Lissajous method

The two signals are connected to the horizontal and the vertical inputs of the oscilloscope. When the ratio of the two frequencies is a rational number, the pattern on the screen will not move. The unknown ratio of the two frequencies can be calculated according to Fig. 5.13 as follows:

$$\frac{f_y}{f_x} = \frac{h}{v}, \tag{5.25}$$

where f_x and f_y are the frequencies of the signals which are connected to the horizontal and vertical inputs of the oscilloscope, and h and v are the lengths of the horizontal and vertical intersections, respectively, on the screen.

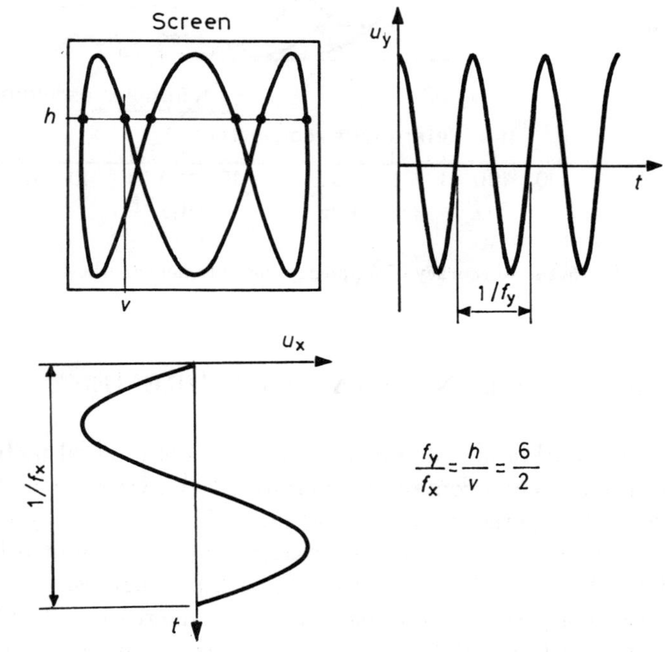

Fig. 5.13 Lissajous pattern

b) Synchronization method

The two signals are connected to the vertical input and to the trigger input of the oscilloscope. When the figure stops, the ratio of the two frequencies is a rational number. If it moves to right, the frequency of the vertical input signal is slightly higher than the frequency of the trigger input signal. If it moves to left, the frequency of the vertical input signal is slightly smaller.

Fig. 5.14 summarizes the characteristics of the above mentioned methods (Bagley, 1967).

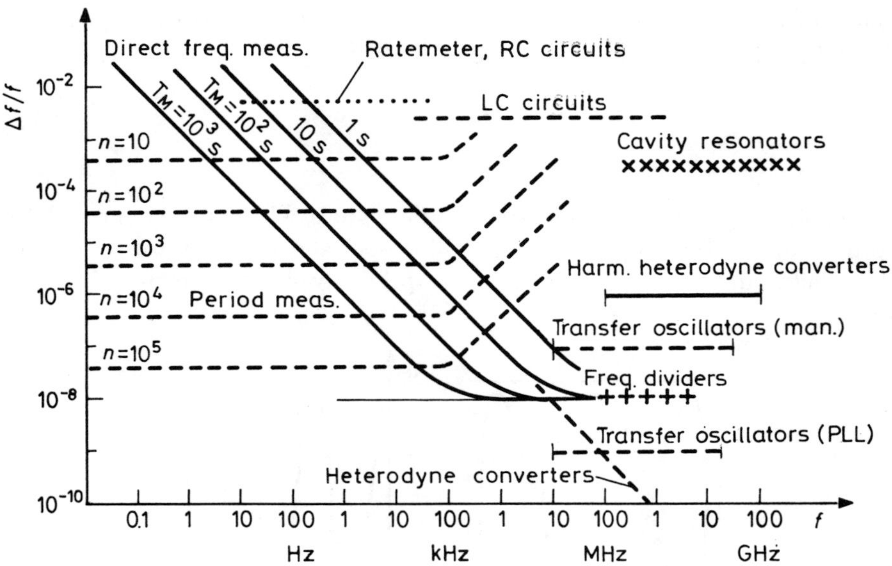

Fig. 5.14 Accuracy of frequency measurement methods

5.4 TIME-INTERVAL MEASUREMENT

Fig. 5.15 shows the block and time diagrams of the time-interval meter, based on a counter. The input circuits contain comparators with hysteresis and changeable comparison level. The operation of the instrument is illustrated by the time diagram. The accuracy of the time-interval measurement is determined by ± 1 count ambiguity and by the signal-to-noise ratio of the input signals. To improve the resolution, the frequency of the time base has to be increased. However, this is limited by the speed of the counter ($f_{\mathrm{rmax}} \approx 500\,\mathrm{MHz}$). To increase the resolution, the average time-interval in periodic events or interpolation techniques can be used.

5.4.1 Average Time-Interval Measurement

Measuring the average time-interval, the uncertainty due to noise will decrease by a factor of $1/\sqrt{N}$, where N is the number of the measured time-intervals. (For example, if $f_{\mathrm{r}} = 100\,\mathrm{MHz}$, the uncertainty of a single time-interval measurement is 10 ns; if $N = 10^6$, the uncertainty of the average time-interval measurement will be 10 ps, but the measurement time will be increased by at least a factor of 10^6.)

A significant problem is that the uncertainty of the measurement process will be decreased only if the measured time-intervals are statistically independent of each other. This means that the repeating frequency of the time-intervals is independent of the frequency of the time base generator. In practice the frequency of the time

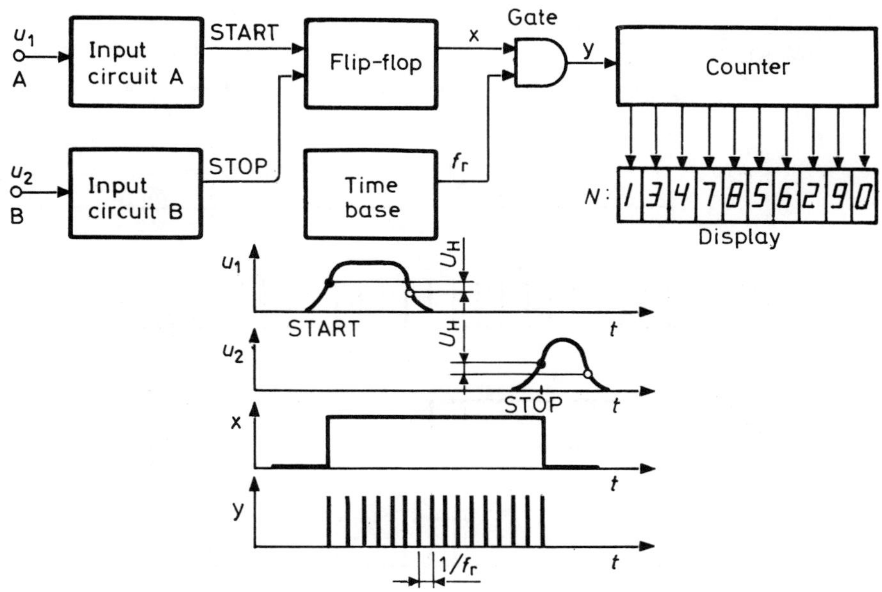

Fig. 5.15 Time-interval measurement

base is modulated by a Gaussian noise in order to avoid unwanted synchronization (Chu, 1974).

5.4.2 Interpolation

The resolution of instantaneous time-interval measurement can be improved by *interpolation techniques*. Fig. 5.16 shows an example.

From the figure it can be seen that in the same time as the start and stop signals two slave oscillators switch on. The frequencies of these oscillators are lower than the frequency of the time base by Δf_0. The impulses of the slave oscillators are counted by two slave counters until there is coincidence between the impulses of the slave oscillators and of the time base.

Knowing the content of the counter (n), of the slave counters (n_1, n_2) as well as the period of the time base (T_r) and of the period of the offset frequency (ΔT_0), the unknown time-interval is calculable:

$$T_x = nT_r + (n_1 - n_2)\Delta T_0 . \tag{5.26}$$

It is concluded that the resolution increases K times when ΔT_0 is equal to T_r/K. The expected accuracy of ΔT_0 is $1/K$, so the permitted uncertainty of the frequency of the slave oscillator is less than $1/K^2$. The accuracy of the equipment is influenced by the speed of the coincidence circuits and of the triggered slave oscillators.

With these interpolation techniques the accomplished resolution is 20 ps (Chu et al, 1978). (In 20 ps the light runs only 6 mm!). There are also other instruments

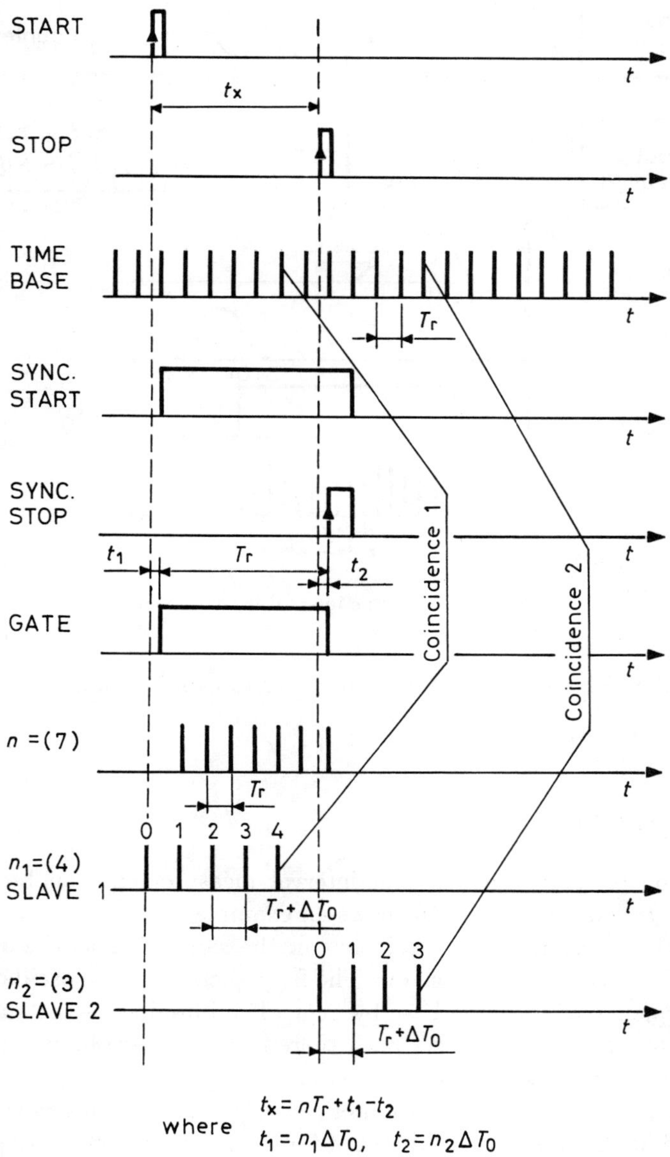

Fig. 5.16 Time-interval measurement with interpolation

where analog interpolator circuits are used to improve the resolution (Sasaki and Jensen, 1980).

5.5 PHASE MEASUREMENT

The relationship of the phase (φ) and the time difference (τ) between the zero crossing points of two sinusoidal signals of the same frequency is as follows:

$$\tau = \frac{\varphi}{2\pi}T_1 = \frac{\varphi}{2\pi f_1} \tag{5.27}$$

where f_1 is the frequency of the two signals.

It can be seen from (5.5) that the τ time difference is equal to the $x(t)$ phase-time. Hence the phase-time can be measured by a time-interval meter, but in daily measurement techniques the phase difference is always measured directly.

In phase measurements special attention should be paid to possible harmonic distortions and noise, because these may introduce severe errors. The effect of noise is discussed in detail in Section 5.3.

5.5.1 Phase Measurement with Counter

Fig. 5.17 shows the block and time diagrams of a phase meter with counter.

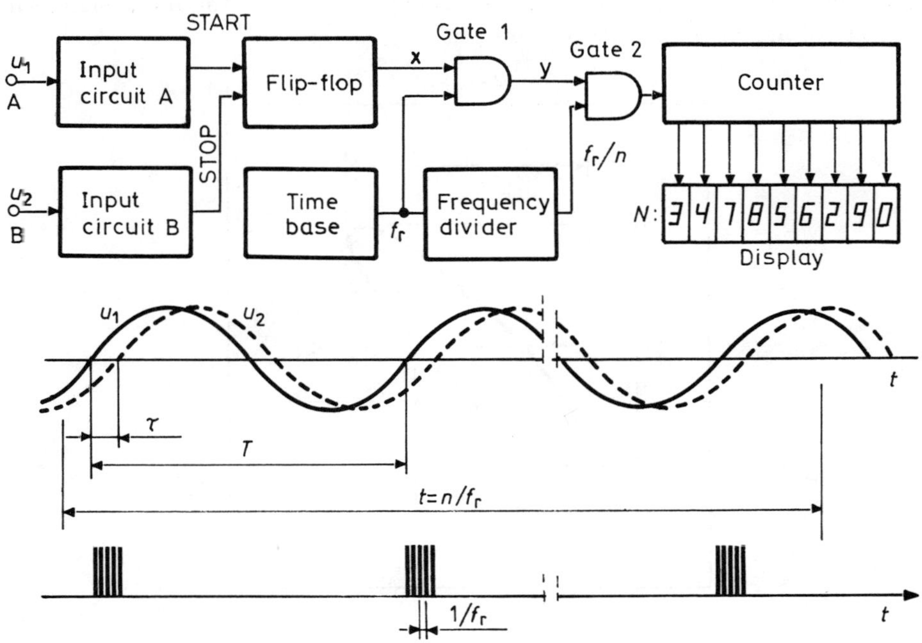

Fig. 5.17 Phase meter with counter

The counter counts the impulse numbers of the periodically appearing τ length impulse packages during the time $t=n/f_r$.

$$N = n\frac{\tau}{T_1}. \tag{5.28}$$

If $n=2\pi m$, the counter will show the phase angle directly in radians:

$$N = m\frac{\tau}{T_1}2\pi = m\varphi. \tag{5.29}$$

The result contains an uncorrectable error, because it is not assured that the measurement time (t) is equal to the whole number of the periods (T_1):

$$\frac{\Delta\varphi}{\varphi} \approx \frac{T_1 - \tau}{t}. \tag{5.30}$$

The error which comes from the ± 1 count ambiguity of the counter is the following:

$$\Delta\varphi_{\text{count}} = \pm 2\pi\frac{f_1}{f_r}. \tag{5.31}$$

This error can be decreased with the increase of the measurement time, if the frequency of the time base is statistically independent from the frequency of the signal. The measurement error is caused also by the noise of the input signals and the offset voltage of the input comparators.

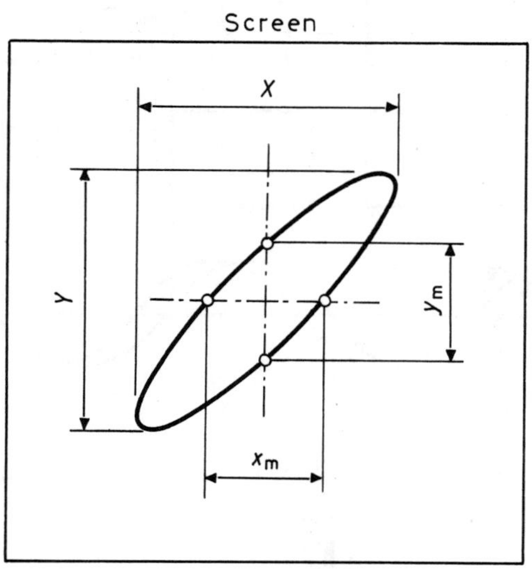

$$\varphi = \text{arc sin}\frac{x_m}{X} = \text{arc sin}\frac{y_m}{Y}$$

Fig. 5.18 Lissajous pattern

5.5.2 Phase Measurement with Oscilloscope

The phase angle can be measured by the Lissajous method. The two signals are connected to the horizontal and vertical inputs of the oscilloscope. The unknown phase angle is calculable according to Fig. 5.18.

The phase can also be measured by a dual beam oscilloscope.

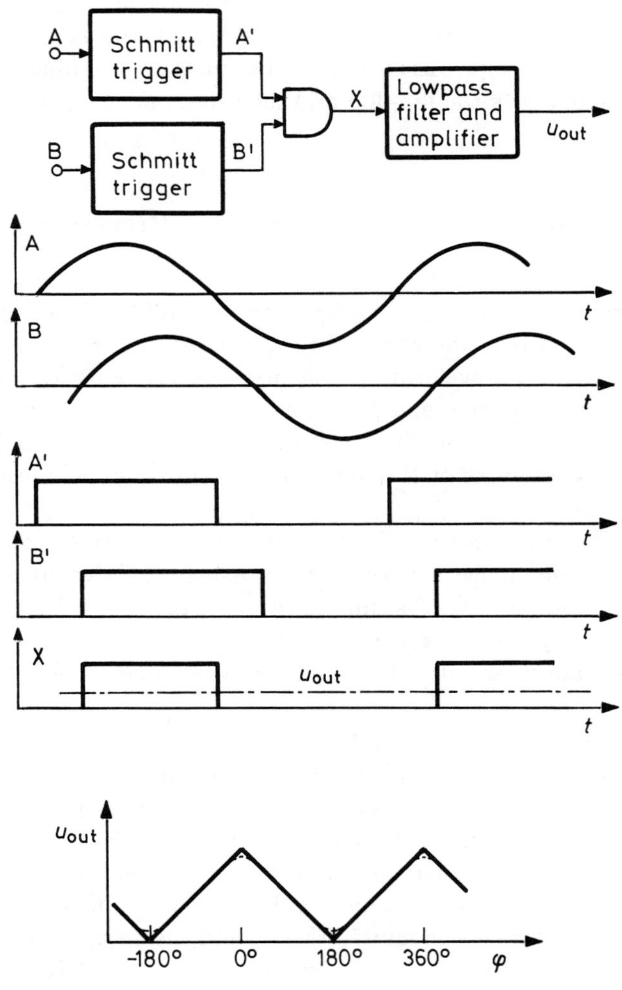

Fig. 5.19 Phase comparator waveforms

5.5.3 Phase Measurement with Phase Detector

Phase detectors are phase-to-voltage converters. They are mostly used in the phase-locked loops (PLL), but their most accurate types are used for phase measurement.

Fig. 5.19 shows the block and time diagrams of a simple phase comparator circuit. The mean value of the impulse train on the output of the AND-gate is proportional to the phase difference.

The main drawback of this phase detector is that its transfer characteristic is linear in the 180° range only. By the division of the frequencies of the square waves by two flip-flops at each input, the phase difference can be measured in the 0–360° range. Edge triggered flip-flops can also be used to measure the phase in the 0–360° range so that one signal sets the flip-flop and the other one resets it (Hager, 1975).

5.5.4 Phase Measurement with Digital Signal Processing

Modern digital signal processors (DSP) offer a new possibility of the measurement of the phase angle between two sine waves of the same frequency, by means of digital processing of the sampled waveforms. Such instruments can even measure time-varying instantaneous frequency and phase, required in the analysis of modulated signals. The algorithms are usually based on the estimation of instants of zero crossings in the time domain, or on the comparison of the discrete Fourier transforms of the two waveforms.

In the *time domain approach* the two sine waves are converted into square waves, and the time difference between zero crossings is determined. The main drawbacks are that the resolution is limited by the sampling rate, and that large phase uncertainties may appear when the input sine waves are distorted by harmonics and additive noise (Wagdy and Lucas, 1985; Mahmud et al, 1988; Mahmud, 1989a).

The *phase progression digitizing technique* is based on continuous counting and sampling at zero crossings only of the signal. The method directly digitizes the phase progression of the signal. The sampling rate must only exceed the Nyquist rate for the modulation, regardless of the carrier frequency, because the reconstruction of the waveform itself is not required, and the information on the phase can already be extracted when this condition is fulfilled. With this method the direct measurement range is between dc and 500 MHz, and, using a pre-scaler or down-converter, the measurement range can be extended to 2 to 18 GHz (Chu, 1989).

In the *frequency domain approach* the real-time bandwidth is a few hundred kHz, because of the need of A/D conversion and FFT. The conversion to the frequency domain very effectively eliminates the influence of harmonics and additive noise (Mahmud, 1989b; Bertocco et al, 1992).

The detailed discussion of the above techniques cannot be given here, these can be found in the literature.

References

Bagley, A. S. (1967): Frequency and Time Measurements. *Handbuch der Physik*, Bd. XXIII, pp. 290–372. Berlin, Springer Verlag.

Barnes, J. A., Chi, A. R., Cutler, L. S., Healey, D. J., Leeson, D. B., McGunigal, T. E., Mullen, J. A., Smith, W. L., Sydnor, R., Vessot, R. F. C. and Winkler, G. M. R. (1971): Characterization of Frequency Stability. *IEEE Trans. on Instrumentation and Measurement*, Vol. IM-20, No. 2, pp. 105–120.

Bertocco, M., Offelli, C. and Petri, D. (1992): Dynamic Behavior of a Digital Phase Estimator. *Record of the IEEE Instrumentation and Measurement Technology Conference*, New York, May 12–14, 1992. 92CH3151-8. pp. 100–105.

Chu, D. C. (1974): Time-Interval Averaging: Theory, Problems and Solutions. *Hewlett-Packard Journal*, Vol. 25, No. 6, pp. 12–15, June 1974.

Chu, D. C. (1989): Phase Digitizing: A New Method for Capturing and Analyzing Spread-Spectrum Signals. *Hewlett-Packard Journal*, Vol. 40, No. 1, pp. 28–35, Feb. 1989.

Chu, D. C., Allen, D. C. and Foster, A. S. (1978): Universal counter Resolves Picosecond in Time Interval Measurements. *Hewlett-Packard Journal*, Vol. 29, No. 8, pp. 2–11, August 1978.

Glaze, D. J., Hellwig, H., Allan, D. W., Jarvis, S. Jr., Wainwright, A. E. and Bell, H. E. (1974): Accuracy Evaluation and Stability of the NBS Primary Frequency Standards. *IEEE Trans. on Instrumentation and Measurement*, Vol. IM-23, No. 4, pp. 489–501.

Hager, J. C. (1975): Edge Triggered Flip-Flops Make 360° Phase Meter. *Electronics*, Vol. 48, No. 17, pp. 100–101.

Hyatt, R. C., Mueller, L. F. and Osterdock, T. N. (1973): A High-Performance Beam Tube for Cesium Beam Frequency Standards. *Hewlett-Packard Journal*, Vol. 24, No. 9, pp. 14–24, Sept. 1973.

Ives, F. H. (1989): Multifunction Synthesizer for Building Complex Waveforms. *Hewlett-Packard Journal*, Vol. 40, No. 1, pp. 52–57, Feb. 1989.

Kartaschoff, P. (1978): Frequency and Time. London, Academic Press, Inc.

Kroupa, V. F. (1973): Frequency Synthesis. London, Charles Griffin.

Lewis, L. L. (1991): An Introduction to Frequency Standards. *Proc. IEEE,* Vol. 79, No. 7, pp. 927-935, July 1991.

Mahmud, S. M. (1989a): Error Analysis of Digital Phase Measurement of Distorted Waves. *IEEE Trans. on Instrumentation and Measurement,* Vol. 38, No. 1, pp. 6–9, Feb. 1989.

Mahmud, S. M. (1989b): High-precision Phase Measurement Using Adaptive Sampling. *IEEE Trans. on Instrumentation and Measurement,* Vol. 38, No. 5, pp. 954–960, Oct. 1989.

Mahmud, S. M., Rusek, A. and Ganesan, S. (1988): A Microprocessor-Based Dual Slope Phase Meter. *IEEE Trans. on Instrumentation and Measurement*, Vol. IM-37, No. 3., pp. 374–378, Sept. 1988.

McNamee, M. D. and Platt, D. L. (1989): A Modular Family of High-Performance Signal Generators. *Hewlett-Packard Journal*, Vol. 40, No. 5, pp. 14–20, Oct. 1989.

Ramsey, N. F. (1987): Quantum Mechanics and Precision Measurements. *IEEE Trans. on Instrumentation and Measurement*, Vol. IM-36, No. 2, pp. 155–157.

Ramsey, N. F. (1991): The Past, Present and Future of Atomic Time and Frequency. *Proc. IEEE,* Vol. 79, No. 7, pp. 921-925, July 1991.

Reinhard, V. (1974): Relativistic Effects of the Rotation of the Earth on Remote Clock Synchronization. *Proc. 6th Annual PTTI Planning Meeting*, 1974, NASA Doc. No. x-814-75-117, pp. 395–424.

Rutman, J. and Walls, F. L. (1991): Characterization of Frequency Stability in Precision Frequency Sources. *Proc. IEEE,* Vol. 79, No. 7, pp. 952-960, July 1991.

Sasaki, G. D. and Jensen, R. C. (1980): Automatic Measurements with a High-Performance Universal Counter. *Hewlett-Packard Journal*, Vol. 31, No. 9, pp. 21–24, Sept. 1980.

Wagdy, M. W. and Lucas, M. S. P. (1985): Errors in Sampled Data Phase Measurement. *IEEE Trans. on Instrumentation and Measurement*, Vol. IM-34, No. 4, pp. 507–509, Dec. 1985.

Wechsler, M. (1989): Characterization of Time Varying Frequency Behavior Using Continuous Measurement Technology. *Hewlett-Packard Journal*, Vol. 40, No. 1, pp. 6–12, Feb. 1989.

Winkler, G. M. R. (1991): Synchronization and Relativity. *Proc. IEEE,* Vol. 79, No. 7, pp. 1029-1039, July 1991.

6

Tools and Methods of Signal Analysis

I. Kollár

Editorial introduction

The purpose of measurement is to extract the maximum amount of information relevant to the investigated phenomenon, in a minimum amount of time and at low cost. This requirement usually implies that the system under test is investigated during the whole available measurement time, and the time signals are analyzed in full length. In other words, ***signal analysis*** is being done. Even in the case of commonly used instruments such as digital voltmeters, where the result is displayed as a single number, usually a time record is processed in order to obtain this. Thus, to be able to determine the error of the result (i.e. the magnitudes of systematic and random errors), the complete signal processing procedure has to be investigated.

Nowadays this tendency is getting even stronger, with the appearance of intelligent, microprocessor-controlled measuring devices. Sophisticated signal processing methods are built into these new instruments, and the user often hopelessly tries to find out what happens inside the instrument itself. However, profound understanding of the measured results is vital for scientists and engineers who intend to use them in their work.

The aim of this chapter is to provide the basics of signal analysis for those who use complex instruments. On the one hand, the most important signal processing methods are going to be discussed, on the other hand, the building blocks and main features of modern 'intelligent' instruments will be treated. Proofs and complicated mathematics will be when possible avoided; instead, explanations based on simple models and illustrations will be given.

The errors of the results of signal analysis will be discussed for the two main signal types: for random signals with continuous spectrum and for sine waves.

6.1 BASIC SIGNAL PROCESSING METHODS

Signal processing means manipulations on so-called signals (usually time functions), defined in Chapter 1. These manipulations may have very different aims, however,

Technology of Electrical Measurements, Edited by L. Schnell
© 1993 John Wiley & Sons Ltd.

the basic principles are usually common and also rather simple. This section deals with these *basic* principles and with the methods that will be used throughout the rest of the chapter.

It is not possible to treat every existing method, since signal analysis or signal processing is already a stand-alone, extensive subject (Bendat and Piersol, 1971; Bendat and Piersol, 1980; Gold and Rader, 1969; Rabiner and Gold, 1975; Oppenheim and Schafer, 1975; Oppenheim et al, 1983; Robinson and Silvia, 1986; Kunt, 1986; Urkowitz, 1983; etc). The attention will be focused on the methods that are already customarily built into the instruments on the market.

For each method the properties of the result (usually the bias and the variance) will be given. The purpose is not only the *analysis* of the methods, but also the *design* of usual signal processing procedures: the necessary instrument settings can be determined on this basis. For example, the necessary record length can usually be calculated from the signal bandwidth and the desired variance reduction. The errors will be evaluated for the two most important types of signal: stochastic signals with continuous spectrum, and sinusoidals with fixed or random phase. Where appropriate, transient signals will also be dealt with.

The application of the error formulas is usually not straightforward. For example, the variance expression of the estimate of the probability density function (6.134) contains the true value of the pdf itself. But this quantity is just to be measured, thus it is unknown! This paradox seems to be unsolvable, and if this were accepted, measurement design would be impossible.

The solution is simple: the variance expression is necessary in order to have an *upper bound* of the error of the estimate. It is generally acceptable that the 'error of the error' is not very small. Thus, the first measurement can be designed on the basis of a rough guess (a priori knowledge, or a quick measurement) of the quantities in the expressions. The result can then be used for checking the validity of this guess; if necessary, the measurement has to be repeated with new parameters. The measurement procedure may become iterative: improved knowledge gained from new measurements helps in the better design of the next one.

In the expressions of the variance (or the bias) often a functional (e. g. the integral value) of a function is present. Since the error need not be exactly known, and usually only uncertain knowledge about these functions is at hand, it would be in vain to exactly evaluate the functional. Instead, approximations are used. Such approximation is e. g. the use of an *equivalent bandlimited white noise* (Bendat and Piersol, 1971, see Fig. 6.1). The true covariance power spectral density function is substituted by the corresponding function of an appropriate bandlimited white noise. The procedure is similar to the definition of the bandwidth of a nonideal bandpass filter. The substitution may yield a very good insight into the behaviour of the estimates, since the variance expressions usually become very simple for the bandlimited white noise.

The equivalent bandwidth can be defined in several ways, and for a given purpose one certain definition should be used. However, theoretical accuracy is generally of no importance: the possible bandwidths are usually close to each other anyway, especially if very selective filters are used. Since the expressions of the

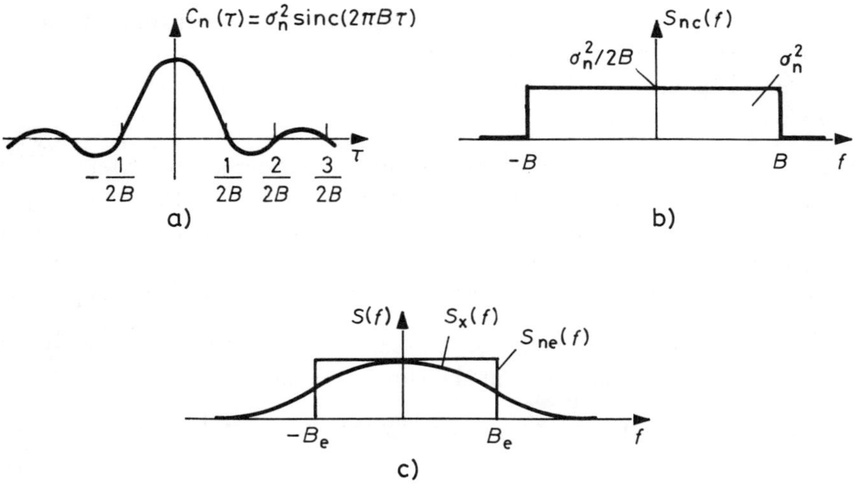

Fig. 6.1 Substitution by an equivalent bandlimited white noise *a)* autocovariance function of the equivalent noise *b)* covariance power spectral density function of the equivalent noise *c)* substitution in the frequency domain

variance are used for approximations (upper bounds) only, the choice of the definition is not decisive. The most common choice is the so-called *noise bandwidth*: the variance σ^2 (the integral value of the covariance power spectral density) and $S_{xc}(0)$ are kept constant in the substitution.

Sometimes not only a scalar quantity but a whole function is to be measured (correlation function, spectrum, probability density function etc). Such measurements are often made by scanning the range of interest; for these cases also the allowable scan rate (the speed of changing the free variable) has to be calculated.

6.1.1 Time Domain and Frequency Domain Methods

Since the pioneering work of Fourier, and especially since the invention of the fast Fourier transform (Cooley and Tukey, 1965), frequency domain analysis (Fourier series expansion, Fourier transform, spectral analysis) has been a powerful tool in the hands of scientists and engineers. Although time domain is very natural and easy to understand, some questions can be more easily answered using frequency domain descriptions. Moreover, easy understanding or visuality are relative notions. Often used models quickly become 'visual': for electrical engineers the concepts of two-sided frequency axis or operational impedances are as visual as the time axis itself.

Time domain and frequency domain have different advantages and disadvantages, they are in several respects complementary to each other. Engineers often prefer frequency domain descriptions, because of the following reasons:

- While the solution of a differential equation needs convolution in the time domain, this convolution is substituted by simple multiplication in the frequency domain. This often makes it possible to explain system behaviour in a visual way.
- It is often possible to decompose signals and noises into different frequency bands. Signal-to-noise ratios can be easily improved in the frequency domain by choosing appropriate bands. Moreover, the energy (power) of periodic signals is concentrated at discrete points in the frequency domain. Thus, frequency domain is very selective with respect to periodic components.
- The attainable dynamic range (80–100 dB) is usually significantly larger in the frequency domain than in the time domain.
- Slight nonlinearities can be more easily recognized and measured by frequency domain methods.
- The FFT is a very powerful tool for fast time domain to frequency domain conversion, and thus e. g. the evaluation of a correlation function is much quicker via the frequency domain than directly in the time domain.

On the other hand, time domain also has important advantages:

- It is very 'natural' to deal with time domain signals (although after some practice, even the two-sided frequency axis becomes very visual).
- Recursive methods often provide on-line calculation possibilities.
- Time varying systems can be modelled more easily in the time domain.
- The transient behaviour of systems can be directly measured in the time domain.
- Time domain digital methods are not very sensitive to the type of signal, while digital frequency domain methods suffer from leakage effects when non-periodic signals are processed.
- Certain nonlinearities (clipping, slew rate etc.) are sometimes more easily recognized in the time domain.

In the following sections both time domain and frequency domain methods will be treated.

6.1.2 Sampling

Modern instruments are usually at least partly *digital*. This means that instead of continuous signals, internally a series of numbers is processed. When this happens, questions like these arise:

- Is it possible that the series of numbers contains exactly the same information as the analog signal?
- How are the important parameters (sampling frequency, bit number) of the conversion to be chosen, in order to maintain the information important to us?

Such questions are answered basically by the *theory of sampling*, discussed in this subsection, and the *theory of quantization*, discussed in the next one. In what follows the sampling theorem will be briefly stated, then the practical consequences, application limitations etc. will be discussed.

To investigate the effect of *sampling*, let us consider an analog signal and its discrete equivalent (Fig. 6.2). It is clear from the figure that when the samples are taken densely enough with respect to the changes of the signal amplitude, the discrete points exhibit the same pattern as the continuous waveform. This statement implies a condition on the sampling rate as a function of the speed of changes of the signal, i. e. of its bandwidth. Consequently, it is obvious to investigate the effect of sampling in the frequency domain.

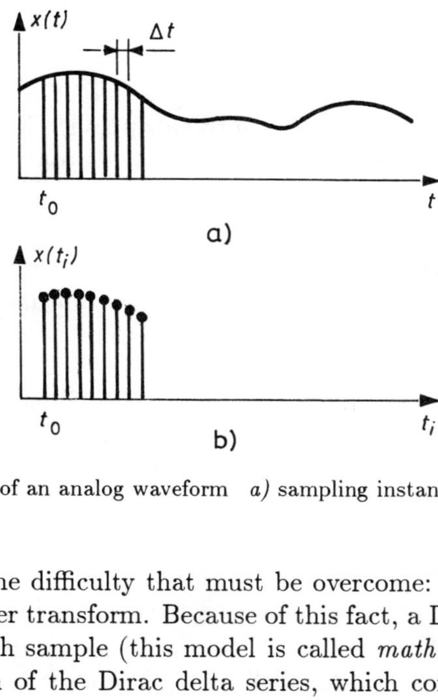

Fig. 6.2 Sampling of an analog waveform *a)* sampling instants *b)* result of sampling

There is only one difficulty that must be overcome: a series of numbers has no (continuous) Fourier transform. Because of this fact, a Dirac delta function used to be attached to each sample (this model is called *mathematical sampling*), and the Fourier transform of the Dirac delta series, which contains exactly the same information as the series of numbers, used to be analyzed. It can be shown that the Fourier transform of the Dirac delta series is a *series of repetitions* of the Fourier transform of the original signal (Fig. 6.3):

$$X_s(f) = \sum_{k=-\infty}^{\infty} X\left(f - \frac{k}{\Delta t}\right), \tag{6.1}$$

where

$X(f)$ is the Fourier transform of the original time function,
Δt is the sampling interval,
t_0 (see Fig. 6.2) is chosen equal to 0.

From (6.1) it is clear that when the repeated spectra do not overlap (do not *alias*), the original spectrum can be cut out, thus the information on the original waveform is kept, too; while when the spectra overlap, the spectrum will be distorted and information is lost. This leads to the sampling theorem (Nyquist, 1928; Shannon, 1949) for time domain signals.

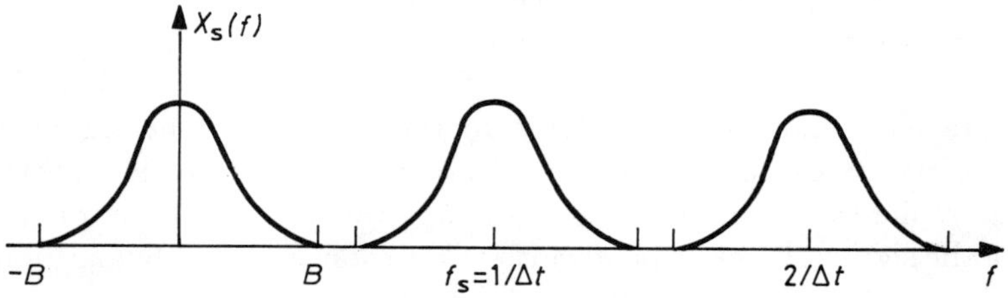

Fig. 6.3 Spectrum of the sampled signal

Sampling theorem I. (for time domain sampling): If a time domain signal is bandlimited, that is, its Fourier transform, $X(f)$, is zero if $|f| \geq B$, no information is lost with sampling frequency $f_s \geq 2B$, that is, the original analog waveform itself can be reconstructed from the samples.

Sometimes, e. g. when using dual-slope A/D converters, see Subsection 2.3.17, not the instantaneous value of the signal is obtained, but its integral mean in given intervals. It can be shown that though (6.1) has to be modified in these cases, the above stated sampling theorem is valid.

By some derivation also the reconstruction algorithm (interpolation formula) can be given:

$$x(t) = \sum_{i=-\infty}^{\infty} x_i \operatorname{sinc}\left(\pi \frac{t - i\Delta t}{\Delta t}\right), \tag{6.2}$$

where $\operatorname{sinc} x = \sin x / x$, and x_0 was taken at $t=0$.

The frequency $2B$ is the so-called *Nyquist rate*, while the frequency, above which the signal may not contain any component ($f_s/2$), is the *Nyquist frequency* (or Nyquist cutoff frequency or Nyquist folding frequency).

This theorem has an equivalent for frequency domain signals, which used to be stated in the following form:

Sampling theorem II. (for frequency domain sampling): If the inverse Fourier transform of a frequency domain signal is time-limited, that is, $x(t)$ is zero outside an interval of length T, no information is lost with sampling density $\Delta f \leq \frac{1}{T}$.

Even if the time domain signal has no Fourier transform — which is the case when treating stochastic signals — a similar theorem applies if the *power spectral density* is bandlimited (Jerry, 1977).

Practical aspects of sampling

Sampling theorem is a very nice one, and is widely used. However, in practical applications a lot of extra aspects arise.

First of all, sampling is usually realized using sample and hold circuits. The timing of sampling is usually somewhat uncertain. This phenomenon is called *timing jitter* or *aperture uncertainty.* The sampled amplitude values may also exhibit some errors, due to the finite transient times of the switches (aperture time), further to noises, imperfections of the tracking circuitry etc (Sheingold, 1986). These errors may often be negligible from the point of view of the measurement goal, but this has to be determined in each case individually.

Timing jitter is investigated e. g. by Souders et al (1990) and Balakrishnan (1962).

One may also notice that (6.2) contains an *infinite* sum. Moreover, it can be proved that this sum contains always an infinite number of nonzero terms: a bandlimited signal cannot be time-limited. In practice, only a finite number of samples can be taken, so (6.2) can never be exactly calculated: only approximate fulfilment of the sampling theorem can be achieved.

If it is accepted that a signal can be *approximately* bandlimited and time-limited at the same time, an interesting expression can be obtained. In this case the number of nonzero samples is

$$N = \frac{T}{\Delta t} \geq 2BT\,. \tag{6.3}$$

This statement is often called *approximate sampling theorem.* It says that with proper sampling $N = 2BT$ samples are enough to describe the signal properly. We will come back to this expression once more, when discussing averaging.

If a signal is processed in a digital way, there is always a risk of violating the conditions of the first sampling theorem. Therefore, an *anti-aliasing filter* is built into the input channel of practically every digital instrument; this limits the bandwidth of the captured signal to half of the sampling frequency. Since real lowpass filters have always a transition band above the passband, the edge of the analyzed frequency band (B) used to be chosen lower than half of the sampling frequency, usually somewhere around $0.4 f_s$ (see Fig. 6.4). However, if the sampling frequency can be chosen significantly higher than the double of the desired bandwidth, a low-order anti-aliasing filter may be sufficient, and the rest (bandlimit reduction and decimation, maybe combined with amplitude and phase correction) can be done by digital filters.

The use of an anti-aliasing filter may introduce unwanted amplitude and phase errors (in the order of 1 dB) in the passband. For the purpose of precision measurements high-quality filters (with small passband errors) are required, or these errors

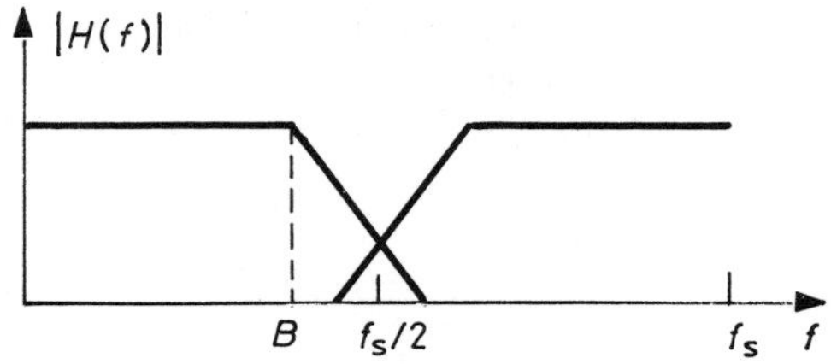

Fig. 6.4 Usual choice of the bandwidth of the anti-aliasing filter

can be compensated for by a carefully designed built-in digital correction filter (see e. g. Kollár et al, 1990).

From the sampling theorem one may come to the conclusion that *oversampling* (sampling over the Nyquist rate) is superfluous: no more information is gained, but storage space is wasted. As was discussed in the previous paragraphs, this is not exactly true, since the conditions of the sampling theorem cannot be exactly met. But this is one aspect only. Let us consider the series of samples in Fig. 6.5a: it seems to consist of two 'folded' sine waves. However, the truth is illustrated in Fig. 6.5b: a sine wave was sampled just above the Nyquist rate.

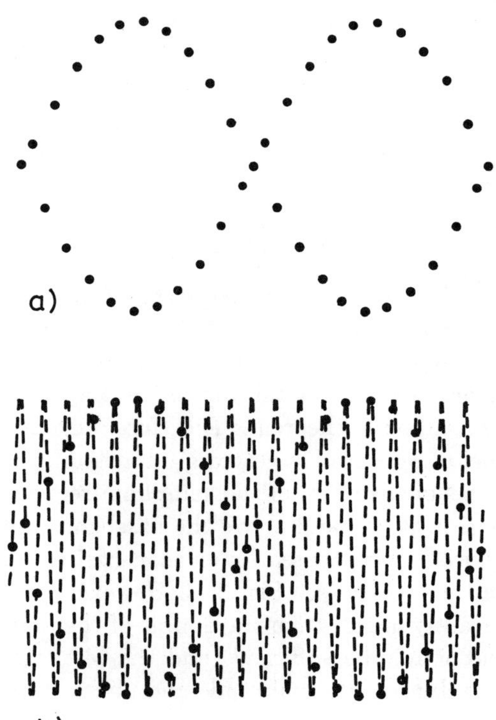

Fig. 6.5 Sampling of a sine wave slightly above the Nyquist rate *a)* series of samples *b)* samples with the original waveform

Seemingly, the human eye has no built-in sinc interpolator! Our brain seeks patterns on each figure, and the simplest pattern is to connect the closest points. Thus, for the visual evaluation of time domain signals, it is usually reasonable to oversample the signal: though later interpolation may be possible, sometimes it is easier to sample a priori with the frequency desirable for visual evaluation. A similar situation occurs when the peak value is sought: it is easier to look for the maximum value in an oversampled signal than to interpolate in the suspected intervals.

Example 6.1

According to the sampling theorem, more than two samples must be taken from a period of a sine wave.

When using a 10-bit A/D converter, in order to ensure that the deviation of the maximum of the samples from the peak value is certainly not larger than 1 LSB, the sampling frequency must be larger than $70f_1$; in order to ensure that subsequent samples do not differ from each other by more than 1 LSB, even at the steepest slope of the signal, $f_s > 6400f_1$ must be assured!

However, when calculating the spectrum — in contrast with time domain demands — it is practical to choose a possibly small sampling frequency, since for a fixed number of samples N, the lower the sampling frequency f_s, the better is the frequency resolution (Fig. 6.6):

$$\Delta f = \frac{1}{T} = \frac{1}{N\Delta t} = \frac{f_s}{N} \,. \tag{6.4}$$

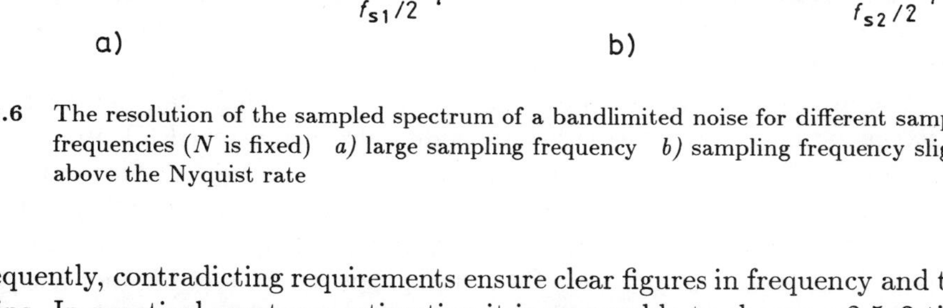

Fig. 6.6 The resolution of the sampled spectrum of a bandlimited noise for different sampling frequencies (N is fixed) *a)* large sampling frequency *b)* sampling frequency slightly above the Nyquist rate

Consequently, contradicting requirements ensure clear figures in frequency and time domains. In practical spectrum estimation it is reasonable to choose a 2.5–3 times higher sampling frequency than the known frequency limit, to ensure adherence to the sampling theorem.

Strangely enough, *undersampling* can also be justified. Here two important cases can be mentioned. The first one is when not all the information is needed from the time record. This is the case e. g. when the mean value of a stochastic signal is to be measured, and it is possible to measure long enough to collect a sufficient number of samples. With undersampling the samples will be independent, which eases the statistical evaluation of the averaging procedure. Though some information is lost indeed, what is necessary is extracted from the signal.

Another reason for undersampling may be the special form of the spectrum. For *bandpass* signals, aliasing into the known signal band can be avoided even if the signal is sampled under the Nyquist rate (see Fig. 6.7). A similar situation occurs when periodic signals are measured: see Subsection 6.2.1 for the principle of sampling oscilloscopes.

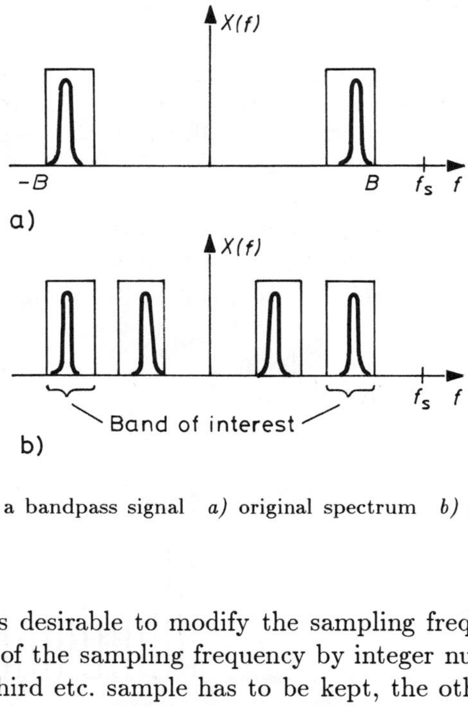

Fig. 6.7 Sampling of a bandpass signal *a)* original spectrum *b)* spectrum of the sampled signal

Sometimes it is desirable to modify the sampling frequency *after* sampling is finished. Division of the sampling frequency by integer numbers is very simple: only every second, third etc. sample has to be kept, the others are to be thrown away. This procedure is called *decimation*. It is to be applied with care, because of the possible aliasing when the signal is not sufficiently bandlimited.

The opposite of decimation is *interpolation*. Eq. (6.2) can always be used, and it can be evaluated at equidistant points also using two FFT-s (see e. g. in Subsection 6.5.2). Though (6.2) gives indeed the exact solution, and it can be proved that having a finite number of samples, the truncated version of (6.2) provides the best approximation of the original waveform in a least squares sense, (6.2) is often not used, because the truncation of a series, given in terms of functions with long oscillations (sinc), results in oscillatory reconstruction. Moreover, if the signal is not exactly bandlimited, around jumps of the original signal the Gibbs phenomenon (overshooting) will show up, which is often undesirable. In Fig. 6.8 such a case is shown: though theoretically a sine wave of infinite length is bandlimited, a fraction of it is not, thus interpolation results in oscillations near to the edges. Thus, usually another approach is followed. From Fig. 6.2 it is straightforward that an ideal lowpass filter is required in $(-f_s/2, f_s/2)$. If desired, a similar lowpass filter can be designed with shorter oscillations (interpolation filter), which is common in digital oscilloscopes.

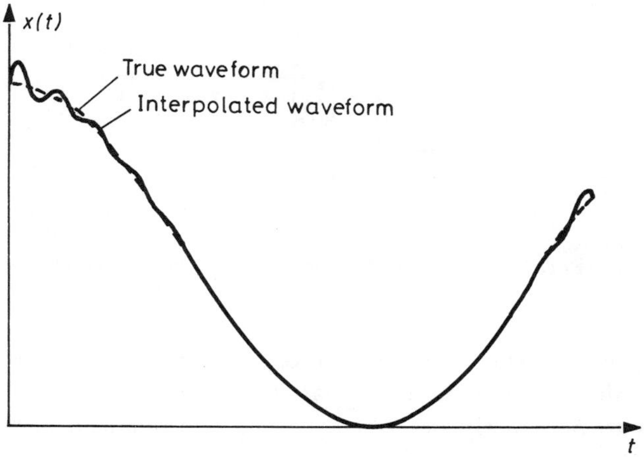

Fig. 6.8 Imperfect reconstruction of a signal, using the interpolation formula

For on-line reconstruction a zero-order hold (DAC) or a higher-order hold can also be used, maybe in combination with an analog lowpass filter to suppress higher frequency contents.

Finally, let us mention an interesting application of the properties of sampling. Since the spectrum appears repeatedly along the frequency axis, sampling may also be used for frequency transposition (modulation): any repetition can be selected by using an appropriately set bandpass filter, thus the signal can be evaluated in another frequency band, e. g. a bandpass one in the baseband. This can be done sometimes even with a sampling frequency lower than the Nyquist rate (*direct decimation*), if the signal is of bandpass nature (Fig. 6.7) or is periodic (Fig. 6.31).

6.1.3 Quantization

Numbers stored in a digital computer are discrete, not only in time but also in amplitude: they are stored in a finite number of bits. Consequently, they are approximations only of the analog values. The nonlinear operation which produces the discrete amplitudes from the continuous ones is called quantization. The transfer characteristic of a so-called uniform quantizer (for which all steps are equal) is shown in Fig. 6.9. x is the input variable, x_q is the output one, x_i, x_{i+1} ... are the input quantization levels, and s_0 is a shifting parameter with respect to the position where an output level is exactly at zero. The special position $s_0 = 0$ is called the *rounding quantizer,* since with steps equal to unity, each sample is rounded to the nearest integer.

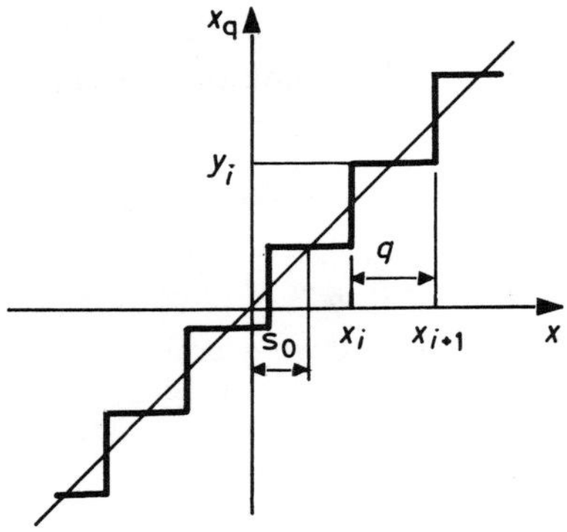

Fig. 6.9 Transfer function of a uniform quantizer

Studying the transfer characteristic of the quantizer, it is obvious that the operation is nonlinear: serious mathematical difficulties may be expected when quantization is to be exactly described, even if the quantum steps are all equal (Dobrowiecki, 1984).

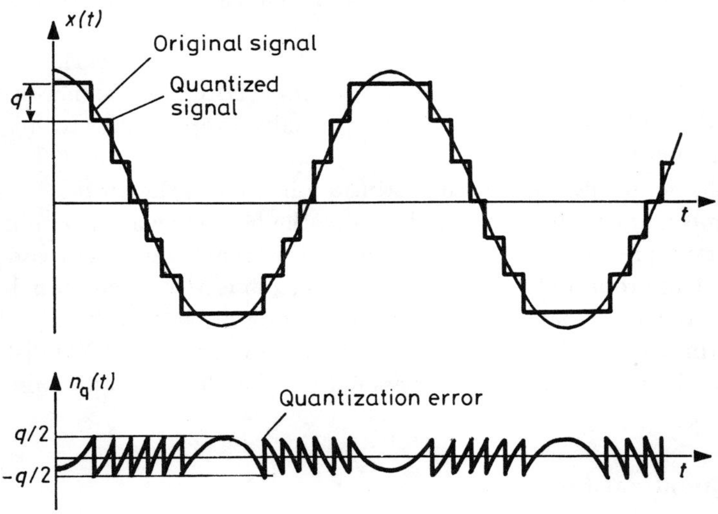

Fig. 6.10 Quantization of a sine wave

Fortunately, in most cases a simple approximate model can be found. Let us consider the quantization of a sine wave (see Fig. 6.10). The quantization error (that is, the difference between the quantized and the original signal) has interesting properties:

- At most of the places the quantization error is 'sawtooth'-like, consequently it has approximately uniform distribution between $(-q/2, q/2)$.

- The form of the quantization error does not resemble the original waveform, since it describes the *local* behaviour of the signal, relative to the next quantum level, and has little to do with the global behaviour.
- The quantization error contains a lot of jumps, thus its spectral content is much wider than that of the original signal. If not sampled too densely (with respect to the density of the jumps), the samples of the error will be apparently independent of each other. The bandwidth of the quantization error is seemingly proportional to the slope of the of the signal ($|\dot{x}(t)|$), and inversely proportional to the quantum size q.

The above observations lead to the so-called *noise model* of quantization (Fig. 6.11): from the statistical point of view, the quantization error may be considered as an additive noise which

- has uniform distribution,
- is uncorrelated with the input signal,
- has a white spectrum (at least when sampled at a sufficiently low rate).

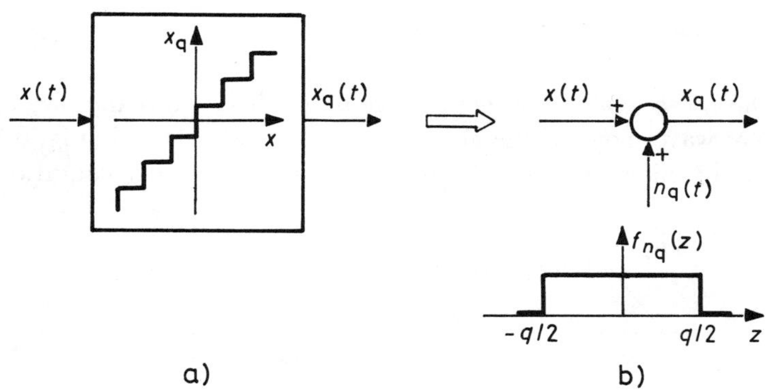

Fig. 6.11 The noise model of quantization *a)* ideal quantizer *b)* statistical model

The noise model as a linear model is very easy to use in the design and the evaluation of signal processing procedures. The signal-to-noise ratio can be easily estimated, and the quantization noise may even be reduced by appropriate filtering.

Sometimes the white spectrum is not required; however, since this means that the measured power spectral density will be modified by an additive constant (white spectrum) only, and, on the other hand, the time domain samples of the error will be uncorrelated with each other, it is reasonable to try to fulfil the conditions of this property.

When quantization is applied in a closed loop, the spectrum of the resulting noise, originating from the quantizer, can be modified, thus concentrating the noise power to frequencies of no interest (noise shaping, see (6.12-4)).

Example 6.2
A sine wave is quantized by a B-bit A/D converter. How large is the signal-to-noise ratio?
Solution:
If the full range of the A/D converter is used,

$$2U_{\mathrm{p}} = 2^B q\,.$$

The power of the signal is $U_{\mathrm{p}}^2/2$, the variance of the quantization noise is $q^2/12$ from the noise model.

$$SNR = 10\log\frac{P_{\mathrm{s}}}{P_{\mathrm{n}}} = 10\log\left(\frac{U_{\mathrm{p}}^2}{2}\frac{q^2}{12}\right) = 10\log(6\cdot 4^{B-1}) = (1.76+6B)\,\mathrm{dB}\,.$$

For $B=10$,

$$SNR = 61.8\,\mathrm{dB}\,.$$

Applicability conditions of the noise model

Obviously, the noise model may not always be applied. For example, when quantizing a square wave, none of the above properties will be valid (Fig. 6.12). Thus, in order to apply this really attractive model, the applicability conditions must be checked.

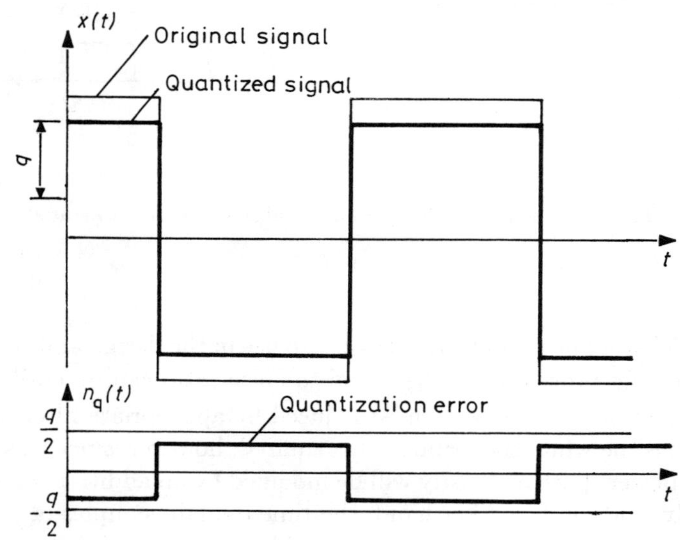

Fig. 6.12 Quantization of a square wave

Unluckily, the analysis of the applicability conditions involves manipulations with the characteristic functions (Fourier transforms of the probability density functions), and tedious derivations lead to the desired results. Therefore, even the quantization theorem will not be stated here. The mathematical details are available in the literature (Widrow, 1956a, 1956b, 1961; Korn, 1966; Sripad and Snyder, 1977; Kollár, 1984b, 1986). Here some heuristic explanations and some rules of thumb will be given.

It is obvious that for the fulfilment of the above conditions, the signal should vary sufficiently with respect to the quantum size. In other words, a small enough quantum size will usually ensure the validity of the noise model. Indeed, in the case of Gaussian signals fulfilment of the condition $q < \sigma$ is enough for the validity of the first two properties.

For signals uniformly distributed between $(-r \cdot q/2\,,\, r \cdot q/2)$, where r is a positive integer, the quantization error will be uniformly distributed. For signals with triangular distribution between $(-r \cdot q\,,\, r \cdot q)$, which is the case e. g. when summing two independent, above defined uniform noises, x and n_{q} will even be uncorrelated (though not independent of each other).

For other signals the signal amplitude should be usually much larger than the quantum size.

If the first two properties hold, the noise model can be used for the evaluation of the moments of quantized signals. For example,

$$\mathcal{E}\{x_{\mathrm{q}}\} = \mathcal{E}\{x + n_{\mathrm{q}}\} = \mathcal{E}\{x\} + \mathcal{E}\{n_{\mathrm{q}}\} = \mathcal{E}\{x\}\,. \tag{6.5}$$

Using this technique, the rules for obtaining the moments of the original signal from the moments of the quantized signal can be given, but care must be taken, because for the calculation of higher-order moments higher-order uncorrelatedness is required. The first four expressions are as follows:

$$\mathcal{E}\{x\} = \mathcal{E}\{x_{\mathrm{q}}\} \tag{6.6a}$$

$$\mathcal{E}\{x^2\} = \mathcal{E}\{x_{\mathrm{q}}^2\} - \frac{q^2}{12} \tag{6.6b}$$

$$\mathcal{E}\{x^3\} = \mathcal{E}\{x_{\mathrm{q}}^3\} - 3\mathcal{E}\{x_{\mathrm{q}}\}\frac{q^2}{12} \tag{6.6c}$$

$$\mathcal{E}\{x^4\} = \mathcal{E}\{x_{\mathrm{q}}^4\} - 6\mathcal{E}\{x_{\mathrm{q}}^2\}\frac{q^2}{12} + \frac{7q^4}{240}\,. \tag{6.6d}$$

These expressions are called *Sheppard corrections.* They show how to estimate moments from quantized observations: the powers of the quantized samples can be averaged, and the moments of the original signal can be determined using (6.6). Since the measurement of moments is the most common task in measurement, Sheppard corrections play an important role in signal processing.

The Sheppard corrections, as stated above, refer to moments of just one single random variable. However, the measurement of the correlation function and of the power spectral density requires the determination of *joint* moments. If the noise model can be applied to both variables, statements similar to those above can also be formulated.

Since the quantization error can often be considered as a *noise*, there is no hope of reconstructing the *exact* input waveform from the quantized one, as happens in sampling. Filters can be designed to approximately reconstruct the original waveform by effectively filtering the noise (Katzenelson, 1962).

There are situations when the applicability conditions cannot be met. In these cases a widely used technique is the application of a *dither*. A dither is an auxiliary signal, added to the input signal, in order to provide the applicability of the noise model (Fig. 6.13). Dithering helps indeed in the fulfilment of the conditions, however, care must be taken, since the variance will usually be increased:

$$\mathrm{var}\{x_\mathrm{q}\} = \mathrm{var}\{x\} + \mathrm{var}\{d\} + \frac{q^2}{12}\,. \tag{6.7}$$

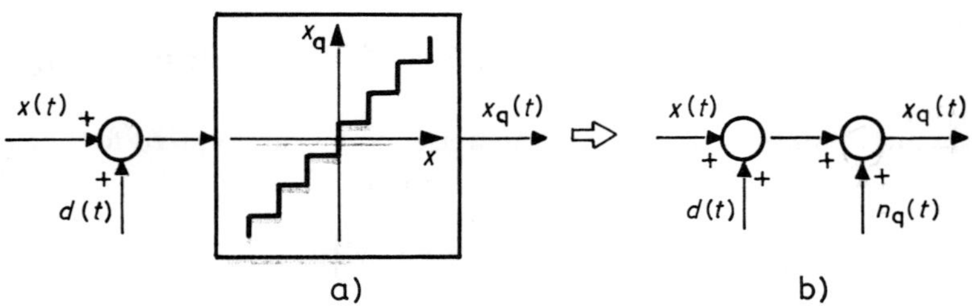

Fig. 6.13 Application of a dither *a)* functional diagram *b)* statistical model

The third property of the quantization noise (white spectrum) has other conditions (Bennett, 1948; Sripad and Snyder, 1977; Claasen and Jongepier, 1981; Kollár, 1986). As was already mentioned above, the sampling frequency should be sufficiently small with respect to the frequency content of the quantization noise:

$$f_s < c\frac{\mathcal{E}\{|\dot{x}(t)|\}}{q}\,, \tag{6.8}$$

where

$$\frac{\mathcal{E}\{|\dot{x}(t)|\}}{q}$$

is the 'average width' of the 'sawteeth' (see Fig. 6.10), and c=0.8−3, depending on the signal form.

This approximate condition can be evaluated by direct measurements, or theoretically, for given special signals. For bandlimited white Gaussian noise an approximate condition is

$$f_s < 9\frac{\sigma_x}{q}B\,, \tag{6.9}$$

where σ_x is the standard deviation of the signal and B is the bandwidth. For a sine wave

$$f_s < 3\frac{U_{\mathrm{p}}}{q} f_1 \,, \tag{6.10}$$

where U_{p} is the peak value, and f_1 is the frequency of the sine wave. In this latter case it should be mentioned that since the quantization noise is periodic, its spectrum is theoretically discrete. However, because of the large number of aliasing components, the power in a finite band, e. g. for

$$\Delta f > \frac{f_{\mathrm{s}}}{3\frac{U_{\mathrm{p}}}{q}} \,,$$

will be more or less independent of the centre frequency. But if sampling is synchronized to the sine wave, the spectrum will be rather peaky. Nevertheless, a high frequency dither will always help to ensure that the quantization error spectrum be white.

The above conditions do not contradict the sampling theorem *applied to the signals* ($f_{\mathrm{s}} > 2B$ or $f_{\mathrm{s}} > 2f_1$), since the values of σ_x/q and U_{p}/q are usually much larger than 1.

Realizations of quantization: A/D conversion and arithmetic rounding

Quantization is performed in two basic operations of digital signal processing: in A/D conversion and in arithmetic rounding.

ADC-s are nonideal elements. The most important types of A/D converters are treated in Subsection 2.3.17.

From the point of view of signal processing, several errors can be considered:

- The input range is limited, larger signals will be clipped.
- The quantization (transition) levels are somewhat uncertain because of internal noises.
- The quantization levels are not exactly at the same distance from each other. This error is characterized by the differential and the integral nonlinearities of the ADC (differential nonlinearity is the maximum deviation of the width of quantization intervals, i. e. the distance between two successive transition levels, from the ideal value; integral nonlinearity is the maximum deviation of the transfer characteristic from the ideal one). However, it must be emphasized that the two kinds of nonlinearities are just two descriptions of the very same phenomenon: inaccurate transition levels.
- Quickly changing signals will result in increased quantization error (the 'effective bit number' decreases).
- etc.

Most of these errors can be decreased or eliminated by careful measurement design. Clipping can be avoided by proper signal conditioning (the input signal should remain within the input range), internal noises are usually independent of the actual quantum level, thus usually act as a small dither, and do not spoil significantly the performance.

Nonlinearity is a more severe problem. This is usually in the order of magnitude of the LSB (least significant bit), and while the above described noise model would theoretically allow a more accurate measurement than the LSB, nonlinearities will not disappear, since signal oscillations around nonideally set quantization levels can never be avoided. There is just one technique to improve ADC performance, and this is again the application of a dither. If a large-amplitude dither is applied, the error of the different quantization levels can be played out against each other, and the ADC nonlinearity can be virtually reduced by some bits (see De Lotto and Paglia, 1986). Dithering may also decrease the effect of an incidental clipping.

Arithmetic rounding causes quite different difficulties. In this case the quantization levels are set exactly, since digital numbers are quantized. But this also means that the quantization error has a discrete distribution, and sometimes cannot be considered as approximately uniformly distributed noise. Moreover, scaling changes the range of the quantization error, floating-point arithmetics produce nonuniformly distributed errors, and cascaded operations may produce correlated errors at the different stages. Thus, arithmetic rounding is a very complex problem which has to be investigated separately for each special case. However, the noise model often gives a rough estimate of the errors.

Finally, two special signal conversion principles will be briefly discussed: sigma-delta modulation which has a more and more important role in precision A/D (and D/A) converters, and stochastic-ergodic conversion. Both techniques produce a discrete-interval, bandwidth-modulated binary signal, which can be processed by digital means.

Sigma-delta modulation

The $\Sigma-\Delta$ modulation principle, first described by Inose et al (1962), has acquired an important role with the development of VLSI technology which made possible its full integration on a single chip. Nowadays already high-resolution, very linear oversampling A/D and D/A converters, based on this principle, are on the market (see Subsection 2.3.17).

The basic idea is very simple (Fig. 6.14). The difference of the input signal $x(t)$ and the output signal $y(t)$ is lowpass filtered and led to a rough quantizer (usually a comparàtor). The resulting binary signal is set to an appropriate level by the one-bit 'D/A converter' (the output range of the converter must be larger than the maximum absolute value of the input signal), and is subtracted from the input signal. The lowpass filter ($H(f)$ in the closed loop has a large gain at low frequencies, and effectively attenuates at high frequencies (it is often an integrator).

It ensures that the comparator will evaluate the average of the difference of the output bit stream and the input signal. The mean value of the binary signal $y(t)$ (measurable using a digital averager, e. g. a lowpass filter) will correspond with high accuracy to the instantaneous value of the relatively slowly changing input signal.

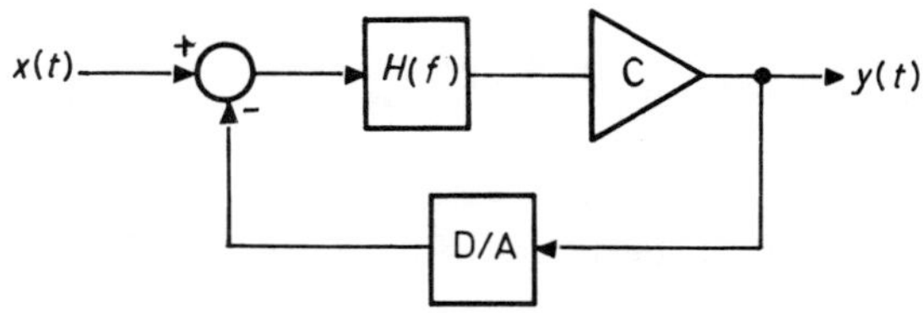

Fig. 6.14 Principle of sigma-delta modulation

For the analysis of the properties of the $\Sigma-\Delta$ modulator, let us analyze the closed loop in Fig. 6.14. Though not completely correct, let us use the noise model (Fig. 6.11) for the comparator, and determine the output signal in terms of the input signal and the noise. Introducing an 'equivalent gain' (g_e) for the comparator,

$$Y(z) = \big(X(f) - Y(f)\big)H(f)g_e + N_q(f)\,, \tag{6.11}$$

where $X(f)$ and $Y(f)$ are the Fourier transforms of the sampled input and output signals, respectively, and $N_q(f)$ is the Fourier transform of the sampled quantization error. After some rearrangement

$$Y(f) = \frac{H(f)g_e}{1+H(f)g_e}X(f) + \frac{1}{1+H(f)g_e}N_q(f)\,. \tag{6.12}$$

It is obvious that because of the oversampling, the power of the useful signal is concentrated at low frequencies, where $H(f)g_e$ is large: here the coefficient of the signal is approximately 1. The noise is practically suppressed at these frequencies; it is large only where the lowpass filter already significantly attenuates. Thus, signal and noise are *separated* in the frequency domain (noise shaping), and an appropriate digital filter can suppress the high frequency noise.

The $\Sigma-\Delta$ modulator has several implementations, since the shape of the transfer function of $H(f)$ can be chosen in different ways (see e. g. Candy and Benjamin, 1981; Norsworthy et al, 1989; Candy and Temes, 1991). There is no space here for a detailed discussion of these possibilities, hence only an interesting result for the case of $H(f)=1/(j2\pi fT_i)$ (simple integrator) will be given instead.

The transfer function of noise shaping is from (6.12)

$$H_{\text{effn}}(f) = \frac{j2\pi fT_i}{j2\pi fT_i + g_e}\,. \tag{6.13}$$

In the band $f \ll 1/(2\pi T_i)$ the power spectral density function of the noise is multiplied by the square of the absolute value of the transfer function:

$$|H_{\text{effn}}(f)|^2 \approx \left(\frac{2\pi fT_i}{g_e}\right)^2. \tag{6.14}$$

The quadratic behaviour shows that for small frequencies the quantization noise can be very well suppressed. Indeed, A/D converters with 18-24 bit resolution can be built using the sigma-delta modulation principle, especially if a higher-order filter is used in the loop.

Stochastic-ergodic conversion

As was shown earlier, uniform dither with the width of one quantum size helps measuring the first moment of the signal. If moreover the signal fulfils the condition

$$|x(t)| \leq \frac{A}{2}, \tag{6.15}$$

where A is the peak-to-peak amplitude of the uniform dither, an interesting observation can be made (Fig. 6.15b): the signal+dither *never surpasses* the levels $\pm A$. This means that only one comparison level is used. Consequently, from the point of view of signal treatment, it is irrelevant how the characteristic of the quantizer looks outside the interval $(-A/2, A/2)$: it can well be a comparator. In other words: if a uniformly distributed random dither is added to a signal $x(t)$, and (6.15) is fulfilled, the quantizer may be substituted by a comparator. This is the basic idea of the stochastic-ergodic converter (SEC, see Wehrmann, 1973; Tumfart, 1976).

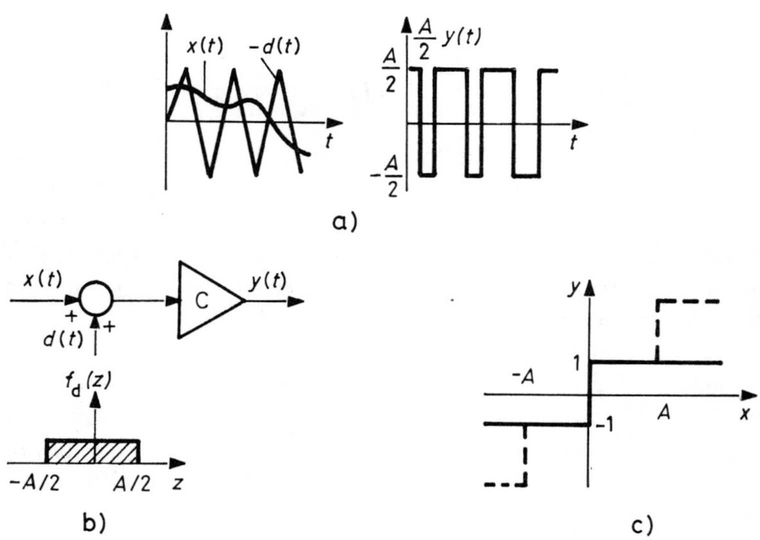

Fig. 6.15 Comparator as a fraction of a uniform quantizer *a)* signal forms *b)* block diagram *c)* transfer characteristic

The basic circuit of the SEC is shown in Fig. 6.15a. The output is a binary signal, usable for digital processing. The time average gives the mean value of the input signal, which is the a property of ergodic processes (hence the name).

In the following paragraphs the basic properties of the binary signal will be investigated.

As was already mentioned, the mean value of the binary signal (multiplied by $A/2$) is equal to the mean value of $x(t)$:

$$\mathcal{E}\{\frac{A}{2}y\} = \mu_x \,. \tag{6.16}$$

The variance can be calculated from the properties of the binary distribution. If the probability of level 1 is p, the mean value is $2p-1$. Thus,

$$p = \frac{1+\frac{2}{A}\mu_x}{2} = 0.5 + \frac{\mu_x}{A} \,. \tag{6.17}$$

$$\mathrm{var}\{\frac{A}{2}y\} = \frac{A^2}{4}4p(1-p) = \frac{A^2}{4} - \mu_x^2 \,. \tag{6.18}$$

If the dither is independent at the examined different time instants, it can be shown that the correlation between the samples is also kept:

$$\mathcal{E}\{\frac{A}{2}y(t)\frac{A}{2}y(t+\tau)\} = \mathcal{E}\{x(t)x(t+\tau)\} \,. \tag{6.19}$$

With these expressions the most important properties of the generated binary signal are given: the properties of concrete estimates based on the SEC will be calculated later.

6.1.4 Averaging and Lowpass Filtering

Measurement results are always corrupted by noise, thus the measured quantities (estimates) have a certain variance. The basic method for decreasing this variance is *averaging*. This subsection deals with the different ways of averaging and with the properties of the averaged quantities, in order to help the design of averaging.

It must be mentioned here that averaging is not the only way of decreasing the variance. In some special situations mean value estimates with smaller variance can also be given, e. g. the arithmetic mean of the minimum and the maximum value in the case of uniformly distributed observation noise. However, averaging is the most general method for variance reduction, because on the one hand averaging is uniformly applicable, and it has similar properties in practically every situation, on the other hand for independent, normally distributed samples the average (the arithmetic mean) is the best estimate of the mean value.

Averaging can be realized both in the continuous and the discrete time domain, and the two realizations are often equivalent to each other (if sampling is done according to the sampling theorem). In the next discussion the two realizations will be treated in parallel, and the conditions of equivalence will also be given.

Simple averaging

The continuous-time method of averaging is *integration* (see Fig. 6.16a). The estimate is unbiased:

$$\mathcal{E}\{\widehat{\mu}_x\} = \mathcal{E}\Big\{\frac{1}{T}\int_0^T x(t)\,\mathrm{d}t\Big\} = \frac{1}{T}\int_0^T \mathcal{E}\{x(t)\}\,\mathrm{d}t = \mu_x\,. \tag{6.20}$$

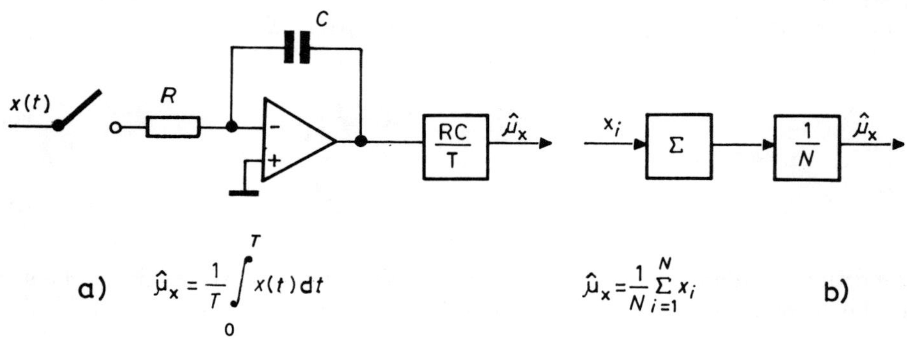

Fig. 6.16 Simple averaging *a)* continuous type *b)* discrete type

The variance can be obtained by a longer derivation (see Bendat and Piersol, 1971). The result is

$$\mathrm{var}\{\widehat{\mu}_x\} = \frac{1}{T}\int_{-T}^{T}\left(1-\frac{|\tau|}{T}\right)C_x(\tau)\,\mathrm{d}\tau\,, \tag{6.21}$$

where $C_x(\tau)$ is the autocovariance function.

This expression is still not too handy for everyday use. However, if the covariance function vanishes for $|\tau|>\tau_0$, where $\tau_0 \ll T$, it is easy to see that the integral can be approximated by the following one:

$$\mathrm{var}\{\widehat{\mu}_x\} \approx \frac{1}{T}\int_{-\infty}^{\infty} C_x(\tau)\,\mathrm{d}\tau = \frac{1}{T}S_{xc}(0)\,, \tag{6.22}$$

where $S_{xc}(f)$ is the covariance power spectral density function, that is, the Fourier transform of the autocovariance function. The above condition is usually fulfilled, since otherwise the sample record is strongly correlated, and averaging is not effective.

For a constant signal, buried in bandlimited white noise with variance σ_x^2 and bandlimit B (see Fig. 6.1b), the variance is obtained in a simple form:

$$\mathrm{var}\{\widehat{\mu}_x\} = \frac{\sigma_x^2}{2BT}. \tag{6.23}$$

This formula is widely used even if the noise in not exactly white: comparing (6.22) and (6.23) it is obvious that they are equivalent if B is substituted by

$$B_{\mathrm{n}} = \frac{\sigma_x^2}{2S_{xc}(0)}, \tag{6.24}$$

the so-called equivalent noise bandwidth (see Fig. 6.1). The necessary integration time can be calculated from the variance of the signal, the desired variance, and the equivalent bandwidth:

$$T > \frac{\sigma_x^2}{2B_{\mathrm{n}}\sigma_{\mathrm{desired}}^2}. \tag{6.25}$$

Example 6.3 Design of signal averaging

Components have to be weighed in a hall, full of running machines. Because of the vibration of the floor, the variance of a simple measurement is too large. Therefore, a damping arrangement has been designed. According to calculations, this arrangement may be well modelled by a first-order linear system, with time constant T_0=0.15 s. The variance decreases indeed, but a reduction is still necessary. Thus the output signal of the scales will be averaged. What is the necessary integration time, if the standard deviation of the measured mass, σ=2 g (measured after the installation of the damping system) is to be decreased to a value of 0.5 g?

Solution:

Let us assume that the noise at the input of the damping arrangement is white, that is, its bandlimit is larger than $1/T_0$. Thus, the power spectral density at the output is

$$S_{xc}(f) \approx S_n|H(f)|^2 = S_n \frac{1}{1+(2\pi f T_0)^2}.$$

The variance of the estimate of the mean is approximately

$$\mathrm{var}\{\widehat{\mu}\} \approx \frac{\sigma_x^2}{2B_nT},$$

where

$$B_n = \frac{\int_{-\infty}^{\infty} S_{xc}(f)\,\mathrm{d}f}{2S_{xc}(0)} = \frac{1}{2}\int_{-\infty}^{\infty} \frac{1}{1+(2\pi f T_0)^2}\,\mathrm{d}f = \frac{1}{4T_0}.$$

The variance of the output of the damping arrangement is to be reduced by a factor of

$$\left(\frac{2\,\mathrm{g}}{0.5\,\mathrm{g}}\right)^2 = 16,$$

that is,

$$2B_nT \approx 16.$$

$$T \approx \frac{16}{2B_n} = 32T_0 = 4.8\,\mathrm{s}.$$

The expression $\frac{\sigma_x^2}{2B_nT}$ may be applied since $T \gg T_0$.

Unfortunately, the approximation (6.22) cannot be applied in the important case of sinusoidal signals. Thus, the averaging of sine waves has to be investigated separately.

Let $x(t) = \mu_x + U_\text{p} \sin(2\pi f_1 t + \varphi)$. The integral mean can be easily calculated:

$$\begin{aligned}\widehat{\mu}_x &= \mu_x + \frac{U_\text{p}}{2\pi f_1 T}\Big(-\cos(2\pi f_1 T + \varphi) + \cos(\varphi)\Big) \\ &= \mu_x + \frac{U_\text{p}}{\pi f_1 T}\sin(\pi f_1 T)\sin(\pi f_1 T + \varphi)\,. \end{aligned} \tag{6.26}$$

The second term is the error of the estimate. This expression can be applied if the phase is fixed. However, if a random phase sinusoidal signal (see Subsection 1.3.1) is measured, and φ is uniformly distributed in $(0,2\pi)$, the estimate is unbiased, and the variance is (see Fig. 6.17):

$$\text{var}\{\widehat{\mu}_x\} = \frac{U_\text{p}^2}{2}\frac{\sin^2(\pi f_1 T)}{(\pi f_1 T)^2} = \frac{U_\text{p}^2}{2}\text{sinc}^2(\pi f_1 T)\,. \tag{6.27}$$

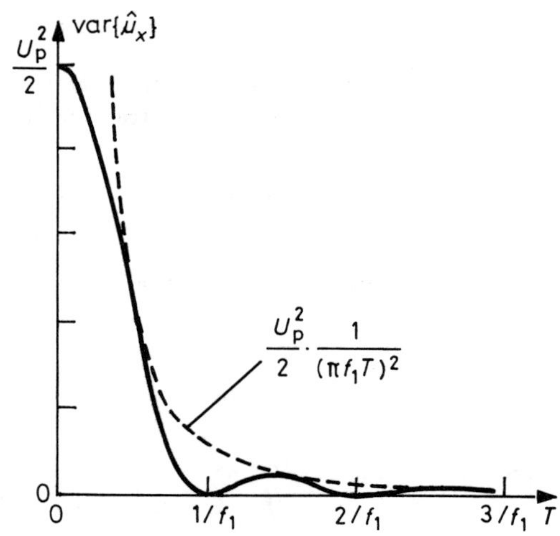

Fig. 6.17 Variance of $\widehat{\mu}_x$ for a random phase sine wave

This variance is small if f_1T is large, and it is zero if f_1T is integer (that is, integer periods are averaged). The 'envelope', that is, the worst-case value, decreases proportionally with the reciprocal of T^2, in contrast to the slower decrease of $\sigma_x^2/(2BT)$.

Similar, though more complex results can be obtained for signals consisting of more than one sine waves. A special problem is *frequency beat* which may occur when the frequencies of two sine waves are close to each other. In such a case the integration time has to be chosen much longer than the reciprocal of the frequency difference, in order to achieve efficient variance reduction.

There is one very important case that still has to be treated, the *mixed* signal:

$$x(t) = \mu_x + U_\mathrm{p} \sin(2\pi f_1 t + \varphi) + n(t)\,. \tag{6.28}$$

Since the individual terms are statistically independent of each other, the variances add up:

$$\mathrm{var}\{\widehat{\mu}_x\} \approx \frac{U_\mathrm{p}^2}{2} \mathrm{sinc}^2(\pi f_1 T) + \frac{\sigma_n^2}{2B_\mathrm{n}T}\,. \tag{6.29}$$

The *discrete* equivalent of the continuous-time estimate is the one illustrated in Fig. 6.16b:

$$\widehat{\mu}_x = \frac{1}{N} \sum_{i=1}^{N} x_i\,. \tag{6.30}$$

This estimate is unbiased too, and the variance can be easily calculated:

$$\mathrm{var}\{\widehat{\mu}_x\} = \frac{1}{N} \sum_{i=-(N-1)}^{N-1} \left(1 - \frac{|i|}{N}\right) C_x(i\Delta t)\,. \tag{6.31}$$

(6.31) can be well approximated in two extreme cases. If Δt is very small (oversampling), the sum will approximate the integral (6.21), thus its value will correspond to the continuous case. If Δt is large, i. e. $C_x(i\Delta t)$ vanishes for $i = 1, 2\ldots$, the well known formula is obtained:

$$\mathrm{var}\{\widehat{\mu}_x\} = \frac{\sigma_x^2}{N}\,. \tag{6.32}$$

Eq. 6.31 can be written in a form, similar to (6.32), if the 'effective (or equivalent) number of samples' is introduced: with

$$N_\mathrm{eff} = \frac{N}{\frac{1}{\sigma_x^2} \sum_{i=-(N-1)}^{N-1} \left(1 - \frac{|i|}{N}\right) C_x(i\Delta t)}\,, \tag{6.33}$$

the variance is

$$\mathrm{var}\{\widehat{\mu}_x\} = \frac{\sigma_x^2}{N_\mathrm{eff}}\,. \tag{6.34}$$

N_eff has a very straightforward meaning: this is the number of independent samples which contain the very same information as the sample record just processed. It it obvious that

$$N_\mathrm{eff} \le 2B_\mathrm{n}T\,, \tag{6.35}$$

and the equation is valid if the sampling theorem is fulfilled.

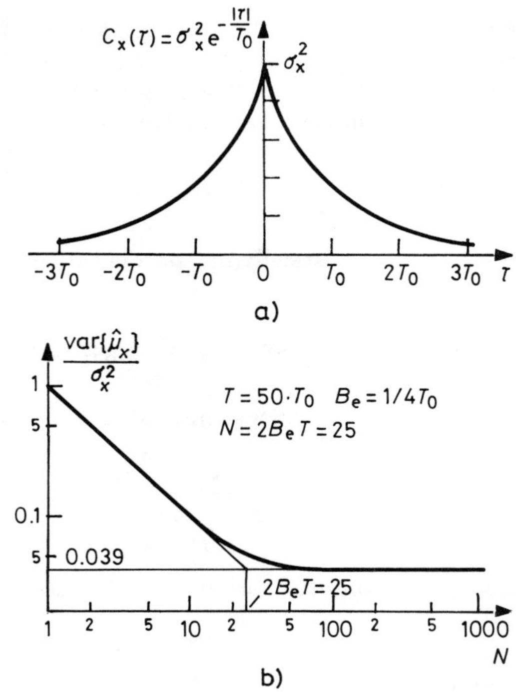

Fig. 6.18 The variance of discrete averaging as a function of N, for fixed observation time T *a)* autocovariance function of the input signal *b)* variance of the arithmetic mean

The behaviour of the variance for a constant observation time $T = N\Delta t$ can be illustrated on a log-log diagram (see Fig. 6.18).

It is easy to see that the variance is σ_x^2 for $N = 1$, and decreases with $1/N$ until it reaches the value $1/(2B_nT)$. The two straight asymptotic lines cross each other at $N = 2B_nT$, which is the same as (6.3). This means that a very simple rule can be given: if the signal is undersampled, (6.32) is to be used; when it is oversampled, (6.23). Discrete averaging is practically equivalent to the continuous one if the signal is sampled according to the sampling theorem: it is irrelevant with respect to the information content, how much larger is the sampling frequency than the Nyquist rate.

For random phase sinusoidal signals the decrease of the variance (that is, of the power) can be described by the value of the equivalent frequency domain transfer function at f_1:

$$P_{\text{out}} = |H(f_1)|^2 P_{\text{in}} \,. \tag{6.36}$$

As will be shown later (Eq. 6.59), this equivalent transfer function is

$$H(f_1) = \frac{\sin \pi f_1 N \Delta t}{N \sin \pi f_1 \Delta t}\,, \tag{6.37}$$

thus the variance is given by

$$\mathrm{var}\{\widehat{\mu}_x\} = \frac{U_\mathrm{P}^2}{2}\left(\frac{\sin\pi f_1 T}{N\sin\pi f_1\Delta t}\right)^2. \qquad (6.38)$$

Table 6.1 summarizes the expressions of the variances, obtained for the different cases.

Table 6.1 Variance of simple mean value estimates. For f_s large, the asymptotic values are given.

	f_s small	f_s large
Random noise	$\frac{\sigma_x^2}{N}$	$f_\mathrm{s} > 2B_\mathrm{n}$ $\frac{\sigma_x^2}{2B_\mathrm{n}T}$
Random phase sine wave	$\frac{U_\mathrm{P}^2}{2}\left(\frac{\sin\pi f_1 N\Delta t}{N\sin\pi f_1\Delta t}\right)^2$	$f_\mathrm{s} \gg 2f_1$ $\frac{U_\mathrm{P}^2}{2}\mathrm{sinc}^2\pi f_1 T$

The disadvantage of simple averaging methods is that the estimate is obtained after the total measurement time only. This can be overcome in the discrete case by *recursive averaging*:

$$\widehat{\mu}_{x,0} = 0,$$

$$\widehat{\mu}_{x,i} = \frac{1}{i}x_i + \frac{i-1}{i}\widehat{\mu}_{x,i-1} = \widehat{\mu}_{x,i-1} + \frac{1}{i}(x_i - \widehat{\mu}_{x,i-1}), \quad i = 1, 2\ldots \qquad (6.39)$$

This recursive averaging also has some problems: i increases without a limit, thus, when implemented, the algorithm must be regularly restarted, in order to avoid numerical overflows. This inconvenience will be overcome by the so-called exponential averaging, to be described in the next subsection.

Exponential (forgetting) averaging

Simple averaging is often substituted by the so-called exponential (or sometimes: forgetting) averaging.

The *analog solution* is shown in Fig. 6.19a. Let us assume that the signal $x(t)$ was fed to the input of the energy-free filter at $t=0$. The frequency transfer function and the impulse response of the filter are:

$$H(f) = \frac{1}{1+j2\pi fK}; \quad h(t) = \frac{1}{K}e^{-\frac{t}{K}}, \quad t > 0. \qquad (6.40)$$

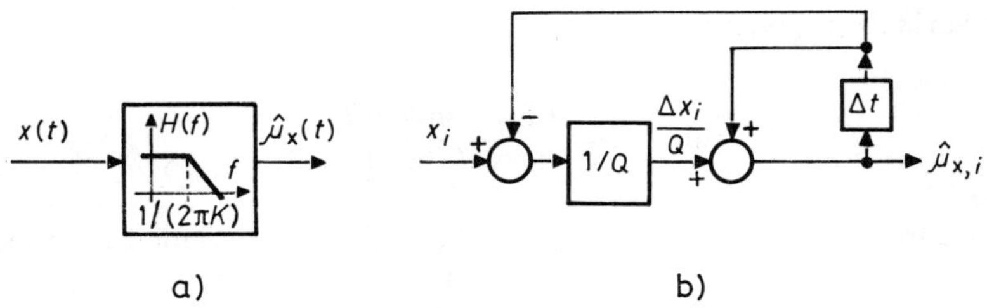

Fig. 6.19 Exponential averaging *a)* analog solution *b)* discrete solution

It is obvious that the expected value of the output signal of the filter is an exponential function of time:

$$\mathcal{E}\{\widehat{\mu}(T_{\rm m})\} = \mu_x(1 - e^{-\frac{T_{\rm m}}{K}}). \tag{6.41}$$

The estimate is distorted, and with an increasing measurement time $T_{\rm m}$, this distortion tends to zero.

The variance can also be obtained by a rather lengthy derivation, and the expression is also complicated, thus will not be treated here. Rather, a more useful expression will be given for the stationary case ($T_{\rm m} \gg K$). If this inequality holds, the output signal of the filter is nearly stationary, and the power spectral density function of the output may be investigated:

$$\mathrm{var}\{\widehat{\mu}(T_{\rm m})\}\Big|_{T_{\rm m}\gg K} \approx C_{\widehat{\mu}}(0) = \int\limits_{-\infty}^{\infty} S_{\widehat{\mu}c}(f)\,\mathrm{d}f = \int\limits_{-\infty}^{\infty} \left|\frac{1}{1+j2\pi fK}\right|^2 S_{xc}(f)\,\mathrm{d}f\,, \tag{6.42}$$

where $S_{xc}(f)$ is the covariance power spectral density function of $x(t)$.

For a bandlimited white noise, the integral can be evaluated:

$$\mathrm{var}\{\widehat{\mu}(T_{\rm m})\}\Big|_{T_{\rm m}\gg K} \approx \int\limits_{-B}^{B} \frac{1}{1+(2\pi fK)^2}\,\frac{\sigma_x^2}{2B}\,\mathrm{d}f = \frac{\arctan(2\pi BK)}{\pi/2}\cdot\frac{\sigma_x^2}{4BK}.$$

When $2\pi BK \gg 1$, the first term of the right side approximately equals 1, and the second term gives the variance:

$$\mathrm{var}\{\widehat{\mu}(T_{\rm m})\}\Big|_{T_{\rm m}\gg K} \approx \frac{\sigma_x^2}{4BK}. \tag{6.43}$$

It can be observed that for $T_{\rm m} \gg K$ the variance is determined only by the time constant K (beside the parameters of the process $x(t)$), and it does not depend on $T_{\rm m}$. If (6.43) is compared to (6.23), it can be stated that from the point of view of variance, the use of the first-order lowpass filter is equivalent to an integration of duration $T = 2K$. However, the measurement time T_{m} must be larger than $2K$: exponential averaging results in a certain information loss.

A consequence of the other assumption $(2\pi BK \gg 1)$ is that the bandlimit of the lowpass filter has to be smaller than the bandlimit of the input signal. If this were not the case, averaging would not modify the spectrum, that is, the variance would remain the same.

When the noise is not a bandlimited white one, but its spectrum is smooth, and $2\pi B_{\mathrm{e}} K \gg 1$, that is, the bandwidth of the lowpass filter is much smaller than that of the noise, it is easy to see that the variance can be well approximated by

$$\mathrm{var}\{\widehat{\mu}(T_{\mathrm{m}})\}\Big|_{T_{\mathrm{m}} \gg K} \approx \int_{-\infty}^{\infty} \frac{1}{1 + (2\pi f K)^2}\, S_{xc}(0)\,\mathrm{d}f = \frac{S_{xc}(0)}{2K}. \tag{6.44}$$

Finally, let us mention that the bias can be eliminated by multiplying the estimate by

$$\frac{1}{1 - e^{-\frac{T_{\mathrm{m}}}{K}}},$$

yet this possibility has no real importance, since this correction is usually not significant (if $T_{\mathrm{m}} > 4K$, this factor is <1.019), and T_{m} can usually be chosen large enough.

Similar results can be quickly obtained also for higher-order lowpass filters. The length of the transients should be determined from the impulse response; it is in the order of magnitude of c/B_0 for settling within 2%, where c is 10–100, depending on the shape of the filter, and B_0 is the filter bandwidth. With a derivation similar to the above one, in the case of an ideal lowpass filter, the following expression of the variance can be obtained:

$$\mathrm{var}\{\widehat{\mu}(T_{\mathrm{m}})\}\Big|_{T_{\mathrm{m}} \gg K} \approx \sigma_x^2 \frac{B_0}{B}, \quad \text{if} \quad B_{\mathrm{s}} < B\,. \tag{6.45}$$

The variance is equal to $\sigma_x^2/(2BT)$ if $B_0 = 1/2T$.

Since for equal bandwidths of the first-order and the ideal lowpass filters the variance reductions are in the same order of magnitude, it is not reasonable to apply higher-order lowpass filters, since their transient behaviour is much worse.

It may be interesting to compare the frequency domain transfer functions of the 'boxcar integrator' with integration length T (see Eq. 6.61), of a first-order lowpass filter and that of an ideal lowpass filter, with parameters providing the same variance of the mean value estimate of a bandlimited white noise (Fig. 6.20).

Example 6.4

The above described variance reduction is often applied in data acquisition channels. If the A/D converter is quick enough, the signal may be oversampled. The variance of the quantization error is $q^2/12$. The bandwidth of the sampled series can now be reduced to the signal bandwidth by using a selective digital lowpass filter. If the sampled quantization noise is white (e. g. if an appropriate dither is applied, possibly with its power outside the signal band), the noise reduction is $2B_0/f_s$. The filtered signal may be decimated, in order to decrease the number of samples to be stored.

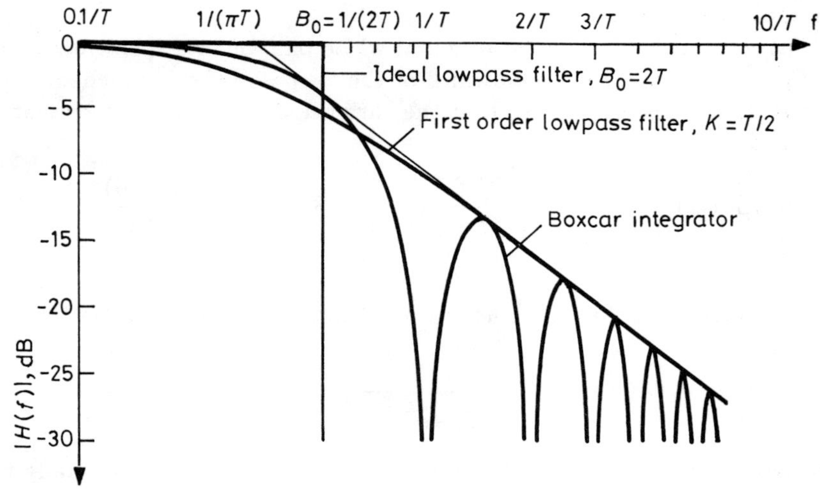

Fig. 6.20 Transfer functions of averaging procedures with the same output variance

For sinusoids the analysis is very simple. The variance of the instantaneous value of a random phase sine wave is $U_{\mathrm{p}}^2/2$, this is multiplied by the square of the attenuation of the filter at the given frequency.

The *discrete equivalent* of analog lowpass filtering is recursive averaging with constant weight (Fig. 6.19b):

$$\begin{aligned}\widehat{\mu}_{x,0} &= 0, \\ \widehat{\mu}_{x,i} &= \widehat{\mu}_{x,i-1} + \frac{1}{Q}(x_i - \widehat{\mu}_{x,i-1}) \quad i = 1, 2, \ldots; \quad Q > 1 . \end{aligned} \tag{6.46}$$

Eq. 6.46 has a very remarkable form: the new estimate is presented as the previous one plus a correction term, which usually has a small weight. A similar approach is followed in the formulation of the Kalman filter (Jazwinski, 1970).

To demonstrate the equivalence of the analog and discrete algorithms, let us suppose that the condition of the sampling theorem is fulfilled, and let us transform the differential equation of the lowpass filter (6.40) into a difference equation:

$$x(t) = \widehat{\mu}(t) + K\frac{d\widehat{\mu}(t)}{dt}\,, \tag{6.47}$$

$$x_i \approx \widehat{\mu}_i + K\frac{\widehat{\mu}_i - \widehat{\mu}_{i-1}}{\Delta t}\,, \tag{6.48}$$

$$\widehat{\mu}_i \approx \widehat{\mu}_{i-1} + \frac{1}{1+\frac{K}{\Delta t}}(x_i - \widehat{\mu}_{i-1}) . \tag{6.49}$$

Thus, (6.46) is obtained with $Q=1+K/\Delta t$. It can also be seen that the approximation of the derivative can be improved by the use of both forward and backward differences, and this improvement yields $Q \sim 0.5 + K/\Delta t$. The difference equation approximates the differential equation well if Δt is sufficiently small, that is, if the sampling is done according to the Nyquist sampling theorem. As will be shown later, (6.46) may be applied also when Δt is large (see (6.53)), but the equivalence will not be true any more.

Let us determine now the explicit form of (6.46):

$$\widehat{\mu}_N = \frac{1}{Q}\sum_{i=1}^{N} x_i \left(\frac{Q-1}{Q}\right)^{N-i} . \tag{6.50}$$

From this expression the mean and the variance of $\widehat{\mu}_N$ can be determined.

$$\mathcal{E}\{\widehat{\mu}_N\} = \mu_x\left(1-\left(\left(1-\frac{1}{Q}\right)^Q\right)^{\frac{N}{Q}}\right) \approx \mu_x(1-e^{-\frac{N}{Q}}). \tag{6.51}$$

The bias is small if $N \gg Q$.

To determine the variance, two cases have to be distinguished. If sampling is performed above the Nyquist rate, the discrete algorithm is equivalent to the analog lowpass filter, and the variance is similar to the one given in (6.43):

$$\mathrm{var}\{\widehat{\mu}(T_\mathrm{m})\}\Big|_{\substack{N\gg Q \\ 2\pi B(Q-0.5)\Delta t \gg 1}} \approx \frac{\sigma_x^2}{4B(Q-0.5)\Delta t} . \tag{6.52}$$

This expression justifies again the equivalence $K \sim (Q-0.5)\Delta t$.

The equivalent integration time is $T=2K\sim(2Q-1)\Delta t$. In other words, this is the record length necessary for obtaining the same variance with simple averaging. Consequently, the number of necessary samples, taken with the same sampling frequency, is $N_\mathrm{e}=2Q-1$. However, because of the above mentioned oversampling, the effective number of samples (the number of independent samples, which provide the same variance when the arithmetic mean is calculated) is less:

$$N_\mathrm{eff}=(2Q-1)\frac{2B}{f_\mathrm{s}} < 2Q-1 .$$

The algorithm (6.46) can be used also in the case of uncorrelated samples (sampling below the Nyquist rate), though this is not equivalent any more to the analog solution. Because of undersampling a part of the information is lost. The variance can be calculated as the sum of a geometric series:

$$\mathrm{var}\{\widehat{\mu}_N\} = \frac{\sigma_x^2}{Q^2}\,\frac{1-\left(\frac{Q-1}{Q}\right)^{2N}}{1-\left(\frac{Q-1}{Q}\right)^2} \approx \sigma_x^2\,\frac{1-e^{-\frac{2N}{Q}}}{2Q-1} \approx \frac{\sigma_x^2}{2Q-1} . \tag{6.53}$$

Because of the above mentioned information loss, this variance must be larger than the one given in (6.52). Indeed, it is larger if $2B\Delta t > 1$.

Let us notice that this result is similar to the one obtained for the analog case: exponential averaging is equivalent to a simple averaging with sample number $N_e \approx 2Q - 1$. However, in the case of exponential averaging more samples have to be processed to obtain this variance.

To obtain the variance for the case of random phase sinusoidal signals, the value of the transfer function at f_1 is to be calculated. For $N \to \infty$ the z-domain transfer function can be calculated:

$$\widehat{\mu}_{x,i} = \frac{Q-1}{Q}\widehat{\mu}_{x,i-1} + \frac{1}{Q}x_i, \tag{6.54}$$

$$\widehat{\mu}(z) = \frac{Q-1}{Q}z^{-1}\widehat{\mu}(z) + \frac{1}{Q}X(z), \tag{6.55}$$

$$\widehat{\mu}(z) = \frac{1}{Q} \cdot \frac{1}{1 - \frac{Q-1}{Q}z^{-1}}X(z) = H(z)X(z). \tag{6.56}$$

The transfer function is obtained from $|H(z)|^2$ with $z^{-1} = e^{-j2\pi f\Delta t}$:

$$\mathrm{var}\{\widehat{\mu}\} = \frac{U_\mathrm{P}^2}{2}\left|\frac{1}{Q} \cdot \frac{1}{1 - \frac{Q-1}{Q}e^{-j2\pi f_1\Delta t}}\right|^2. \tag{6.57}$$

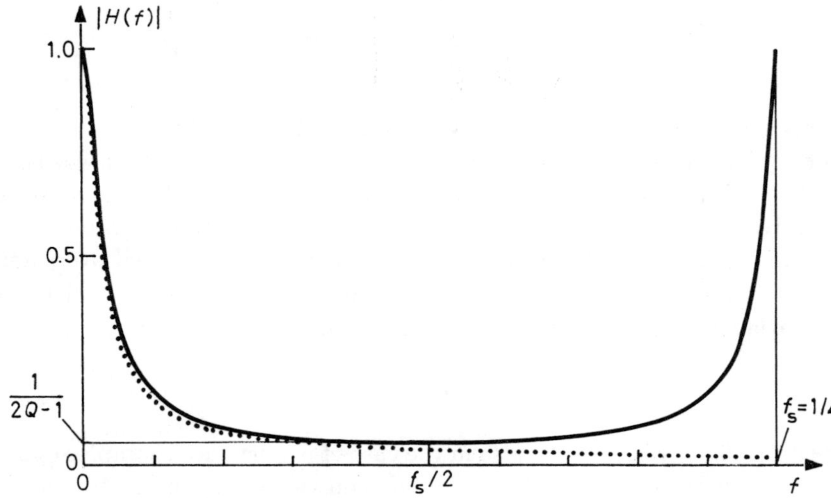

Fig. 6.21 Equivalent transfer function of exponential averaging for $Q=10$

The behaviour of this variance is illustrated in Fig. 6.21. For $f_1 \ll 1/\Delta t$,

$$|H(f)|^2 \approx \frac{1}{1 + (2\pi f Q\Delta t)^2}. \tag{6.58}$$

The other name of this type of averaging is *forgetting* averaging since earlier values are taken into account with a decreasing weighting. This decrease has an exponential shape (similar to (6.51)), explaining the name *exponential* averaging.

Since this kind of averaging can follow slow changes of the mean, it is applicable in nonstationary cases, too.

Moving averaging and signal enhancement

Moving averaging and signal enhancement are special averaging techniques. *Moving averaging* (the result is also called *running sample mean*) means that samples are averaged within a small neighbourhood of a time instant to obtain a point of the waveform (see Fig. 6.22a). Since in the frequency domain time delay corresponds to multiplication by a complex exponential, it is easy to determine the equivalent transfer function, because a geometric series is to be summed. If N is odd:

$$H(f) = \frac{1}{N} \sum_{i=-\frac{N-1}{2}}^{\frac{N-1}{2}} e^{-j2\pi i f \Delta t} = \frac{1}{N} e^{j\pi f (N-1)\Delta t} \frac{e^{-j2\pi f N \Delta t} - 1}{e^{-j2\pi f \Delta t} - 1}$$

$$= \frac{1}{N} \frac{e^{-j\pi f N \Delta t} - e^{j\pi f N \Delta t}}{e^{-j\pi f \Delta t} - e^{j\pi f \Delta t}} = \frac{\sin \pi f N \Delta t}{N \sin \pi f \Delta t}, \tag{6.59}$$

where Δt is the sampling interval (see Fig. 6.22b). The result is the same for N even.

Using this procedure, the averaging length $N\Delta t$ is usually chosen to a value much smaller than the time length of the variations in the signal, thus the width of the equivalent frequency transfer function is much larger than the bandlimit of the signal, that is, at the signal frequencies the attenuation is negligible. Meanwhile, high frequency noise and periodic disturbances of frequency $k/N\Delta t$ ($k \neq pN$) can be effectively suppressed.

Moving averaging can be used for the processing of both periodic and transient signals, corrupted by high frequency noise.

For $N \to \infty$ (keeping $T = N\Delta t$ constant) the so-called boxcar integration is obtained. An estimate of the mean value is produced at every instant:

$$\widehat{\mu}(t_0) = \frac{1}{T} \int_{t_0 - \frac{T}{2}}^{t_0 + \frac{T}{2}} x(t)\,\mathrm{d}t\,. \tag{6.60}$$

It is easy to see that the equivalent transfer function can be obtained from (6.59) with $\Delta t \to 0$ and $N\Delta t = T = \text{const}$:

$$H(f) = \frac{\sin \pi f T}{\pi f T}\,. \tag{6.61}$$

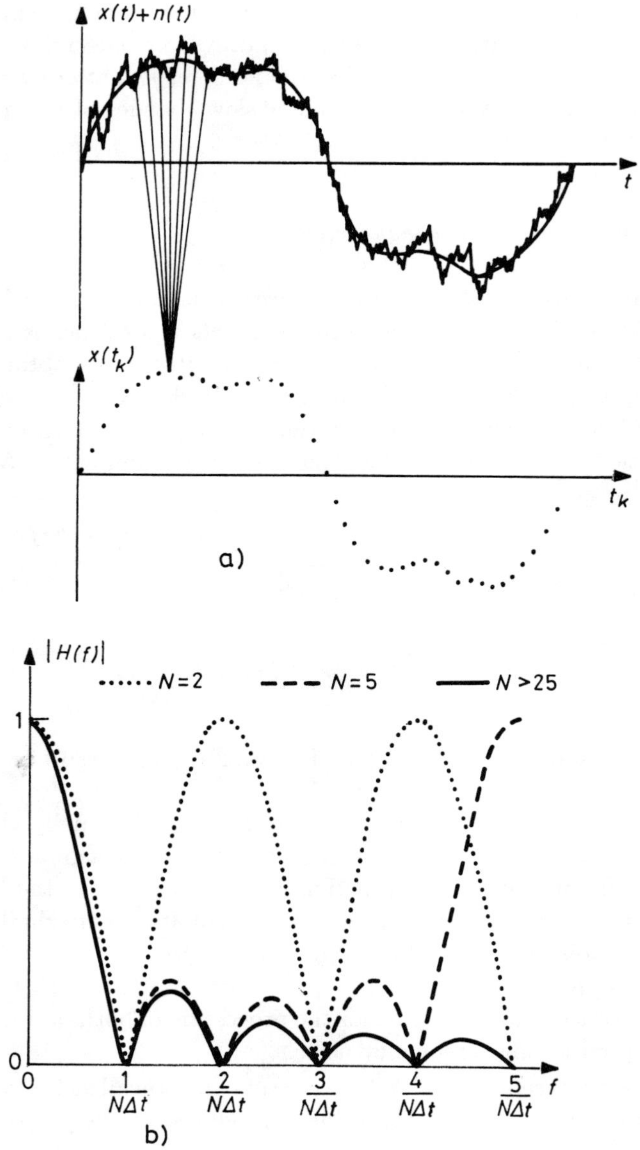

Fig. 6.22 Moving averaging *a)* averaging in the time domain *b)* the equivalent transfer function

A useful approximation of the bias can be derived for the continuous case by developing $x(t)$ into a Taylor series around t_0:

$$\mathrm{b}\{x_{\mathrm{av}}(t_0)\} = \frac{1}{T} \int_{t_0-\frac{T}{2}}^{t_0+\frac{T}{2}} x(t_0) + x'(t_0)(t - t_0)$$

$$+\frac{x''(t_0)}{2}(t-t_0)^2+\ldots\,\mathrm{d}t-x(t_0)\approx\frac{x''(t_0)}{24}T^2\,. \qquad (6.62)$$

The variance can be calculated using the results obtained for simple averaging, or the equivalent transfer function can be applied:

$$\mathrm{var}\{x_{\mathrm{av}}\}=\int_{-\infty}^{\infty}|H(f)|^2S_{\mathrm{n}}(f)\,\mathrm{d}f\,.$$

Signal enhancement means averaging of samples taken from the same positions of a repetitive waveform, buried in noise (Fig. 6.23a). To realize this, trigger pulses, synchronized to the signal, are necessary. Using this trigger, the signal can be sampled properly, that is, in the case of a periodic signal the sampling frequency is exactly equal to the frequency of the fundamental harmonic. It is easy to see that the attenuation is 0 dB at every multiple of the fundamental frequency, i. e. the signal itself will not be attenuated at all. The equivalent transfer function is the same as above, the only difference is that instead of Δt the signal period length T is to be written:

$$H(f)=\frac{\sin\pi fNT}{N\sin\pi fT}\,. \qquad (6.63)$$

A typical application is shown in Fig. 6.23c. The transfer function $H(f)$ of the noisy network is to be determined. The trigger can easily be generated from the known excitation signal.

The variance can be similarly determined as for the moving average.

Example 6.5

The signal-to-noise ratio of a noisy periodic signal is $SNR=-10\,\mathrm{dB}$. The noise is a bandlimited white noise with $B=2\,\mathrm{kHz}$. The power of the periodic signal is $P_{\mathrm{s}}=0.1\,\mathrm{V}^2$, the period length is $T=100\,\mathrm{ms}$. The noisy signal is sampled, $N=1000$ samples are taken from a period, and moving averaging is performed, by calculating the arithmetic mean of $N_{\mathrm{m}}=20$ neighbouring points. How large is the improvement in the *SNR*? What is the width of the 95% confidence intervals of the instantaneous values of the signal?

Solution:

The sampling frequency is

$$f_{\mathrm{s}}=\frac{1}{NT}=10\,\mathrm{kHz}\,.$$

The noise is oversampled, thus the resulting variance is not σ_{n}^2/N, but

$$\sigma_{\mathrm{out}}^2=\frac{\sigma_{\mathrm{n}}^2}{\frac{2B}{f_{\mathrm{s}}}N_{\mathrm{m}}}=\frac{\sigma_{\mathrm{n}}^2}{8}\,.$$

The signal-to-noise ratio is improved by $10\log 8=9\,\mathrm{dB}$.

The variance of the averaged samples is

$$\sigma_{\mathrm{out}}^2=\frac{\sigma_{\mathrm{n}}^2}{8}=\frac{P_{\mathrm{s}}}{10^{\frac{SNR}{10}}}\cdot\frac{1}{8}=0.125\,\mathrm{V}^2\,,$$

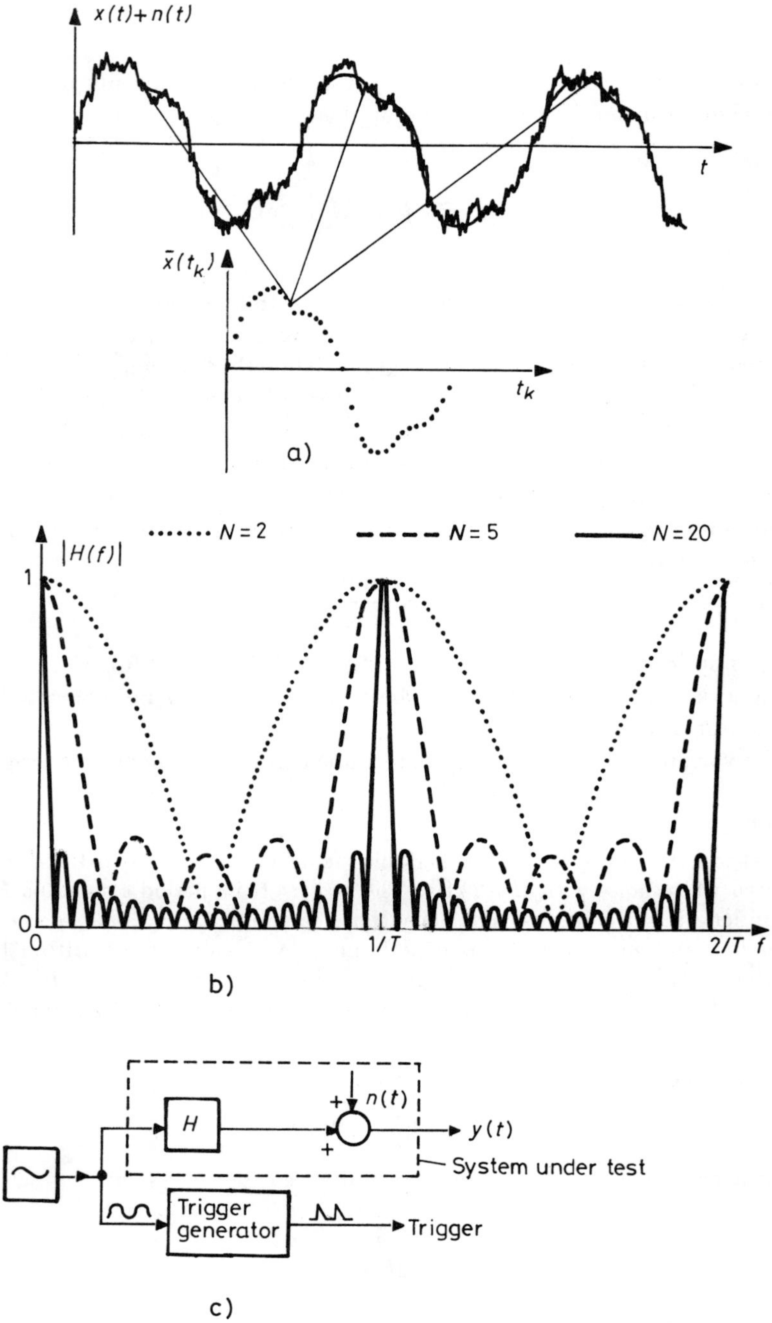

Fig. 6.23 Signal enhancement *a)* averaging in the time domain *b)* the equivalent transfer function *c)* application example

and the width of the confidence intervals is from the normal distribution

$$\pm 1.96\sqrt{0.125}\,\mathrm{V} \approx \pm 0.69\,\mathrm{V}.$$

It is worth mentioning that signal enhancement can also be applied for a not strictly periodic signal (e. g. a repetitive transient signal), if a synchronous trigger can be generated. However, the concept of the equivalent transfer function is not directly applicable in this case.

6.1.5 Fourier Transform

As was discussed in Subsection 6.1.1, frequency domain methods have outstanding importance in engineering. In this subsection the methods of the transformation between time and frequency domain, i. e. realizations of the Fourier transform, will be treated.

The Fourier transform of absolute integrable functions is the following integral:

$$X(f) = \mathcal{F}\{x(t)\} = \int_{-\infty}^{\infty} x(t)e^{-j2\pi ft}\,\mathrm{d}t\,. \tag{6.64}$$

This transform generally has a unique inverse pair:

$$x(t) = \mathcal{F}^{-1}\{X(f)\} = \int_{-\infty}^{\infty} X(f)e^{+j2\pi ft}\,\mathrm{d}f\,. \tag{6.65}$$

If the variable ω is used instead of $2\pi f$, a factor $1/2\pi$ has to be introduced into one of the above expressions.

The properties of the Fourier transform are well known (see e. g. Papoulis, 1962; Bracewell, 1965), they will not be enumerated here. Instead, the relationship between the above *continuous* Fourier transform and the *discrete* one will be discussed.

The discrete Fourier transform (DFT) and its inverse are defined for discrete signals as:

$$X_k = \sum_{i=0}^{N-1} x_i e^{-j2\pi\frac{ik}{N}}, \quad x_i = \frac{1}{N}\sum_{k=0}^{N-1} X_k e^{j2\pi\frac{ik}{N}}, \quad i,k = 0\ldots N-1\,. \tag{6.66}$$

The discrete Fourier transform can be investigated on its own in the discrete domain, however, it is also an approximation of (6.64) in $(0,T)$:

$$X(f) \approx \int_{0}^{T} x(t)e^{-j2\pi ft}\,\mathrm{d}t \approx \sum_{i=0}^{N-1} x(i\Delta t)e^{-j2\pi fi\Delta t}\Delta t\,. \tag{6.67}$$

This is also the Fourier transform of the result of mathematical sampling in (0,T).

The result is still continuous. In order to discretize it, the appropriate frequency points should be chosen. It seems reasonable to select N points also in the frequency domain. Since the Fourier transform of a sampled series is periodic with the period length $f_s = 1/\Delta t$ (see Subsection 6.1.2), the distance between the frequency points has to be

$$\Delta f = \frac{f_s}{N} = \frac{1}{N\Delta t} = \frac{1}{T}\,. \tag{6.68}$$

Let us notice that with this choice

$$\Delta t \Delta f = \frac{1}{N}\,. \tag{6.69}$$

Using this expression, (6.67) becomes

$$X(k\Delta f) \approx \sum_{i=0}^{N-1} x(i\Delta t)e^{-j2\pi\frac{ki}{N}}\Delta t\,. \tag{6.70}$$

For sake of simplicity, Δt used to be chosen 1, and the samples used to be denoted by x_i and X_k, respectively. With these notations the discrete Fourier transform is obtained. It is obvious that approximation was used at three important steps:

- truncation of the improper integral to (0,T),
- discretization in the time domain,
- discretization in the frequency domain.

The errors of the discrete Fourier analysis of continuous-time signals follow from the above approximations, and will be treated in Subsection 6.7.7.

It is interesting to note that there are cases where there is no loss of information during the above discretization procedure. If a periodic signal contains a limited number of harmonics only, and an integer number of periods is measured, and the sampling frequency is higher than double the frequency of the highest harmonic, the proper spectrum is obtained, except for the finite numbers at the places of the Dirac delta functions.

The DFT has a very regular structure, which can be used for significant acceleration of the calculations. In what follows, the possibilities of this acceleration will be briefly presented.

Fast Fourier transform (FFT)

The most common means of fast computation of the DFT is the FFT (fast Fourier transform, Cooley and Tukey, 1965). The basic ideas can be illustrated as follows.

Let us introduce the notation

$$W_N = e^{-j\frac{2\pi}{N}}, \tag{6.71}$$

that is, the Nth root of 1. With this notation the DFT is

$$X_k = \sum_{i=0}^{N-1} x_i W_N^{ik}. \tag{6.72}$$

Although there are N^2 combinations of i and k, the coefficients of x_i have N different values only, because $W_N^N = 1$, thus

$$W_N^{ik} = W_N^{(ik \bmod N)}. \tag{6.73}$$

Let us suppose now that N is even. In this case the time series $\{x_i\}$ can be decimated, by forming two 'subseries', $\{y_r\}$ and $\{z_s\}$:

$$\begin{array}{cccccccccc} x_0, & x_1, & x_2, & x_3, & x_4, & x_5, & x_6, & x_7, & \ldots \\ y_0, & & y_1, & & y_2, & & y_3, & & \ldots \\ & z_0, & & z_1, & & z_2, & & z_3, & \ldots \end{array}$$

The original DFT can be split into two parts:

$$X_k = \sum_{i=0}^{N-1} x_i W_N^{ik} = \sum_{r=0}^{N/2-1} y_r W_N^{2rk} + \sum_{s=0}^{N/2-1} z_s W_N^{(2s+1)k}$$

$$= \sum_{r=0}^{N/2-1} y_r W_{N/2}^{rk} + W_N^k \sum_{s=0}^{N/2-1} z_s W_{N/2}^{sk} = Z_k + W_N^k Y_k. \tag{6.74}$$

As seen from (6.74), for $k = 0 \ldots N/2 - 1$ the N-point DFT has been expressed by two $N/2$-point DFT-s. Let us notice now that the $N/2$-point DFT-s are periodic with period length $k_{\mathrm{p}} = N/2$, and

$$W_N^k = -W_N^{k-N/2}. \tag{6.75}$$

This means that the full original DFT can be simply expressed in terms of the two half-length ones. While for the N-point DFT about N^2 complex multiplications were necessary, the use of the two half-length DFT-s reduces this to about

$$\left(\frac{N}{2}\right)^2 + \left(\frac{N}{2}\right)^2 + \frac{N}{2} \ll N^2. \tag{6.76}$$

The above procedure can be repeated if $N/2$ is still even. If N is a power of 2, the reduction is striking: instead of $2N^2$ real multiplications only $2N\log_2 N$ real

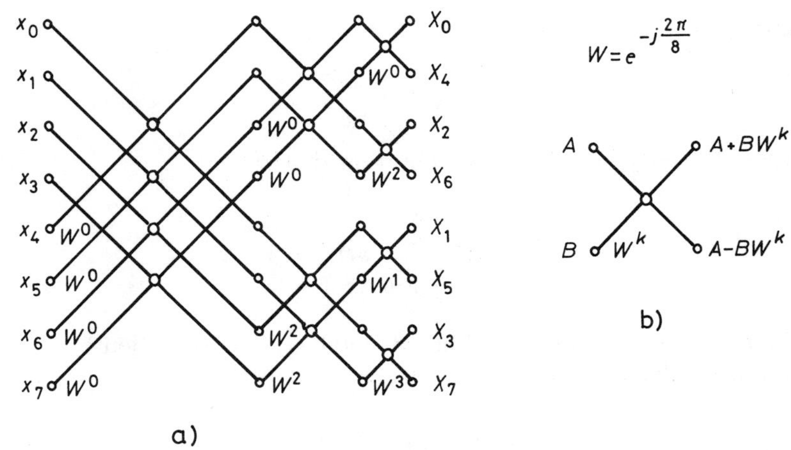

Fig. 6.24 Radix-2 decimation-in-time FFT for $N=8$ *a)* data flow graph *b)* butterfly

multiplications are necessary (see Fig. 6.24), which is for $N=1024$ about 100 times less than $2N^2$!

The above described procedure leads to the so-called radix-2 DIT (base-2, decimation in time) FFT, illustrated in Fig. 6.24. The basic operation (calculation of X_k from Y_k and Z_k) is illustrated beside the flow diagram: it is called *butterfly*.

The fast Fourier transform has a lot of variants. First of all, there is another possibility for decimation at the other end of the flow chart: DIF (decimation in frequency) FFT algorithms can also be developed. The base of the FFT may also be different from 2; the reduction of computations may rely on a prime factorization of N (Kolba and Parks, 1977).

As illustrated in Fig. 6.24, the output sequence is not in the natural order of indices, as a consequence of successive decimations. However, in-place FFT-s can also be developed, at the cost of a somewhat more complex flow diagram.

There is no space in this book for a detailed discussion of all the possibilities of the FFT: the reader is referred rather to the literature (Brigham, 1974; Nussbaumer, 1982).

The FFT is not the only way of evaluating the DFT. Another possibility is transcription into a convolution form.

Chirp z-transform

Let us consider the following identity:

$$2ik = i^2 + k^2 - (k-i)^2 \,. \tag{6.77}$$

Substituting this into the expression of the DFT, the following formula is obtained:

$$X_k = \sum_{i=0}^{N-1} x_i W_N^{ik} = e^{-j\pi\frac{k^2}{N}} \sum_{i=0}^{N-1} \left(x_i e^{-j\pi\frac{i^2}{N}} \right) e^{+j\pi\frac{(k-i)^2}{N}} \,. \tag{6.78}$$

Seemingly, the operations of this expression are based on the values $e^{-\frac{i^2}{N}}$. The terms of this series can be considered as samples of a swept complex exponential (see Fig. 6.25), called a *chirp* function, after its sound when its real or imaginary part is led to a loudspeaker. The expression (6.78) is the chirp z-transform form of the DFT. Its evaluation consists of three steps:

- multiplication of the input series by the chirp function,
- discrete convolution of the result with the chirp function,
- another multiplication with the chirp function.

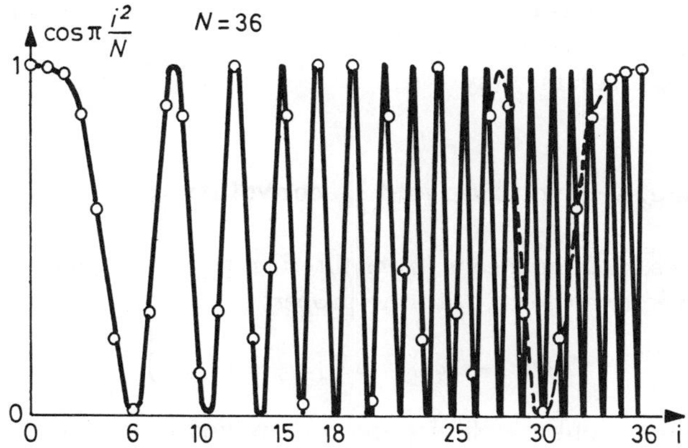

Fig. 6.25 The real part of the chirp function

Eq. (6.78) can be directly evaluated using CCD elements (see Subsection 6.5), but can also be used for the fast calculation of the DFT with arbitrary point length, since the convolution can be evaluated by three base-2 FFT-s (FFT-s of the two series extended by zeros, multiplication, and inverse FFT, see Rabiner and Gold, 1975).

Winograd Fourier transform algorithm (WFTA)

An interesting alternative to the FFT is based on the work of Winograd (1976, see also Silverman, 1977). This algorithm is based on the factorization of the transformation matrix of the DFT in the following form: $\mathbf{W}=\mathbf{S}\cdot\mathbf{C}\cdot\mathbf{T}$, where $\mathbf{C}$ is a diagonal matrix of size M, and $\mathbf{S}$ and $\mathbf{T}$ are incidence matrices (their elements are all equal to 0, -1 or +1). An example for $N=3$ is shown in Fig. 6.26.

It is obvious that having such a factorization with a small M value, the number of necessary multiplications is drastically reduced. Unfortunately, there is no straightforward way to find the factorizations, and the structure of the algorithm

$$\begin{bmatrix} W_3^0 & W_3^0 & W_3^0 \\ W_3^0 & W_3^1 & W_3^2 \\ W_3^0 & W_3^2 & W_3^1 \end{bmatrix} = \underbrace{\begin{bmatrix} 1 & 0 & 0 \\ 1 & 1 & 1 \\ 1 & 1 & 1 \end{bmatrix}}_{N \times M} \underbrace{\begin{bmatrix} 1 & 0 & 0 \\ 0 & \cos\frac{2\pi}{3}-1 & 0 \\ 0 & 0 & j\sin\frac{2\pi}{3} \end{bmatrix}}_{M \times M} \underbrace{\begin{bmatrix} 1 & 1 & 1 \\ 0 & 1 & 1 \\ 0 & 1 & 1 \end{bmatrix}}_{M \times N}$$

$$W_3 = e^{-j\frac{2\pi}{3}} = \cos\frac{2\pi}{3} - j\sin\frac{2\pi}{3}$$

Fig. 6.26 Winograd factorization for N=3

is usually not too regular. As a consequence, what is gained in the reduction of the computational need, is usually lost in the organization of the calculation cycles. Thus, the Winograd transform is usually not much faster than the FFT; it is used in special purpose programs only.

6.1.6 Filtering, Convolution and Deconvolution

For most engineers, filtering means *linear* filtering, that is, an operation on the time series or the Fourier transform, with the property

$$\mathcal{T}\{ax_i + by_i\} = a\mathcal{T}\{x_i\} + b\mathcal{T}\{y_i\}. \tag{6.79}$$

Although sometimes nonlinear filters are also applied (e. g. the *median* filter, see Bovik et al, 1983; Lee and Kassam, 1985), in the vast majority of applications linear filters are used, because of several reasons:

- the theory of linear operations is well elaborated,
- linearity allows a design, independent of the signal levels,
- a lot of components behave linearly in the allowable range,
- etc.

Filters are usually applied for the modification of the behaviour of signals. They are often designed in the frequency domain, because there the effect of a filter can be described as multiplication with its transfer function, thus noise suppression, band-selectivity etc. have straightforward meanings.

Filters can be applied in the continuous time domain (analog filters) or in the discrete time domain (discrete-time filters). Although a lot of problems can be solved only by analog filters, the importance of the discrete ones, especially of the digital ones, is growing rapidly. Digital signal processing has no drift, offset, or aging problems, and advanced integrated circuits offer a lot of computing power at reasonable speed. On the other hand, errors of quantization, arithmetic rounding, under- and overflows are special problems of digital tools.

Digital convolution can be executed directly, or by multiplication in the transformed domain, making use of the efficiency of the FFT or similar transforms (Agarwal and Cooley, 1977; Nussbaumer, 1982). In the case of the FFT there is one important aspect that must be considered: the result is *circular convolution*, i. e. the two time series are convolved in such a manner, that the samples, which 'step out' from before the other record, reappear at the other end of the shifted record (Fig. 6.27). Another explanation may also be given on the basis of the periodicity of the Fourier transform of a sampled series.

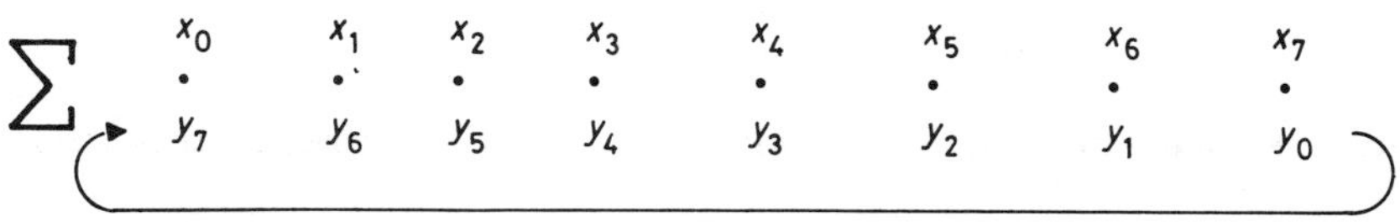

Fig. 6.27 Circular convolution

The design of digital filters is extensively treated in the literature (Rabiner and Gold, 1975; Gold and Rader, 1969; Oppenheim and Schafer, 1975; etc).

The effect of every linear dynamic system on a signal can be modelled by filtering. The time domain description of this effect is *convolution*. The aim of signal processing in measurement is often to get rid of this signal deterioration. This operation, which can be referred to as inverse filtering, which is a frequency domain concept, is called *deconvolution* in the time domain. Thus, deconvolution is one of the key issues in precision measurement.

Unfortunately enough, deconvolution is usually ill-posed. The cause of this fact can be easily understood in the frequency domain. While filtering means multiplication by the transfer function, its inverse is division by it. However, transfer functions generally have small-amplitude parts (stopbands, frequency band over the bandlimit). In these bands the measured signal has to be divided by a very small number. Since some measurement noise is always present in every frequency band, this noise can be unacceptably amplified by this division, making the result of inverse filtering or deconvolution useless. Consequently, a trade-off has to be found between noise amplification and signal deterioration.

Deconvolution is extensively treated in the literature (Mendel, 1983; Nahman, 1986; Tikhonov and Arsenin, 1972; Widrow and Stearns, 1985). There are different approaches for the deconvolution of transient and stochastic signals.

6.1.7 Modulation and Demodulation

Modulation and demodulation are basic operations in telecommunications. However, with the extensive use of frequency domain methods in measurement, they have growing importance also in this field. Two simple cases will be briefly discussed here: mixing and coherent demodulation, since these will be directly used later in this book. For other methods, see e. g. Peebles (1976).

The simplest case of modulation is mixing: the signal is multiplied by a sine wave (or a general periodic signal).

The effect of mixing can be easily studied in the case of signals which have Fourier transform.

$$\mathcal{F}\{x_{\mathrm{m}}(t)\} = \mathcal{F}\{x(t)\cos(2\pi f_1 t + \varphi)\} = \mathcal{F}\{x(t)\} \star \mathcal{F}\{\cos(2\pi f_1 t + \varphi)\}$$
$$= X(f) \star \Big(0.5e^{j\varphi}\delta(f - f_1) + 0.5e^{-j\varphi}\delta(f + f_1)\Big)$$
$$= 0.5e^{j\varphi}X(f - f_1) + 0.5e^{-j\varphi}X(f + f_1)\,. \tag{6.80}$$

Thus, mixing results in frequency splitting: the original spectrum is repeated around $\pm f_1$.

The effect is the same in the case of stochastic signals, but the derivation is somewhat more involved, and will not be included here. However, some conceptual problems must be discussed. The signal $x(t)\cos(2\pi f_1 t + \varphi)$ is not stationary[1] for fixed phase φ, thus the power spectral density function is not directly applicable. Nevertheless, if a random phase sine wave is measured, that is, the phase φ can be considered as uniformly distributed random variable in $(0,2\pi)$, the mixed signal will also be stationary, and the PSD will be (if the split spectra do not overlap):

$$S_{x\mathrm{m}}(f) = 0.25S_x(f - f_1) + 0.25S_x(f + f_1)\,. \tag{6.81}$$

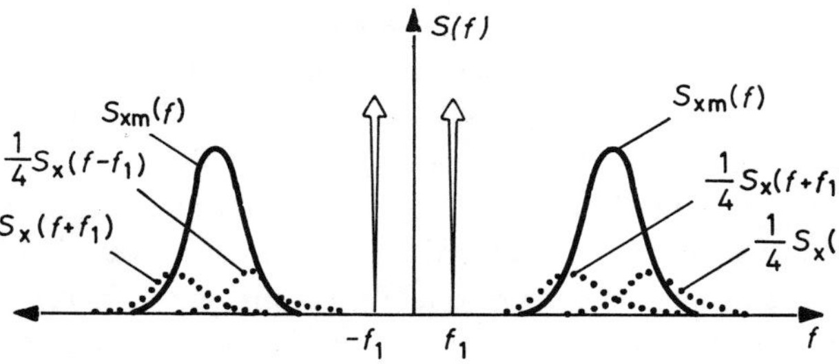

Fig. 6.28 The effect of mixing by a sine wave

As it was shown in Subsection 6.1.1, sampling is also a special kind of mixing: multiplication by a Dirac delta series. Since the Fourier transform of a periodic Dirac delta series is a frequency domain Dirac delta series, the Fourier transform of the input signal is repeated at each frequency component (see Eq. 6.1).

The mixed signal can be restored e. g. by *synchronous* or *coherent demodulation.* It is *repeated modulation* by a signal of the same frequency: the result is

$$x_{\mathrm{md}}(t) = x(t)\cos(2\pi f_1 t + \varphi_{\mathrm{m}})\cos(2\pi f_1 t + \varphi_{\mathrm{d}})$$

[1]The properties of this signal change periodically with period length $T_1 = 1/f_1$, therefore this process is sometimes referred to as *cyclostationary.*)

$$= x(t)\big(0.5\cos(\varphi_{\rm m} - \varphi_{\rm d}) + 0.5\cos(2\pi 2f_1t + \varphi_{\rm m} + \varphi_{\rm d})\big)\,, \tag{6.82}$$

$$X_{\rm md}(f) = 0.5\cos(\varphi_{\rm m} - \varphi_{\rm d})X(f) + 0.25e^{j(\varphi_{\rm m}+\varphi_{\rm d})}X(f - 2f_1)$$
$$+0.25e^{-j(\varphi_{\rm m}+\varphi_{\rm d})}X(f + 2f_1)\,, \tag{6.83}$$

The high frequency components can be filtered out, and the central part kept. However, it can be noticed that the phases have an extremely important influence on the result: the two phases must be equal (the modulating cosines must be synchronized), otherwise the demodulated signal may even disappear. The situation is similar in the case of stochastic signals. Because of this reason, mixing to zero frequency is to be performed with extreme care.

A question that quickly arises after the above considerations sounds like this: is it not possible to move the spectrum of a signal *without spectrum splitting*?

Spectrum splitting is a consequence of the two Dirac delta functions in the Fourier transform of the sine wave. To avoid splitting, a signal with just one Dirac delta function in the spectrum is necessary. There is a signal like this: the complex exponential. Such a complex-valued signal can be handled as follows.

Using Euler's formula,

$$e^{j(2\pi f_1 t+\varphi)} = \cos(2\pi f_1t + \varphi) + j\sin(2\pi f_1t + \varphi)\,. \tag{6.84}$$

Signal processing can be performed in two parallel channels: in the 'real' one modulation is done by a cosine, in the 'imaginary' one by another one, shifted by 90°. Both signals are true signals, but if considered as real and imaginary parts of a complex one, the desired spectrum shifting is performed without splitting. This procedure is called complex modulation (or demodulation).

6.1.8 Special Techniques

The aim of this section was to summarize the basics of signal processing. Though the most important signal processing methods were indeed summarized, it has to be mentioned that a lot of methods exist, which are not yet customarily built into instruments but have important applications in the solution of special measurement problems. Though in difficult situations the reader is referred to the literature of signal processing (e. g. Bendat and Piersol, 1971, 1980; Crochiere and Rabiner, 1983; Jazwinski, 1970; Oppenheim and Schafer, 1975; Rabiner and Gold, 1975; Robinson and Silvia, 1986; etc), some characteristic problems and solutions will be briefly mentioned here.

Preprocessing of the data record

Standard signal processing methods are usually based on certain assumptions, like stationarity or absence of gross reading errors. When these assumptions are not met, the data record is to be treated prior to the application of standard procedures.

One of the most often used methods is the *detection and removal of outliers.* These are such measured points, which are corrupted by large local disturbances, usually due to electromagnetic interference, or temporary malfunction of the data acquisition channel. Such outliers can be detected by statistical tests. They may be removed by human intervention or by median filtering (see later).

Nonstationarity often means a sudden or slow *level shifting* or a linear or nonlinear *trend,* superimposed on the signal. Such trends may be detected by appropriate tests. *Trend removal* is important in many applications.

Signal processing is usually most effective when the spectrum is close to a white one. If a priori knowledge about the spectrum of the measured signal is at hand, the signal may be prefiltered, in order to 'prewhiten' the spectrum before signal processing.

Data preprocessing often includes the *qualification of data,* as tests for periodicities, stationarity, normality, whiteness etc.

Detection of weak signals in noise

A common problem in many applications is to determine whether a given signal is present in a noisy sample record or not. This problem usually requires the use of detection theory: the answer yes or no has to be given with a possibly small error probability. However, the purpose is usually to process the data record in such a way that the answer becomes obvious (e. g. a *large* peak is present in the spectrum, with small relative variance, which indicates the presence of a sine wave).

If the signal is sinusoidal, or a periodic one, there is a theoretically straightforward solution: by decreasing the bandwidth in the power spectral density measurement, the signal-to-noise ratio can be arbitrarily improved. However, this also means that the observation time must be adequately long, and sufficient processing capacity must be at hand. These requirements cannot always be met. In such cases special methods may be applied, as Pisarenko harmonic decomposition, Prony's method, AR or ARMA model fitting etc. (See e. g. Kay and Marple, 1981; Thomson, 1982; Haykin, 1983.)

In the general case, when an arbitrary (but a priori given) waveform is to be detected, a so-called *matched filter* (see e. g. Peebles, 1976) can be applied. It can be shown that for a signal $s(t)$, a filter with the impulse response $h(t) = s(t_0 - t)$ will provide optimal detection properties, if the noise is white. Thus, it may be said that the filter is *matched* to the signal.

Orthogonal transforms

Fourier series expansion, regularly used in signal analysis, does nothing more than expresses a given signal in terms of an orthonormal set of signals:

$$\left\{e^{j2\pi k f_1 t}\right\}, \quad k = 0, \pm 1, \pm 2, \ldots \tag{6.85}$$

This signal representation has a very useful intuitive meaning (harmonic components). However, this is not the only orthogonal series expansion (or transform) that has important applications in measurement. In principle every complete orthogonal set of functions can be used for this purpose. The best known sets of this kind are the Haar- and the Walsh-functions. There is also another possibility for stochastic signals: sample functions can be very effectively characterized by the first terms of the series expansion on the set of the eigenfunctions of the stochastic process itself: this is called Karhunen-Loève expansion.

But what is the purpose of a series expansion, especially if this has no direct intuitive meaning? There are two main motivations:

- Certain kinds of signals can be well described by a few of the first terms of a given type of the series expansion. This gives a possibility of data compression: storage with small memory usage.
- Some coefficients of a series expansion may be very sensitive to certain features of the signal: the calculation of these coefficients helps feature extraction. For example, Walsh functions are often used in the investigation of nonidealities of A/D converters.

Optimal filtering

Filter design usually starts from the specification of the transfer function. However, this specification may be rather difficult to determine, if the spectra of the useful signal and the noise are overlapping. In this case a trade-off has to be found between noise and signal distortion. Such a compromise is usually found by minimizing a given cost function. This approach, that is, obtaining a signal processing procedure by minimizing (or maximizing) an objective function is called *optimal* measurement. The filters designed for optimal filtering in a least squares sense are the Wiener filter and the Kalman filter (Jazwinski, 1970).

A common difficulty is that the desired quantity often cannot be directly measured, its value has to be estimated from the output signal of the system. There are standard methods for continuously estimating the state variables of a dynamic system from the output signal, these are called observers, and are related to Kalman filters.

Nonlinear filtering

As already mentioned, the theory of linear filtering is well elaborated, and these filters have advantageous properties, first of all the superposition principle can be applied. There are, however, situations when nonlinear filters prove to be more effective, e. g. when dealing with data corrupted by impulsive noise, or when the signal contains abrupt changes, or when the noise is 'heavy-tailed', i. e. the tails of the probability density function vanish very slowly (Bovik et al, 1983; Lee and

Kassam, 1985), or when the amplitude of the noise is definitely limited, e. g. uniform noise is corrupting the measurements. However, the theory of nonlinear filtering is neither easy to use, nor well elaborated, thus it is advisable in very special situations only, when linear filters cannot help (e. g. for the suppression of impulsive noise). A typical example is the so-called *median filtering.* In the simplest realization each point is substituted by the median of all the points in a small environment.

Model-based signal processing (Parametric signal processing)

Signal processing methods can be either nonparametric or parametric (see Subsection 1.2.4). In the first case no a priori information is used about the structure of the signal or of the system under test. Commercial measuring devices usually take a nonparametric approach, since by this the user need not provide extensive information on the internal structure of a dynamic system. However, the parametric approach can prove to be much more effective, at the cost of the required a priori information (Candy, 1986; 1988). If a parametric model (that is, a mathematical equation with some free parameters to be determined) of the system or the signal is available, this model can be fitted to the measured data. Typical examples are linear and polynomial regressions, when a straight line or a polynomial is fitted in a least squares sense to the measured data, respectively, or the use of matched filters, usually applied for the detection of transient waveforms. When sine waves are measured by means of spectrum analyzers, the results are also interpreted on the basis of a signal model.

A very general parametric model of signals is the so-called ARMA (autoregressive – moving average) model. When applying this model, it is usually assumed that the discrete-time signal is the output of a hypothetical digital filter, driven by white noise. When using this model, first the parameters of the filter are determined, and the required features (for instance, the spectrum) are determined from the parametric model.

The filter can also be sought in other forms, such as lattice form etc.

When a model of a system is sought, the procedure is usually called *system identification* (Eykhoff, 1974; Goodwin and Payne, 1977; Sorenson, 1980; Norton, 1986; Ljung, 1987; Söderström and Stoica, 1989; Schoukens and Pintelon, 1991).

Homomorphic signal processing

There are rather special techniques among the tools of signal processing in measurement that cannot be dealt with in this book, but should at least be mentioned. An example is *homomorphic* signal processing. Roughly, the Fourier transform of the logarithm of the time domain signal or of its spectrum (the so-called *cepstrum*) is investigated. The result can be used for demodulation (the logarithm makes addition from multiplication), or for the processing of echoic data (Robinson and Silvia, 1986).

6.2 MEASUREMENT OF TIME RECORDS

The most 'natural' way of signal processing is to examine time records. The tools that help this activity are the instruments that visualize them (oscilloscopes, recorders, plotters etc), and those which store them (tape recorders, data loggers, transient recorders etc). These two functions are often combined in a single instrument (e. g. storage oscilloscopes). In this section a short overview of the possible solutions will be given.

6.2.1 Oscilloscopes

The 'golden rule' of measurement is: if possible, check the results of any instrument, using an oscilloscope. Indeed, oscilloscopes are the basic instruments of every electrical laboratory.

Classical oscilloscopes can display on the built-in cathode ray tube (CRT) any repetitive waveform for which a usable synchronization signal (trigger) can be generated.

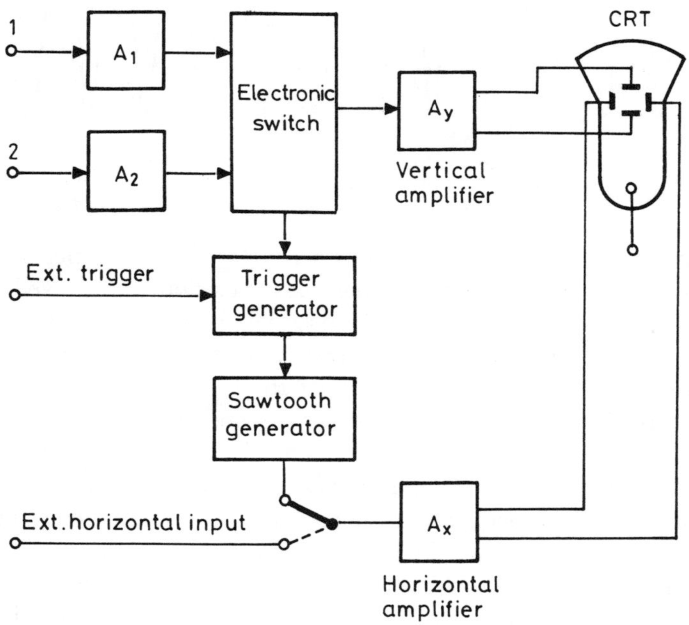

Fig. 6.29 Block diagram of a classical oscilloscope

The bandlimit of the oscilloscope depends on the speed of the CRT and the input amplifier. Modern oscilloscopes can display signals up to 1 GHz. Using non-storage CRT-s, the lowest allowable signal frequency is determined by the physiology of the human eye: signals under 20 Hz usually vibrate in an intolerable manner.

The read-off accuracy is limited by the linearity of the static deflection and size of the screen, to about 1%, which can be somewhat improved by numerical displaying of the cursor position.

Nowadays oscilloscopes are often equipped with different additional modules, like digital voltmeter, rms voltmeter, conditional trigger generator, time/frequency meter, logic state analyzer, spectrum analyzer etc. These modules can be used to increase the read-off accuracy, and make surprisingly complex measurements possible.

Sampling oscilloscope

The frequency limit of oscilloscopes can be significantly increased in the case of repetitive waveforms, using the principle of *equivalent sampling* (Fig. 6.30).

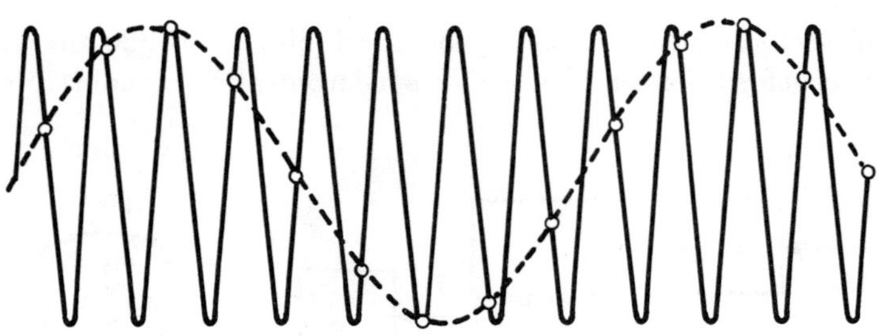

Fig. 6.30 Principle of equivalent sampling

Only one sample is taken from every period (or every kth period) of the high frequency periodic signal, at slowly sliding phase positions. Displayed sequentially, these samples will show the shape of the original signal. However, the CRT may work at a much lower speed than the frequency of the signal: it is the sample and hold circuit only that must work quickly enough.

It is interesting to throw a glance at the spectrum of the sampled signal (Fig. 6.31).

Obviously, the sampling theorem is violated: there are harmonics well above $f_S/2$. However, since the spectrum of the useful periodic signal contains lines only, with a carefully chosen sampling frequency the transformed lines will not coincide, and the desired spectral shape will appear, transformed to lower frequencies.

The frequency limit of sampling oscilloscopes can be 10–50 GHz. A disadvantage is noise sensitivity: since the signal shape is put together from samples of different signal periods, sampling jitter and additive noise may cause visible 'trembling' of the displayed waveform.

There is another principle that can be used in sampling oscilloscopes: *random sampling,* where samples are taken randomly of both the signal and the time base, thus samples are displayed not sequentially, but in random manner (see Mulvey, 1970; Frye and Nahman, 1964). Random sampling imposes theoretically no fre-

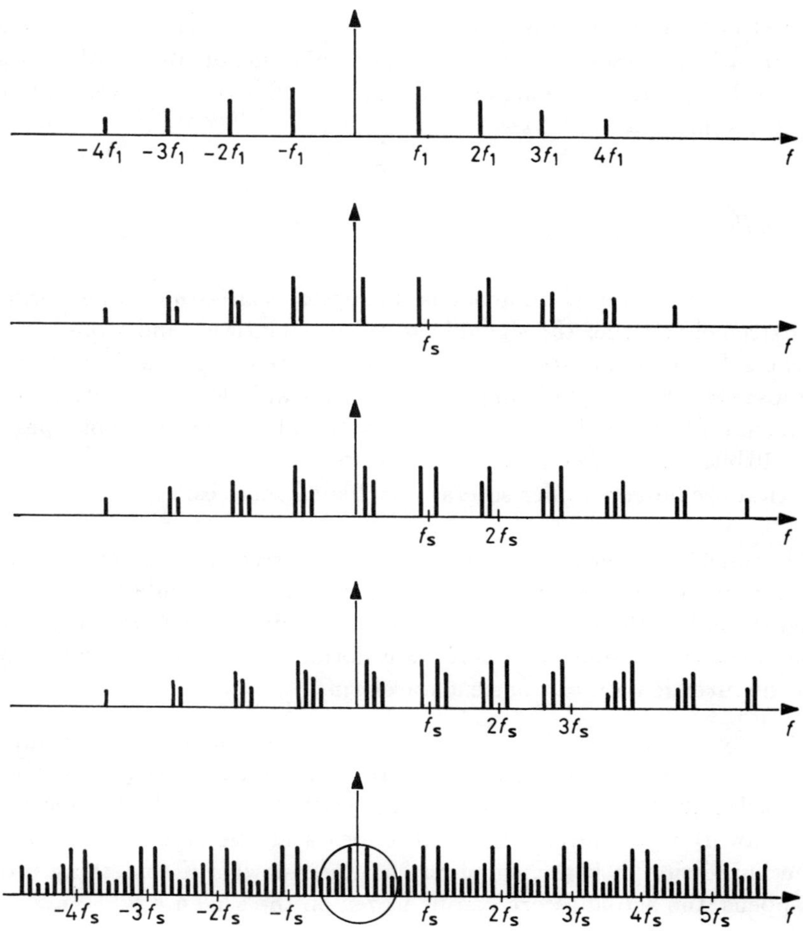

Fig. 6.31 The effect of equivalent sampling in the frequency domain

quency limitation, however, the frequency limit is in the same order of magnitude as above, because of the bandwidth of the input channels and the finite width of the sampling pulses.

Digital oscilloscopes also work with signal samples, however, they will be treated later in this subsection, separately.

Storage oscilloscope

Although digital oscilloscopes also store data records, the term *storage oscilloscope* is generally used for equipment based on storage cathode ray tubes. These are special CRT-s which are able of holding the shape of even one single plot. They can be used for the study of nonrepetitive (transient) and low frequency (<10 Hz) signals. The lower frequency limit is about 0.1 mHz, because in writing mode the screen has limited capability of keeping the information. The writing speed can reach 2500 cm/μs, which allows the measurement of transients of 50–100 MHz bandwidth.

With the appearance of high-speed digital oscilloscopes the storage CRT gradually loses its importance.

Digital oscilloscope

The most flexible type of oscilloscope is the *digital oscilloscope*. The working principle is extremely simple: the signal is sampled, quantized, and stored in a digital memory; displaying and postprocessing are performed on the stored waveform. Digital oscilloscopes can sample from practically arbitrarily low frequencies to the GHz range, the amplitude resolution is 6–8 bits for high frequency oscilloscopes and can reach 12–16 bits for low frequency (audio range) ones.

Digital oscilloscopes offer several very handy services:

- It is possible to study the waveform *before* the trigger pulse (pretriggering). In this mode the oscilloscope continuously samples and circularly stores the input signal, and at the trigger pulse, or after an additional delay, stops sampling: the memory contains the desired waveform. This feature is especially useful in the investigation of transient waveforms.
- A slowly varying signal can be monitored continuously. In this 'rolling' mode the actual contents of the memory are continuously displayed on the screen, at each sampling instant the record is shifted to the left, the oldest sample is thrown away, and the new one appears at the right. The screen gives the impression as if the signal was scanned slowly. If something interesting happens, the actual record can be frozen on the screen.
- A reference waveform can be stored in the memory, and used for comparisons with the actually measured one.
- The stored waveform can be easily documented on a printer or a plotter.
- The principle of equivalent sampling can be used for the investigation of high frequency periodic signals.
- If the signal is sampled over the Nyquist rate, the point density can be increased by interpolation (see Subsection 6.1.2).
- High frequency signal components (like impulses or oscillatory slopes) can often be detected by a special-purpose circuitry at the input (e. g. glitch detection).
- Complex signal processing methods can be built into the oscilloscope, such as FFT, transfer function measurement, integration and differentiation, signal enhancement, digital filtering etc.

6.2.2 Signal Recording and Storage

Classical methods of signal recording are the use of strip chart signal recorders with automatic compensators, which transversally move a pen on a slowly advancing paper strip, and the use of T-Y or X-Y recorders. The advantage is robustness and the immediate hardcopy obtained, the disadvantage is that signal processing cannot be directly applied. The bandlimit is determined by the speed of the mechanical system moving the pen (<10 Hz), the accuracy is not better than 0.1%.

The bandlimit can be significantly increased in the so-called *transient recorders*. From outside, these resemble conventional X-Y recorders, however, internally they do not directly plot the signal, but sample and store it with the speed of the electronic circuitry, and plotting is performed from the memory.

Analog data recording is generally done by using tape recorders. Instrumentation tape recorders can store signals of several channels in parallel. Their higher frequency limit is about 1 MHz, the lower limit is 0 Hz, if frequency modulated signal recording is used.

Analog signal recording is gradually being substituted by digital equipment (data loggers). The main reason is that once a signal is in digital form, it can be stored without degradation, moreover, microprocessors offer a cheap way of different kinds of digital signal processing. Portable recorders also give the possibility of preprocessing and data compression, increasing the efficiency of storage.

A special type of digital signal recorder is the so-called *transient analyzer*. This is a digital signal recorder with additional built-in display and data processing capabilities, and like this, it is very similar to a digital oscilloscope (see above).

The key element of every digital signal recorder is the A/D converter. A/D converters are discussed in Subsection 2.3.17.

6.3 MEASUREMENT OF THE MEAN VALUE

The usual way of measuring the mean value of a given signal is *averaging*. Averaging is treated in detail in Subsection 6.1.4, thus for the usual solutions the bias and variance are already given. Also, the implementations are straightforward (integrator, lowpass filter, digital summer etc). In this subsection a special mean value measuring device will be discussed only: the one based on a stochastic-ergodic converter (see also Subsection 6.1.3).

The arrangement is shown in Fig. 6.32. The output of the comparator is sampled and summed by the up/down (U/D) counter. A is the peak-to-peak amplitude of the dither (see Fig. 6.15), N is the number of counts.

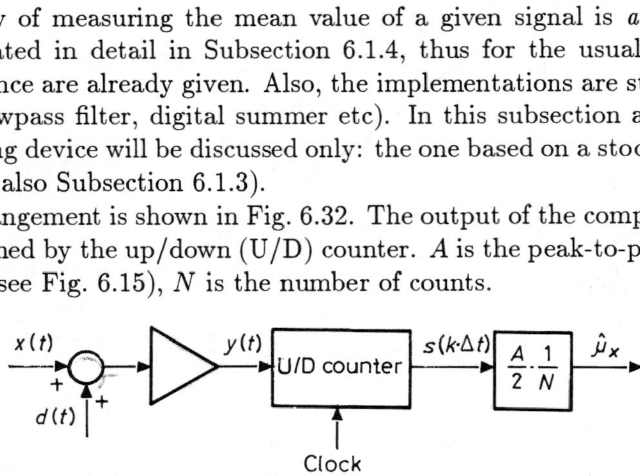

Fig. 6.32 Mean value measurement using a stochastic-ergodic converter

From the results given in Subsection 6.1.3 it follows that the mean estimate is unbiased:

$$\mathcal{E}\{\widehat{\mu}_x\} = \mu_x \,. \tag{6.86}$$

By a longer derivation also the variance of the estimate of the mean value can be determined on the basis of (6.16–19):

$$\operatorname{var}\{\widehat{\mu}_x\} = \frac{1}{N}\Big(\frac{A^2}{4} - \mu_x^2 - \sigma_x^2\Big) + \frac{1}{N}\sum_{i=-(N-1)}^{N-1}\Big(1-\frac{|i|}{N}\Big)C_{xx}(i\Delta t)\,. \tag{6.87}$$

The second term can be recognized as the variance of the result of conventional discrete averaging (6.31). The first term is nonnegative since $|x(t)| \le A/2$, thus (6.87) cannot be smaller than (6.31), as is expected.

The variance of the estimate as a function of N, with fixed measurement time $T = N\Delta t$, is illustrated in Fig. 6.33 for a random signal.

The behaviour of the variance can be explained in an intuitive way: for small values of N the samples are practically independent, that is, the variance is inversely proportional to N, while for dense sampling the variance approximates the continuous case.

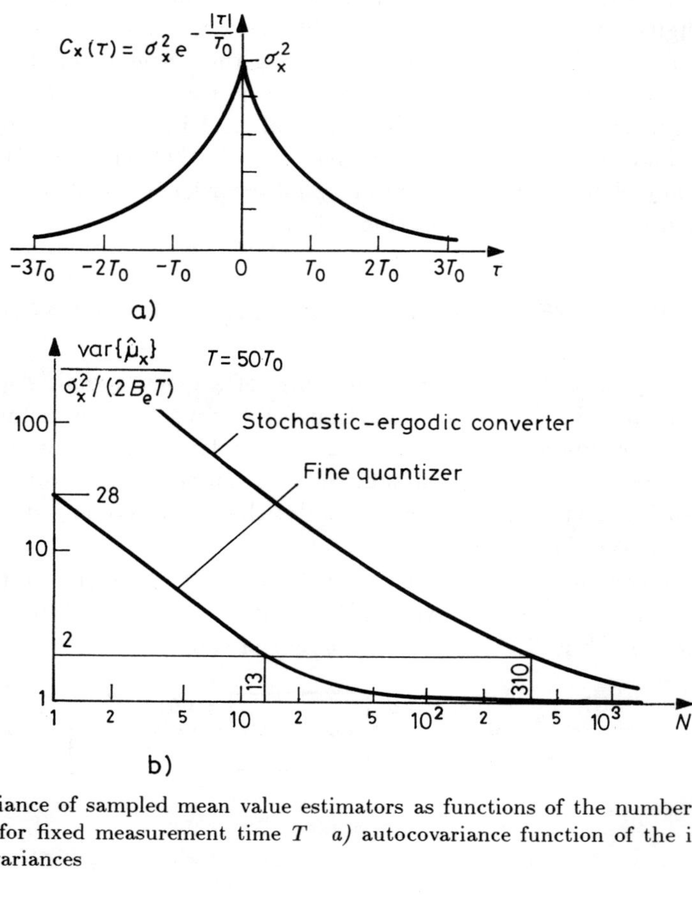

Fig. 6.33 Variance of sampled mean value estimators as functions of the number of samples N, for fixed measurement time T *a)* autocovariance function of the input signal *b)* variances

The crosspoint of the two asymptotic lines is at

$$N = 2BT\frac{\frac{A^2}{4} - \mu_x^2}{\sigma_x^2}, \tag{6.88}$$

where B is the equivalent bandwidth of the input signal. This observation leads to a very interesting statement: the equivalent bandwidth of the output of the comparator is

$$B_e = B\frac{\frac{A^2}{4} - \mu_x^2}{\sigma_x^2}. \tag{6.89}$$

This expression, obtained in an indirect way, would be difficult to derive directly.

The above expressions can be used for the design of an arrangement for mean estimation, or for the determination of the properties of the output. The circuitry shown in Fig. 6.32 is cheap and simple, the accuracy is determined by the accuracy of the distribution of the uniformly distributed noise, and the length of the counter.

6.4 MEAN SQUARE VALUE AND VARIANCE MEASUREMENT

The mean square value, $\mathcal{E}\{x^2(t)\} = \psi_x^2$, is nothing other than the *power* of the signal, and has to be measured quite often. For example, in order to design the measurements, the variance is often necessary, which is in direct relationship with the power:

$$\text{var}\{x\} = \sigma_x^2 = \psi_x^2 - \mu_x^2. \tag{6.90}$$

The measurement of ψ_x^2 means the measurement of the mean value of $x^2(t)$, thus the results of Section 6.3. can be directly used. On the other hand, since

$$\mathcal{E}\{x^2(t)\} = R_{xx}(0) \quad \text{and} \quad \sigma_x^2 = C_{xx}(0), \tag{6.91}$$

mean square value and variance measurements are special cases of correlation measurement, the results are in strong relationship with those of Section 6.5.

If the estimate is

$$\widehat{\psi_x^2} = \frac{1}{T}\int_0^T x^2(t)\,\mathrm{d}t, \tag{6.92}$$

applying the results of Subsection 6.1.4 (6.20), it is easy to see that this estimate is unbiased.

A simple expression can be obtained for Gaussian signals, using the equality

$$\mathcal{E}\{x_1x_2x_3x_4\} = \mathcal{E}\{x_1x_2\}\mathcal{E}\{x_3x_4\} + \mathcal{E}\{x_1x_3\}\mathcal{E}\{x_2x_4\}$$
$$+\mathcal{E}\{x_1x_4\}\mathcal{E}\{x_2x_3\} - 2\mu_1\mu_2\mu_3\mu_4. \tag{6.93}$$

It is easy to show that if $x(t)$ is a zero mean Gaussian process, the autocovariance function of $x^2(t)$ is $2C_x^2(\tau)$, thus the power spectral density has to be convolved by itself. Generally this autoconvolution increases the bandwidth. However, surprisingly enough, for bandlimited white Gaussian noise the *noise bandwidth* does not change (see Fig. 6.34).

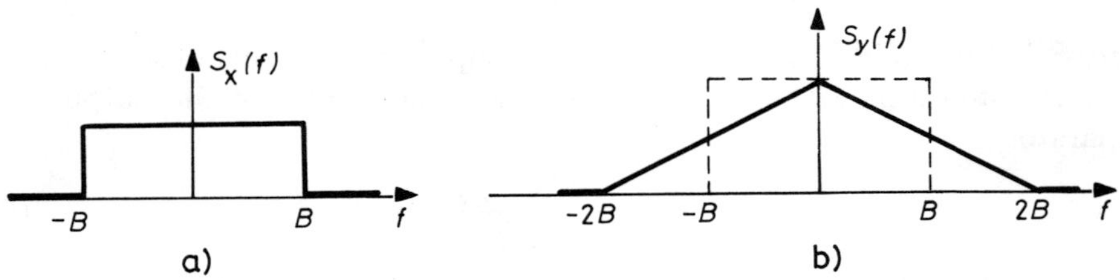

Fig. 6.34 Equivalent bandwidth of $x^2(t)$ *a)* original spectrum *b)* spectrum of $y(t)=x^2(t)$ and the spectrum of the equivalent white noise

From (6.23–24) it follows that

$$\mathrm{var}\{\widehat{\psi_x^2}\} \approx \frac{2C_x^2(0)}{2B_\mathrm{n}T} = \frac{\sigma_x^4}{B_\mathrm{n}T}\,. \tag{6.94}$$

Strictly speaking, in (6.94) the so-called *statistical bandwidth* should be applied:

$$B_\mathrm{s} = \frac{\left(\int_{-\infty}^{\infty} S_{x\mathrm{c}}(f)\,\mathrm{d}f\right)^2}{2\int_{-\infty}^{\infty} S_{x\mathrm{c}}^2(f)\,\mathrm{d}f} = \frac{\sigma_x^4}{2\int_{-\infty}^{\infty} S_{x\mathrm{c}}^2(f)\,\mathrm{d}f}\,, \tag{6.95}$$

but since this is usually approximately equal to B_n, the latter may be used.

For Gaussian signals, the exact expression can be calculated also for the more general case $\mu_x \neq 0$:

$$\mathrm{var}\{\widehat{\psi_x^2}\} = \frac{2}{T}\int_{-T}^{T}\left(1-\frac{|\,t\,|}{T}\right)\left(C_x^2(t) + 2\mu_x^2 C_x(t)\right)\mathrm{d}t. \tag{6.96}$$

If the width of $C_x(t)$ is much smaller than T,

$$\mathrm{var}\{\widehat{\psi_x^2}\} \approx \frac{2}{T}\int_{-\infty}^{\infty}\left(C_x^2(t) + 2\mu_x^2 C_x(t)\right)\mathrm{d}t. \tag{6.97}$$

Evaluating this expression for the bandlimited white noise,

$$\mathrm{var}\{\widehat{\psi_x^2}\} \approx \frac{1}{BT}(\sigma_x^4 + 2\mu_x^2\sigma_x^2)\,. \tag{6.98}$$

It is interesting to note that if $\mu_x = 0$, the *relative variance* is

$$\epsilon_{\mathrm{v}}^2\{\widehat{\psi_x^2}\} \approx \frac{1}{BT}\,. \tag{6.99}$$

The properties of the estimate have to be determined for the sinusoidal case separately. Expressions (6.26–27) can be used for the signal

$$y(t) = \left(\mu_x + U_{\mathrm{p}}\sin(2\pi f_1 t + \varphi)\right)^2 = \mu_x^2 + 2\mu_x U_{\mathrm{p}}\sin(2\pi f_1 t + \varphi) + \frac{U_{\mathrm{p}}^2}{2}\left(1 - \cos(4\pi f_1 t + 2\varphi)\right). \tag{6.100}$$

If a random phase sine wave is investigated, the expected value is

$$\mathcal{E}\{\widehat{\psi_x^2}\} = \mu_x^2 + \frac{U_{\mathrm{p}}^2}{2}\,, \tag{6.101}$$

the variance is

$$\mathrm{var}\{\widehat{\psi_x^2}\} = 2\mu_x^2 U_{\mathrm{p}}^2 \mathrm{sinc}^2(\pi f_1 T) + \frac{U_{\mathrm{p}}^4}{8}\mathrm{sinc}^2(2\pi f_1 T)\,, \tag{6.102}$$

and the relative variance is for $\mu_x = 0$

$$\epsilon_{\mathrm{v}}^2\{\widehat{\psi_x^2}\} = \frac{1}{2}\mathrm{sinc}^2(2\pi f_1 T)\,. \tag{6.103}$$

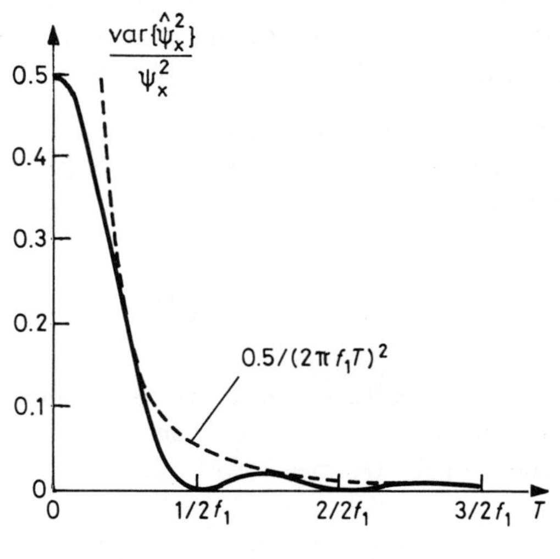

Fig. 6.35 Relative variance of $\widehat{\psi_x^2}$ for a random phase sine wave

If two or more sine waves are present in the signal, the variance can be derived in a similar manner. An incidental frequency beat causes similar problems as in the measurement of the mean value (see Subsection 6.1.4). The remedy is also the same: the integration time has to be much longer than the reciprocal of the frequency difference, in order to achieve efficient variance reduction.

The mixed case (sine+noise) requires special considerations. The estimate

$$\widehat{\psi_x^2} = \frac{1}{T}\int_0^T \Big(\mu_x^2 + 2\mu_x U_\mathrm{p}\sin(2\pi f_1 t+\varphi) + 2\mu_x n(t)$$

$$+U_\mathrm{p}^2\sin^2(2\pi f_1 t+\varphi) + 2U_\mathrm{p}\sin(2\pi f_1 t+\varphi)n(t) + n^2(t)\Big)\,\mathrm{d}t \tag{6.104}$$

contains five terms that cause variance. Fortunately, these are mutually uncorrelated. The variance is:

$$\mathrm{var}\{\widehat{\psi_x^2}\} \approx 4\mu_x^2\frac{U_\mathrm{p}^2}{2}\mathrm{sinc}^2(\pi f_1 T) + 4\mu_x^2\frac{\sigma_n^2}{2BT}$$

$$+\frac{U_\mathrm{p}^4}{8}\mathrm{sinc}^2(2\pi f_1 T) + 4\frac{U_\mathrm{p}^2}{2}\mathrm{sinc}(\pi f_1 T)\frac{\sigma_n^2}{2BT} + \frac{\sigma_n^4}{BT}\,. \tag{6.105}$$

Using the results of Section 6.3 and Subsection 6.1.4, the variance of simple discrete and of exponential averaging can also be easily determined.

The measurement of the mean square value requires squaring. This is inconvenient in the analog case, since the dynamic range of the available squaring circuits is small (40–50 dB), and their frequency band is also limited. A solution will be given in Subsection 6.7.3, here only two important special cases will be mentioned:

1. If the signal is pure sinusoidal with zero mean, the mean square value can be determined from an arbitrary measured parameter (peak value, mean absolute value etc, see Subsections 2.4.3-4):

$$\mathcal{E}\{\xi^2\} = \frac{U_\mathrm{p}^2}{2} = \frac{\pi^2}{2}\mathcal{E}^2\{|\,\xi\,|\} = 2\pi^2\mathcal{E}^2\{\xi^+\}\,. \tag{6.106}$$

2. If the signal is Gaussian with zero mean value, the measurement of the mean square value can be substituted by the measurement of the mean absolute value (or the mean of the rectified wave):

$$\mathcal{E}\{\xi^2\} = \frac{\pi}{2}\mathcal{E}^2\{|\,\xi\,|\} = 2\pi\mathcal{E}^2\{\xi^+\}\,. \tag{6.107}$$

6.5 MEASUREMENT OF CORRELATION FUNCTIONS

The definition of correlation is (see Subsection 1.3.1):

$$R(\tau) = \mathcal{E}\{x_1(t)x_2(t+\tau)\}\,. \tag{6.108}$$

The measurement of the correlation function usually means some kind of averaging of terms like $x_1(t)x_2(t+\tau)$. Let us first consider this expression.

From (6.108) it is obvious that the expected value of this product is equal to the correlation value, that is, the product itself as an estimate is unbiased.

For the calculation of the variance the evaluation of fourth-order moments is necessary. This is generally impossible in simple terms. However, for Gaussian signals, using (6.93) the result can be quickly obtained:

$$\mathrm{var}\{x_1(t)x_2(t+\tau)\} = (1+r^2)\sigma_1^2\sigma_2^2 + \mu_1^2\sigma_2^2 + \mu_2^2\sigma_1^2 + 2\mu_1\mu_2 r\sigma_1\sigma_2\,, \tag{6.109}$$

where σ_1 and σ_2 are the standard deviations of x_1 and x_2, respectively, and r is the correlation coefficient.

If $\mu_1 = 0$ and $\mu_2 = 0$, the above expression simplifies to:

$$\mathrm{var}\{x_1(t)x_2(t+\tau)\} = (1+r^2)\sigma_1^2\sigma_2^2\,. \tag{6.110}$$

It is easy to see that this variance is the largest if $r \approx \pm 1$, and for autocorrelation measurement

$$\mathrm{var}\{x_1(t)x_1(t+\tau)\} \le \mathrm{var}\{\widehat{\psi^2_{x_1}}\}\,, \tag{6.111}$$

while for small values of r the left side becomes half of this value. Consequently, the rule of thumb is quite simple: for $r \approx \pm 1$ the variance can be approximated by the variance of the mean square value estimate, while for small values of r the variance will be half of this. In the latter case the relative variance can be very large: for small values of r, a very long measurement time will be necessary in order to obtain a reasonable estimate of the correlation function.

If the above products are averaged in time, the general expression of the correlation measurement is obtained.

$$\widehat{R}_{12}(\tau) = \frac{1}{T}\int_0^T x_1(t)x_2(t+\tau)\,\mathrm{d}t\,. \tag{6.112}$$

For Gaussian signals, using (6.21) and (6.93), the general expression is:

$$\mathrm{var}\{\widehat{R}_{12}(\tau)\} = \frac{1}{T}\int_{-T}^{T}\left(1-\frac{|t|}{T}\right)\Big(R_{11}(t)R_{22}(t) + R_{12}(t+\tau)R_{12}(t-\tau)\Big)\,\mathrm{d}t - 2\mu_1^2\mu_2^2\,. \tag{6.113}$$

This general expression can be evaluated for any special signal. If $\mu_1 = \mu_2 = 0$ and $C_1(\tau)$ and $C_2(\tau)$ both disappear for $\tau > \tau_0$, where $\tau_0 \ll T$,

$$\mathrm{var}\{\widehat{R}_{12}(\tau)\} \approx \frac{1}{T} \int_{-\infty}^{\infty} \Big(C_{11}(t)C_{22}(t) + C_{12}(t+\tau)C_{12}(t-\tau) \Big)\,\mathrm{d}t \leq \frac{\sigma_1^2 \sigma_2^2}{BT} . \quad (6.114)$$

For large values of τ the second term in the integral is usually negligible, and the variance is approximately

$$\mathrm{var}\{\widehat{R}_{12}(\tau)\} \approx \frac{\sigma_1^2 \sigma_2^2}{2BT} . \quad (6.115)$$

It is interesting to note that this variance value is independent of the value of $R_{12}(\tau)$.

The random phase sinusoidal case and the mixed case (sine+noise) will not be treated here, on the basis of the above discussions the reader can derive the expressions for these signals, or use (6.102) or (6.105) as upper limits of the variance.

There is an additional very important aspect of correlation measurement that did not arise in the measurement of the mean square value: the correlation (or the covariance) function has to be investigated for different values of τ, that is, a *function* has to be estimated, not a scalar value only. This means that also the covariance of the estimates for different τ values has to be considered.

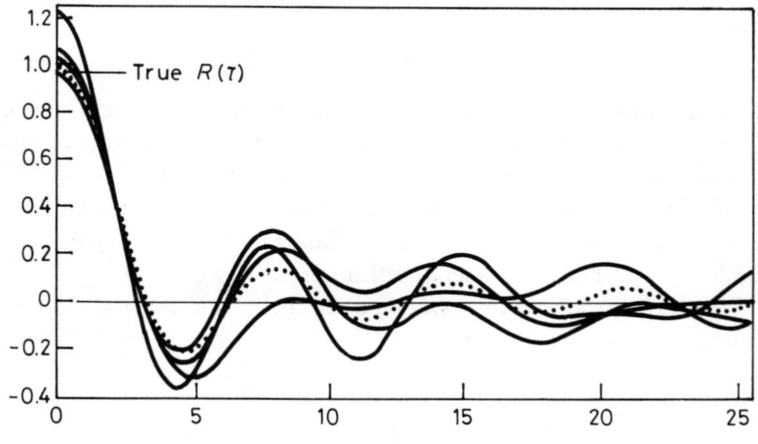

Fig. 6.36 Estimated correlation function of a bandlimited white noise, $B = 160$ Hz, $T = 410$ ms, $BT = 66$.

The more or less exact treatment of these covariances requires lengthy calculations, not to be presented here. The most important result is that when calculated from the *same* sample record, the correlation estimates for only slightly different values of τ strongly depend on each other. The consequence is that the variance of the correlation function estimate will not cause wild variations in the estimated functions, but smooth 'waves' will appear instead, that can be extremely misleading (see Fig. 6.36).

Because of this phenomenon, special care must be taken when evaluating correlation plots. A possibility is to plot also the confidence bounds around the measured function, either calculated from the above variances, or determined from several successive experiments (see Subsection 1.6.2). Another solution may be to superpose a few measured correlation function plots (see Fig. 6.37). As the 'fake' function parts will be different from measurement to measurement, they can be easily recognized.

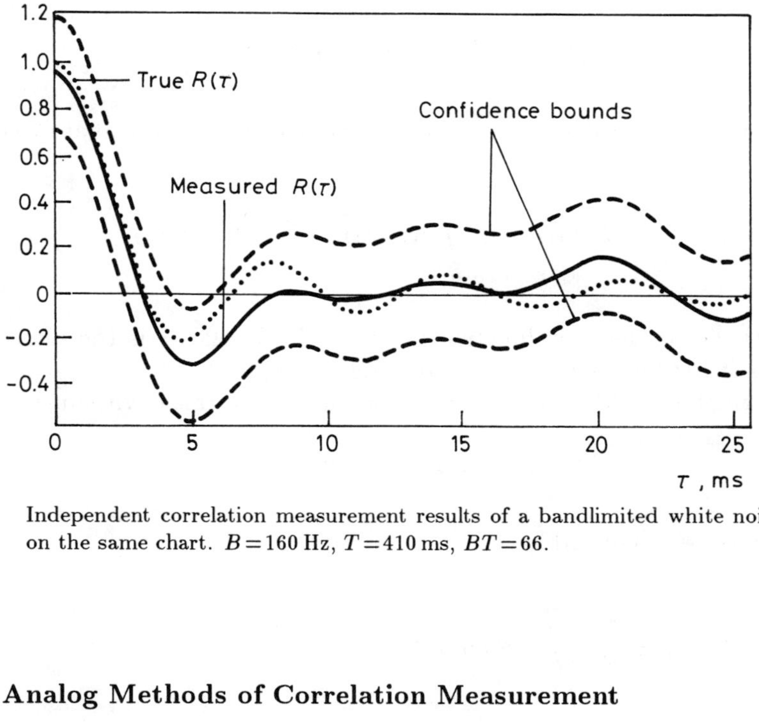

Fig. 6.37 Independent correlation measurement results of a bandlimited white noise, plotted on the same chart. $B=160$ Hz, $T=410$ ms, $BT=66$.

6.5.1 Analog Methods of Correlation Measurement

The basic block diagram of analog correlation measurement is shown in Fig. 6.38. The arrangement realizes the following estimate:

$$\widehat{R}_{12}(\tau) = \frac{1}{T} \int_0^T x_1(t-\tau)x_2(t)\,\mathrm{d}t\,, \tag{6.116}$$

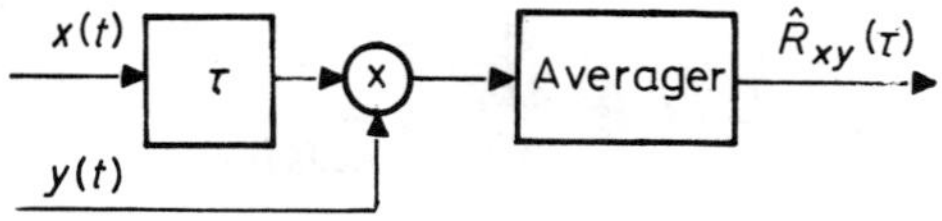

Fig. 6.38 Analog correlator

which is equivalent in statistical sense to (6.108), but needs a delay block instead of the 'predictor' (for $t+\tau$).

The usual solution for the scanning of the correlation function is to slowly vary the value of τ, while a lowpass filter as averager produces the estimate. A simple rule for determining the upper limit of the *scan rate* is as follows.

The expected value of the signal, arriving at the input of the averager (lowpass filter) is $R(\tau(t))$, where $\tau(t)$ is the time function of the changing delay, usually $\tau(t)=ct$. It is easy to see that if the bandwidth of $R(\tau)$ is B_e, the bandwidth of $R(ct)$ is cB_e. The bias will be small if the lowpass filter does not seriously modify $R(\tau(t))$, that is, its bandwidth is large enough:

$$\frac{1}{2\pi K} \gg cB_\mathrm{e} \qquad \text{or} \qquad c \ll \frac{1}{2\pi B_\mathrm{e} K}\,. \tag{6.117}$$

The bias of the scanned autocorrelation measurement is approximately equal to

$$\mathrm{b}\{\widehat{R}(\tau_0)\} = \frac{1}{K}\int_0^{t_1} R(\tau(t))e^{-\frac{t_1-t}{K}}\,\mathrm{d}t - R(\tau_0)\,, \tag{6.118}$$

where $\tau(t)=ct$ is the time function of the realized delay, c is the scan rate, and $\tau(t_1)=\tau_0$. If $R(\tau)$ is approximately constant in $\tau(t_1-4K)<\tau<\tau(t_1)=\tau_0$, the bias will be negligible. If $R(\tau)$ can be well approximated in this environment of τ_0 as

$$R(\tau) \approx R(\tau_0) + R'(\tau_0)\Delta\tau\,,$$

the bias may be calculated as

$$\mathrm{b}\{\widehat{R}(\tau_0)\} \approx \frac{1}{K}\int_{t_1-4K}^{t_1} R'(\tau_0)c(t_1-t)e^{-\frac{t_1-t}{K}}\,\mathrm{d}t \approx KR'(\tau_0)c\,. \tag{6.119}$$

The maximum allowable value of c can be determined from this expression, too.

Considering the variance, (6.114) and (6.115) may be applied with $T=2K$, but it must be assured that the scan rate is not too high, that is, the 'independent points' of $R(\tau)$ are all measured for a time at least a few times K. The density of these 'independent points' is in the order of the Nyquist rate, $2B_\mathrm{e}$, where B_e is the equivalent bandwidth of the covariance function. The condition may be

$$c = \frac{\mathrm{d}\tau(t)}{\mathrm{d}t} \ll \frac{\frac{1}{2B_\mathrm{e}}}{2K} = \frac{1}{4B_\mathrm{e}K}\,, \tag{6.120}$$

which is practically the same as (6.117).

Realization of the time delay

Time delay is difficult to realize because there is no simple electrical component or circuit which would have the transfer function

$$H(f) = e^{j2\pi f\tau} . \tag{6.121}$$

Some practical solutions are given below.

Artificial delay line

In a narrow frequency band (in the MHz range) it is possible to approximate (6.121) using discrete components, for example applying an LC network (this solution is used e. g. in TV sets). However, the bandwidth and the accuracy is limited. For low frequency applications other solutions are required.

Tape recorder

The 'classical' way of delaying is the use of a *tape recorder* (Fig. 6.39).

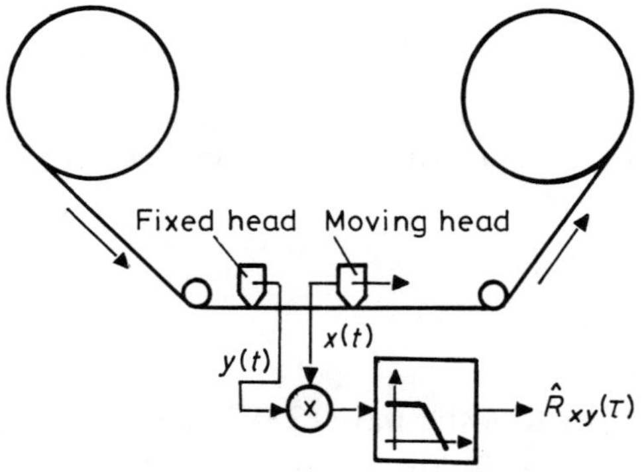

Fig. 6.39 Correlation measurement using a tape recorder

First the signals are recorded. When playing them back, the recorded signals are read by a fixed and a movable head. The delay depends on the actual distance of the heads and the recording speed (*not* the playback speed). With a recording speed of 18 cm/s and a head distance 1–10 cm, the realizable delay is 0.05–0.5 s.

Though the frequency range of tape recorders is somewhat higher than the audio band, this delay range limits the usable frequency range of correlation measurement below $1/(2 \cdot 0.05)\,\text{s} = 10\,\text{Hz}$.

Charge coupled devices

Between 1 kHz and 1 GHz charge coupled devices (CCD-s) can be used for signal delay. These are analog shift registers; the working principle is shown in Fig. 6.40. The 'potential traps', generated by the three-phase clock signal (illustrated for the time instants T_1, T_2, T_3), drift the charge (electrons) step by step to the right. The charge in each trap is proportional to a sample of the signal, thus the CCD works in discrete time domain. The delay depends on the clock frequency and the number of stages: $\tau = N/f_c$, thus the delay can be easily regulated. Assuming 32–512 stages, the realized delay may be 30 ns – 1 s.

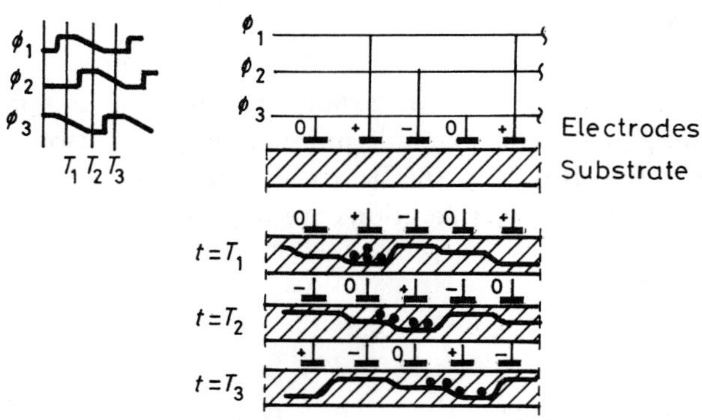

Fig. 6.40 Working principle of charge coupled devices

It is interesting to note that correlation estimation is more effective if all available samples are used at one time. This is implemented in special integrated CCD correlators.

Surface acoustic wave devices

At high frequencies surface acoustic wave (SAW) devices can be used for the realization of the delay (Butler, 1980; see Fig. 6.41).

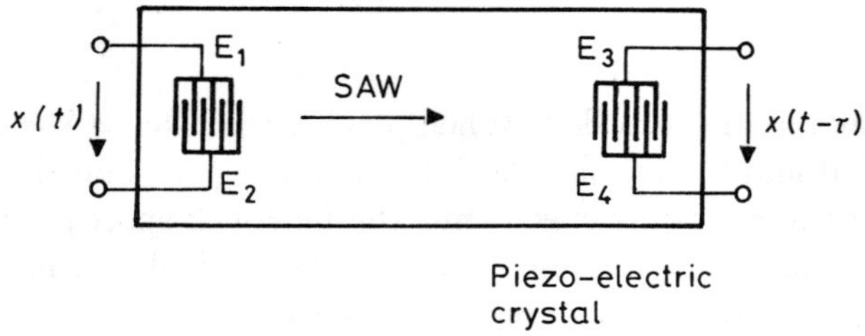

Fig. 6.41 Realization of delay with surface acoustic waves

The electrical signal is transformed into mechanical waves with the help of the electrodes E_1, E_2, placed onto the surface of a piezo-electric crystal. The mechanical waves travel along the surface at a speed of about 1000 m/s, i. e. the delay on 10 cm is 1 ms. The sensor electrodes are similar to the excitation electrodes.

If several electrodes are placed into the path of the travelling waves, different delay values can be realized, and the correlation function can be evaluated at different points. Moreover, with the help of small capacitances, integrated onto the surface of the crystal, a wave can be 'frozen', and cross correlation can be measured, for which the multiplication is realized using a nonlinearity.

SAW devices can also be directly used for correlation measurement, see Subsection 6.7.6 for the realization of convolution.

6.5.2 Digital Measurement of Correlation

Nowadays most correlators, used at low or medium frequencies, are *digital* ones. This means that the input signals are quantized both in time and frequency, and the digital samples are used in the estimation.

Two basic groups of correlators may be distinguished. The first one is the group of *on-line* correlators, which immediately process the arriving samples, and store correlation estimates only. The hybrid sign correlator and the polarity coincidence correlator are typical examples. The other group contains *off-line* correlators, which calculate the correlation estimate from a measured record, usually trying to utilize all the information that can be extracted. Indirect correlators are the most typical off-line ones.

General digital correlator

The most straightforward solution is shown in Fig. 6.42. This correlator may be both an on-line one (in this case the delay is realized by temporary storage of earlier samples), or an off-line one.

The correlation function is usually measured at the points $\tau = k\Delta t$, where $k = 0, 1, 2, \ldots$ or $k = 0, \pm 1, \pm 2, \ldots$, but with appropriate timing, $\widehat{R}(\tau)$ can be measured for any value of τ.

The disadvantage of this solution is the limited speed due to the several-bit multiplication and averaging, and also the expensive A/D converter and multiplier circuits. As a consequence, this arrangement is rarely used in correlators.

The arrangement shown in Fig. 6.42 can also be used for measurement of $\widehat{R}(\tau)$ with a significantly higher density than the conversion rate of the A/D converters, using the *vernier* principle (see Fig. 6.43). The two relatively slow A/D converters are working with a slightly different conversion rate. The resolution of the correlation measurement, Δt, can be much smaller than the conversion time of the A/D converters, T_{AD}.

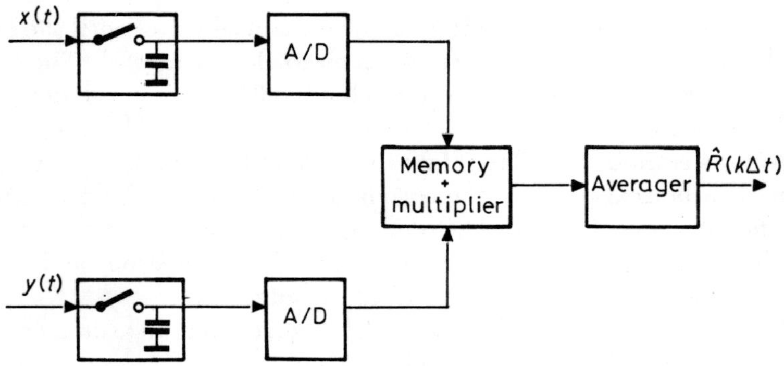

Fig. 6.42 Digital correlation measurement

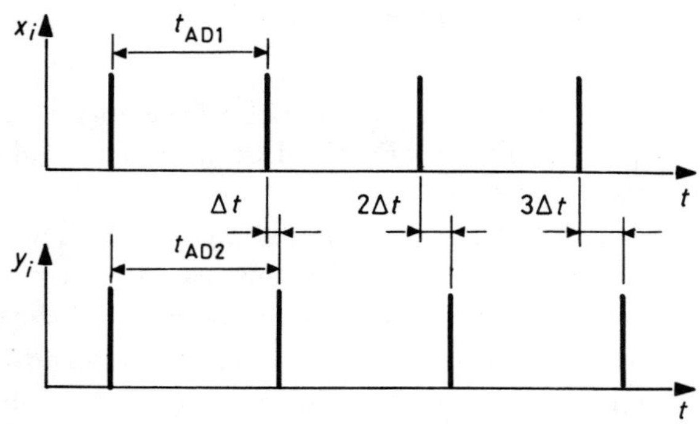

Fig. 6.43 The vernier principle

It is obvious that the bias is zero. The variance of the averaged estimate depends on the sampling strategy. If the averaged products are statistically independent, the variance will be given by (6.109–11), divided by the number of averaged terms N. If the products are statistically not independent, (6.113–5) can be used.

Hybrid sign correlator

Digital multiplication can be avoided if at least one of the signals is a binary one. This is done e. g. in the so-called hybrid sign (or ring or relay) correlators (Fig. 6.44).

There are two input channels: a high-speed, one-bit channel and a low-speed one with an A/D converter. The high-speed channel contains a stochastic-ergodic converter (see Subsection 6.1.3) at the input. The binary signal is sampled and transferred by a shift register. While a sample travels through the shift register, the A/D converter in the other channel converts just one sample.

The output of the A/D converter is stored into a register, and multiplied one by one with the bits coming out from the shift register. It is obvious that by this

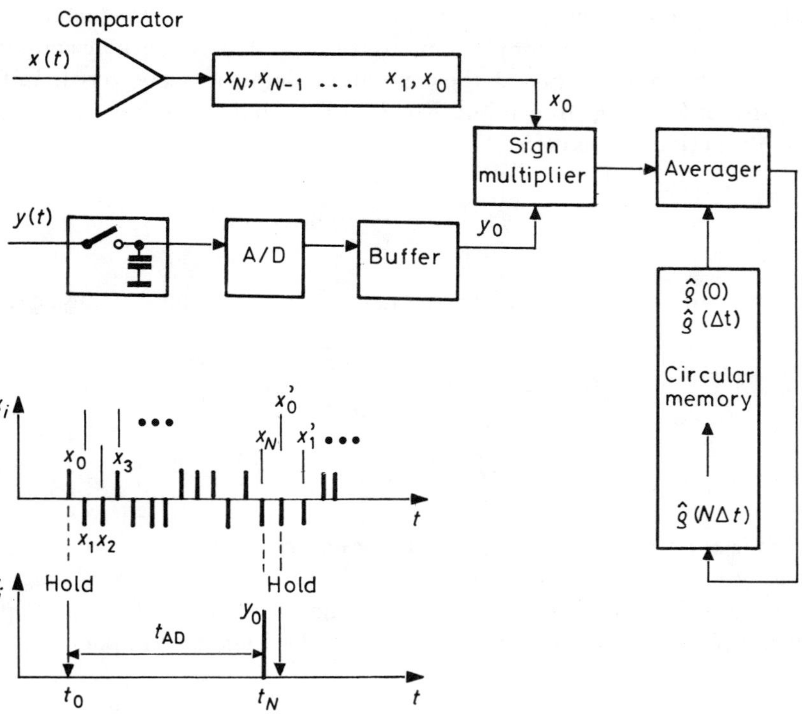

Fig. 6.44 Hybrid sign (relay) correlator

multiplication — which can be done very quickly with the value ± 1 of the binary channel — estimates of the correlation function are obtained at the points $k\Delta t$. These products are averaged to the appropriate correlation estimates, circulating in the memory. When a cycle is finished, the A/D converter is ready with the next sample, and the procedure restarts.

The advantage of the hybrid sign correlator is that it combines the speed of one-bit manipulations with the accuracy of an A/D converter.

It can be shown that the resulting estimate is unbiased:

$$\mathcal{E}\{\widehat{R}_{xy}(\tau)\} = \mathcal{E}\{\frac{A}{2}\widehat{\varrho}(\tau)\} = R_{xy}(\tau)\,, \tag{6.122}$$

where $\widehat{\varrho}(i\Delta t)$, produced in the circular memory, is the average of the output results of the sign multiplier.

The variance of one single product can also be easily calculated, since by squaring the binary signal gives just 1:

$$\mathrm{var}\Big\{x_i y_0 \frac{A}{2}\Big\} = \mathcal{E}\Big\{\big(x_i y_0 \frac{A}{2}\big)^2\Big\} - \mathcal{E}^2\Big\{x_i y_0 \frac{A}{2}\Big\} = \frac{A^2}{4}\psi_y^2 - R_{xy}^2(\tau)\,. \tag{6.123}$$

The variance of the averaged estimate depends again on the interdependence of the averaged products (see above).

If the two signals have jointly normal distribution, even the stochastic-ergodic converter is not necessary: a simple comparator may be used, with ac coupling (that is, the mean value of the signal at the comparator input must be zero). In this case it can be shown (see e. g. Hagen and Farley, 1973) that the shape of the measured correlation function is correct:

$$\mathcal{E}\{\widehat{\varrho}(\tau)\} = \sqrt{\frac{2}{\pi}}\frac{1}{\sigma_x}C_{xy}(\tau)\,. \tag{6.124}$$

The variance can be determined for zero mean Gaussian signals, using again $x_i^2 \equiv 1$:

$$\mathrm{var}\{\widehat{C}_{xy}(\tau)\} = \mathrm{var}\Big\{\sqrt{\frac{\pi}{2}}\sigma_x\widehat{\varrho}_{xy}(\tau)\Big\}$$

$$= \mathcal{E}\Big\{\Big(\sqrt{\frac{\pi}{2}}\sigma_x\widehat{\varrho}_{xy}(\tau)\Big)^2\Big\} - \mathcal{E}^2\Big\{\sqrt{\frac{\pi}{2}}\sigma_x\widehat{\varrho}_{xy}(\tau)\Big\} = \Big(\frac{\pi}{2} - r^2(\tau)\Big)\sigma_x^2\sigma_y^2\,, \tag{6.125}$$

where $r(\tau)$ is the correlation coefficient between $x(t)$ and $y(t+\tau)$.

It may be surprising that the variance is in the same order of magnitude as the one of the high-resolution case (6.110), while the absence of the multiplier and the high sampling rate of the shift register provides high-speed operation.

Polarity coincidence correlator

Another interesting arrangement is the one where both channels contain a stochastic-ergodic converter each (see Fig. 6.45). The multiplication is realized by a coincidence gate (hence the name of the correlator). The averager is an up/down counter, similarly to the mean value measurement using the SEC principle (Section 6.3). The delay is controlled by the clock frequency of the shift register.

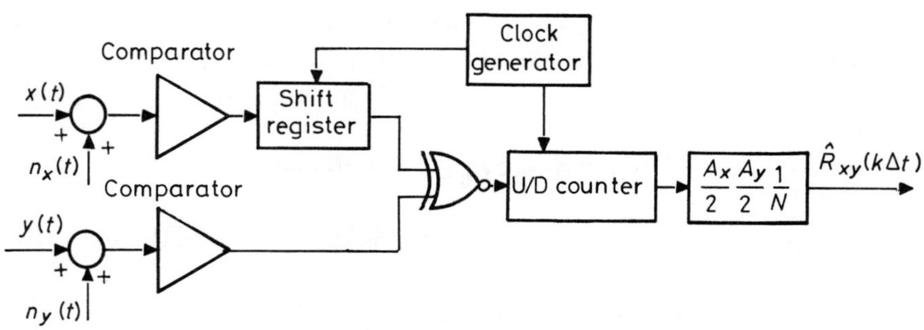

Fig. 6.45 Polarity coincidence correlator

Since the correlation function is to be measured for different values of τ, the correlator either measures $\widehat{R}(\tau)$ one by one for each desired value of τ, or is extended by a circular memory, similar to that in the hybrid sign correlator (Fig. 6.44).

$\widehat{R}(\tau)$ is unbiased again; the variance is for $N=1$

$$\mathrm{var}\{\widehat{R}_{xy}(\tau)\} = \frac{A_1^2 A_2^2}{4} - R_{xy}^2(\tau)\,. \tag{6.126}$$

It is obvious that for sufficiently high values of N the estimation may have the same efficiency as the one based on high-resolution ADC-s.

The SEC-s may also be omitted from the polarity coincidence correlator when measuring zero mean Gaussian processes (Fig. 6.46).

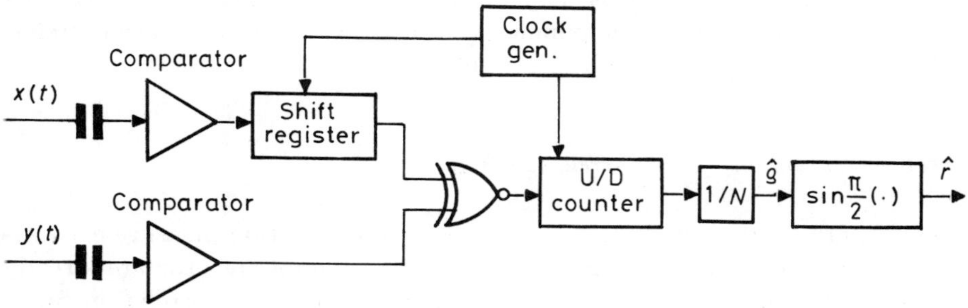

Fig. 6.46 Polarity coincidence correlator for zero mean Gaussian signals

This arrangement measures the normalized covariance function $r(\tau)$. However, in this case the averaged estimate will be biased:

$$\mathcal{E}\{\mathrm{sign}(x)\mathrm{sign}(y)\} = \frac{2}{\pi}\arcsin(r)\,. \tag{6.127}$$

This is the reason why the nonlinear block is applied at the output: if the variance of $\widehat{\varrho}(\tau)$ is small, the output is approximately unbiased.

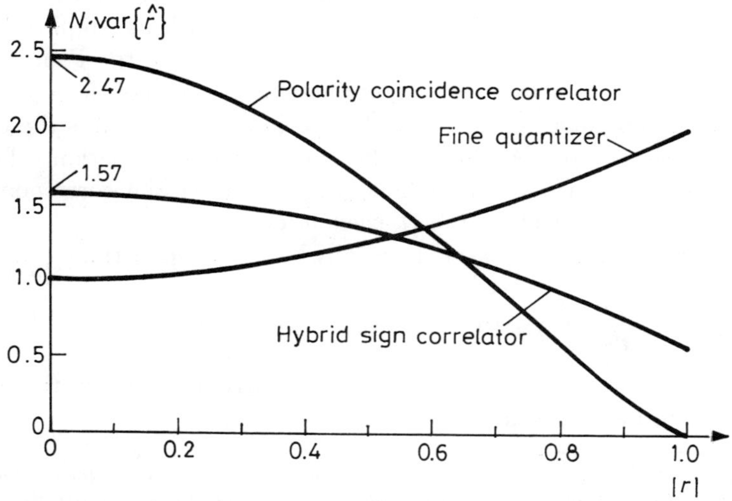

Fig. 6.47 Comparison of the variance of different correlators for zero mean Gaussian signals

From the above unbiasedness, linearizing the nonlinearity and using the properties of binary signals it can be shown (Kollár, 1988) that

$$\mathrm{var}\{\widehat{r}\} \approx \frac{1}{N}\left(\left(\frac{\pi}{2}\right)^2 - \arcsin^2(r)\right)(1-r^2)\,. \tag{6.128}$$

The normalized variances of the above three correlators for zero mean Gaussian signals are shown in Fig. 6.47.

It is interesting to note that the variances are comparable; moreover, the polarity coincidence correlator is the best for $|r| > 0.65$. This fact and speed and low price are the reasons for the popularity of the polarity coincidence correlators, especially in high frequency applications, like radio astronomy.

Indirect measurement of the correlation function

In general-purpose computers the 'natural' way of correlation measurement would be the one shown in Fig. 6.42. However, if the two sample records are already at hand, there is a much more effective way of calculating the correlation function estimate, via the frequency domain. The effectiveness is a consequence of two facts: firstly, correlation is equivalent to a multiplication in the frequency domain, secondly, the DFT can be very effectively calculated via FFT[2].

It may sound paradoxical that while in discrete correlation estimation each sample must be multiplied by each sample, that is, N^2 sample pairs must be multiplied, there is a way to significantly reduce the number of multiplications. The main reason is that not all the products are necessary, but certain *sums of products* only.

Indirect correlation measurement can also be interpreted in the following way. The discrete Fourier transform produces a Fourier series expansion of the sample record. The autocorrelation function of each function in the Fourier series is explicitly known: it is a cosine of the same frequency, with an amplitude equal to the power of the relevant sine wave. Since the sine waves in the Fourier series are orthogonal, the autocorrelation function of the series is a sum of the autocorrelation functions of the individual components. The periodogram contains the coefficients of the Fourier series representation of the autocorrelation function. The inverse discrete Fourier transform of the periodogram evaluates the corresponding time function, from the coefficients of the Fourier series.

Indirect correlation estimation is usually based on the periodogram (see Subsection 6.7.1):

$$\widehat{R}(i\Delta t) = \mathrm{FFT}^{-1}\left\{\overline{\{X(k\Delta f)\}}\{Y(k\Delta f)\}\right\}. \tag{6.129}$$

To decrease the variance (see later), either the periodograms or the correlation estimates need to be averaged.

[2] It must be mentioned that the DFT (via FFT) is not the only effective means of calculation of the correlation function. Polynomial and number theoretic transforms can also be used for fast correlation calculation (Nussbaumer, 1982).

Using the expression of the DFT (6.71–2), the expected value of (6.129) can be calculated:

$$\mathcal{E}\{\widehat{R}(i\Delta t)\} = \left(1-\frac{i}{N}\right)R(i\Delta t) + \frac{i}{N}R((i-N)\Delta t), \quad i = 0,1,2,\ldots,N-1\,. \tag{6.130}$$

The first term is the correlation function itself, multiplied by the Bartlett window (see Subsection 6.7.1), as expected. What is confusing however, is the second term, which is the (-T,0) part of the correlation function, copied onto the 'regular' function (see Fig. 6.48). If $C(i\Delta t)$ is not equal to 0 for $i \geq N/2$, this causes an annoying bias.

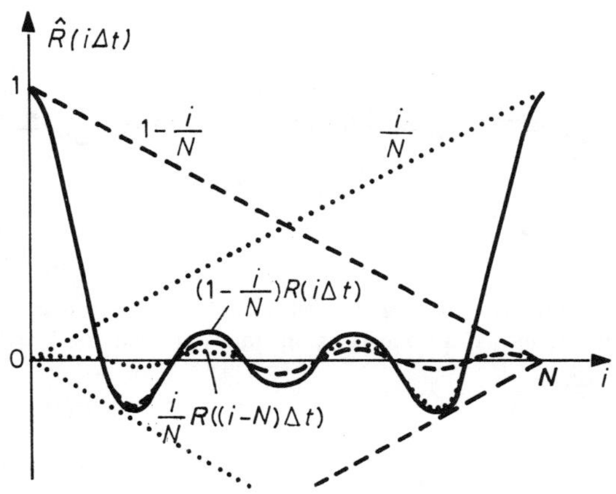

Fig. 6.48 Circular correlation (with $R(0)$ normalized to 1)

Where does this overlapping come from? To answer this question, the steps of data processing have to be carefully considered.

The calculation starts from a sample time record of length T. This record is Fourier transformed via FFT, the result is a discrete series with distances $1/T$. These series can still be transformed back to time domain, resulting in the original records, since the DFT is invertible.

The next step is multiplication, and this is the source of the problem. The time domain equivalent of multiplication is convolution (or correlation), which broadens the signal from width T to $2T$. The time domain equivalent of the sampled spectrum is repeated at distances T. To avoid overlapping, the spectrum of the signal of width $2T$ should have been sampled at points $1/2T$ from each other. This is not the case, therefore time domain overlapping occurs.

The cause is already known, now the remedy must be found. A possibility is to increase the density of frequency domain samples. This can be done by adding zeros to the time records, extending the (virtual) length to $2T$ (zero padding). The resulting correlation estimate is illustrated in Fig. 6.49. Another possibility is to keep the record length, but decrease the width of the useful signal by truncating it to length $T/2$ (see Fig. 6.50).

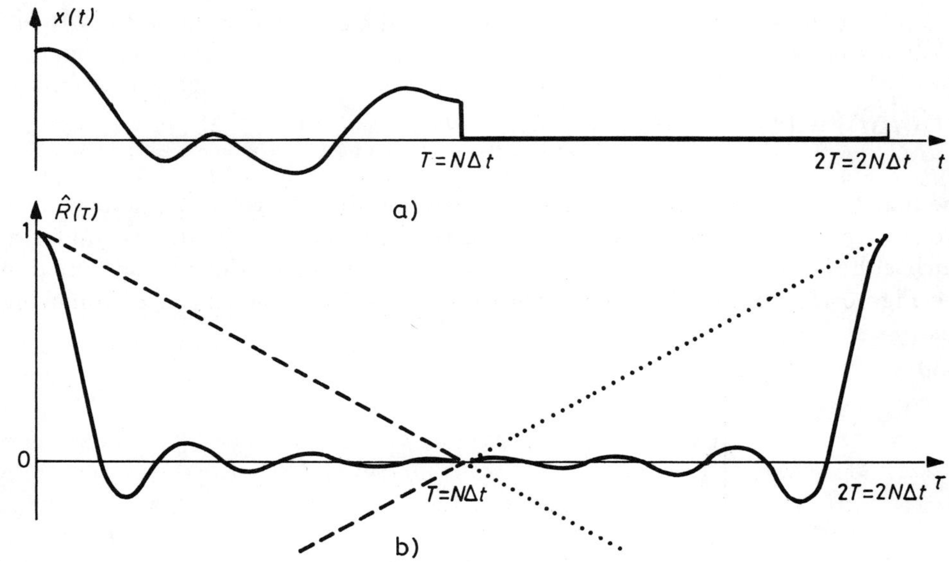

Fig. 6.49 Elimination of the overlapping by adding zeros to the time record *a)* the time record, padded with zeros *b)* autocorrelation function with the Bartlett window

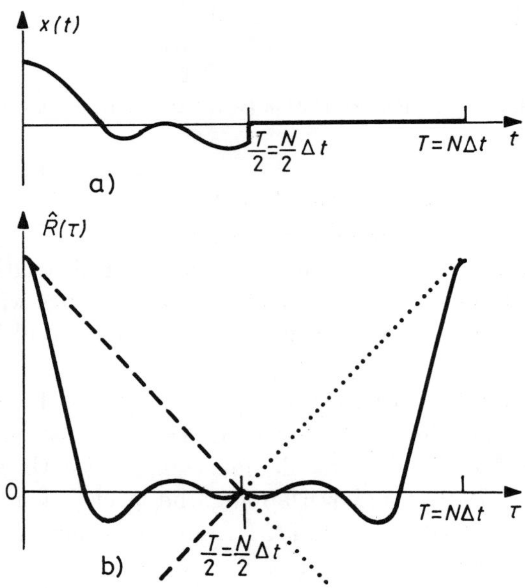

Fig. 6.50 Elimination of the overlapping by truncation of the time record *a)* truncated time record *b)* autocorrelation function with the Bartlett window

The estimate (6.129) is called *circular correlation*; because the algorithm acts as if when shifting one record vs. the other, the leading samples would reappear circularly at the other end of the series. The phenomenon is analogous with circular convolution (Subsection 6.1.6).

Once overlapping is eliminated, the only bias is caused by the Bartlett window. This can be eliminated dividing the estimate by the appropriate samples of the Bartlett window. However, care must be taken, because this division gives rise to large variances, especially near to the edges, where division is done by small values. Therefore, the estimate is usually corrected for $-N/2 \leq i \leq N/2$ only.

This large variance can also be explained in another way: since the record length is fixed, only a small number of multiplied sample pairs is available for large values of $\tau = i\Delta t$.

The autocorrelation function can also be evaluated in the indirect way for lags at $|\tau| > N\Delta t$, if in (6.129) one of the input records is delayed by an appropriate value. Moreover, such estimates for different delay values can be combined in order to eliminate the effect of the Bartlett window (short-time unbiased spectrum estimation, STUSE: Rabiner and Allen, 1979).

The variance can be calculated similarly, as for the analog case, since the signal has to be sampled above the Nyquist rate in order to obtain a usable estimate of the correlation function. If the covariance function disappears at the half-width of the Bartlett window, the variance at $\tau = T/2$ is about the same as the variance at $\tau=0$ (the number of the averaged products is $N/2$, thus (6.115) has to be multiplied by 2), but division by 0.5 increases this value by a factor of 4.

6.6 MEASUREMENT OF PROBABILITY FUNCTIONS

Probability density and distribution functions are measured on the basis of the following expressions:

$$F(x) = \mathrm{P}(\xi < x), \qquad f(x) \approx \frac{\mathrm{P}\left(x - \frac{\Delta x}{2} \leq \xi < x + \frac{\Delta x}{2}\right)}{\Delta x}, \tag{6.131}$$

calculating the relative frequency of the considered event. The estimate of the distribution function is usually unbiased, while the bias of the probability density functions (pdf-s) is equal to the difference between the integral mean and the central value of the pdf in the given interval. If the pdf is smooth, this bias can be calculated from the first terms of the Taylor series expansion of the pdf:

$$\mathrm{b}\{\widehat{f}(x_0)\} = \frac{1}{\Delta x} \int\limits_{x_0 - \frac{\Delta x}{2}}^{x_0 + \frac{\Delta x}{2}} f(x_0) + f'(x_0)(x - x_0) + \frac{f''(x_0)}{2}(x - x_0)^2 + \ldots \, \mathrm{d}x - f(x_0)$$

$$\approx \frac{f''(x_0)}{24} \Delta x^2 \,, \tag{6.132}$$

where $''$ denotes second-order derivation.

For the use of this approximation $f(x)$ must be smooth, otherwise the higher-order terms of the series expansion must also be considered, or this method cannot be applied.

The variance can be calculated for the case of statistically independent indicator variables, since in this case the sum has binomial distribution. In the case of probability distribution functions the variance is

$$\operatorname{var}\{\widehat{F}(x)\} = \frac{F(x)(1-F(x))}{N}. \tag{6.133}$$

For pdf-s the expression is somewhat more complex:

$$\operatorname{var}\{\widehat{f}(x)\} \approx \frac{1}{\Delta x^2}\frac{f(x)\Delta x(1-f(x)\Delta x)}{N} \approx \frac{1}{N\Delta x}f(x). \tag{6.134}$$

For the analog case no simple formula can be given since the indicator variable is obtained from the samples by a strongly nonlinear operation, hence the bandwidth of the indicator function cannot be explicitly given. For bandlimited white noise Bendat and Piersol (1971) suggest, on the basis of simulations,

$$\operatorname{var}\{\widehat{f}(x)\} \approx \frac{0.09}{2BT\Delta x}f(x), \tag{6.135}$$

a value much smaller than the one that can be obtained from (6.134) by using $N \approx 2BT$, valid for the input signal. This verifies the increased bandwidth of the indicator signal.

For the probability distribution function no similar result was published. Since the operator which produces the indicator function is very nonlinear in this case, too, it may be expected that the bandwidth increases. A conservative upper limit of the variance is

$$\operatorname{var}\{\widehat{F}(x)\} < \frac{F(x)(1-F(x))}{2BT}. \tag{6.136}$$

Comparing (6.132) to (6.134) and (6.135) it is interesting to note that contradictory requirements provide small bias and variance. The measurement design recipe may be the following: Δx has to be chosen small enough in order to have small bias, and the variance can be decreased to the desired level by choosing N sufficiently high.

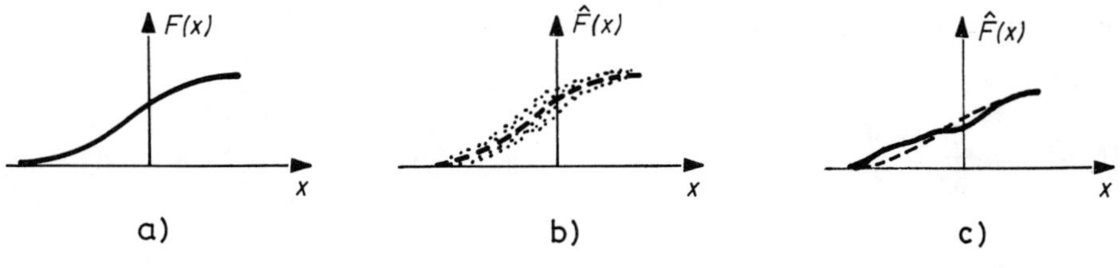

Fig. 6.51 Measured probability distribution functions *a)* the true distribution function *b)* uncorrelated measurement point *c)* strongly correlated measurement points

Typical results of the measurement of the distribution function are shown in Fig. 6.51. Fig. 6.51b illustrates the case when the estimates at different values of x are statistically independent. It is clearly seen that the variance is maximal around $F(x) \approx 0.5$. Fig. 6.51c shows what happens if the neighbouring estimates are highly correlated (e. g. if the distribution estimate is obtained by integrating the pdf estimate). The variance of the probability distribution estimate is not apparent from the plot: the smooth measured function gives the illusion of a good estimate.

In the following paragraphs some typical arrangements will be shown to illustrate the most common solutions in the measurement equipment.

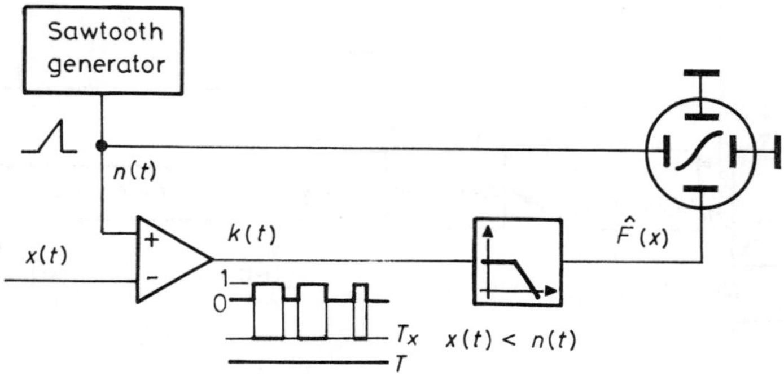

Fig. 6.52 Analog measurement of the probability distribution function

Fig. 6.52 presents the analog way of probability distribution measurement. The 'sawtooth generator' slowly sweeps through the total amplitude range. The comparator compares the actual value of the signal $x(t)$ with the slowly changing reference value. The fill-in factor of the output signal of the comparator is equal to the probability distribution function: this can be estimated by an averager (lowpass filter). The 'sawtooth' and the output of the lowpass filter are led to a CRT or an X–Y recorder. The scan rate can be determined similar to what was done for the measurement of the autocorrelation function (Subsection 6.5.1).

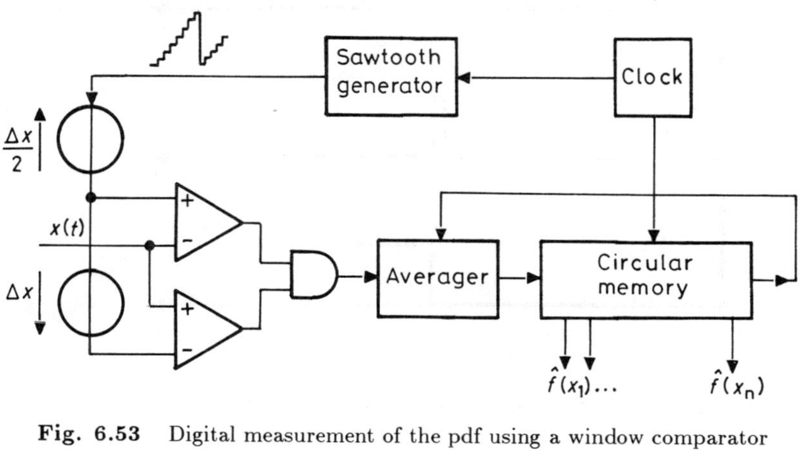

Fig. 6.53 Digital measurement of the pdf using a window comparator

A hybrid pdf meter is shown in Fig. 6.53. The so-called *window comparator* generates the indicator function of the interval $x \pm \Delta x/2$. Let us notice that the integral of the estimated pdf can be different from 1 (if the variance is not small enough). If this is undesirable, the result may be scaled at the end of the measurement.

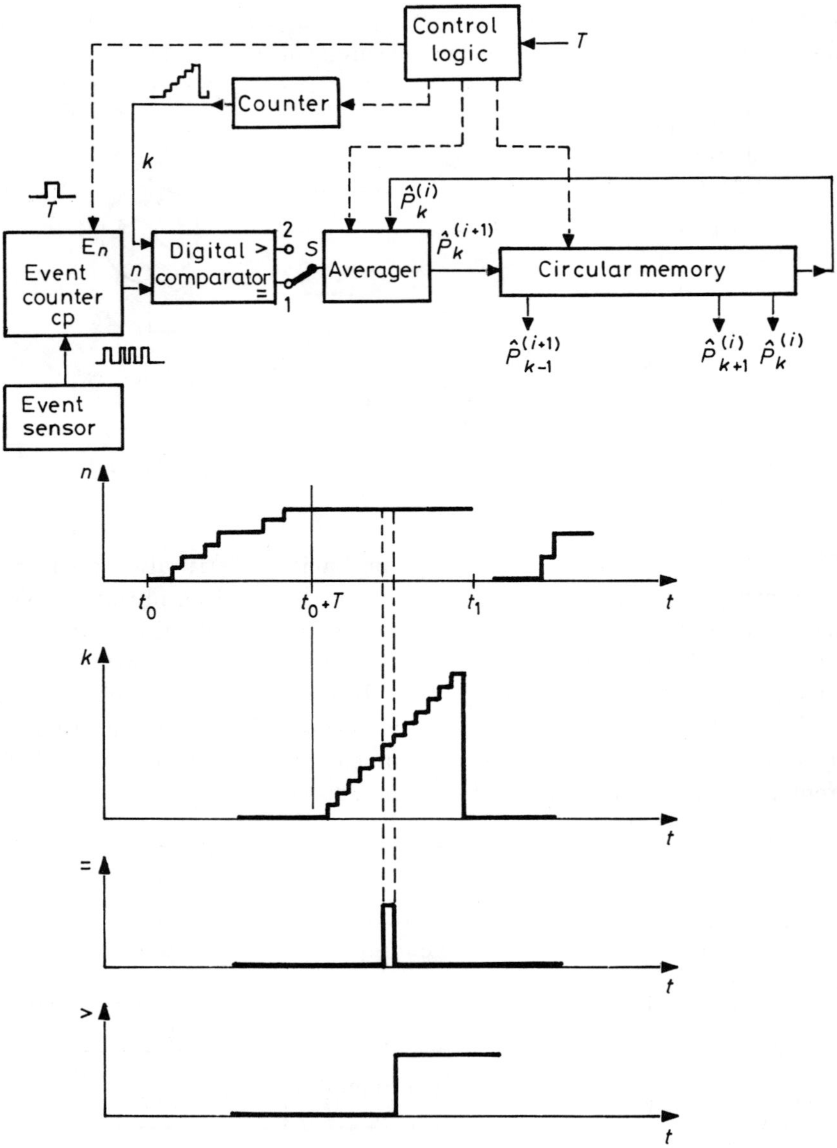

Fig. 6.54 Measurement of event density

A maybe even more simple solution is digital pdf measurement, when the samples are quantized, and simply a histogram is calculated, usually with a more coarse resolution than that of the A/D converter.

Probability functions are measured not only when investigating the amplitude of a random variable, but also when the distribution of events in time is considered. Examples of such events: particle detection, telephone calls, passing vehicles, arrival of customers etc. Two kinds of statistics used to be measured: the number of events in a given time (this is a discrete distribution, often a Poisson one), or the time between successive events (a continuous distribution, usually an exponential one).

A measurement device of the first kind is shown in Fig. 6.54. The measurement consists of two phases. First the event counter counts the events in a given time T. Let the result be n. In the second phase $\widehat{P}_n$ is averaged with 1, the other probability estimates with 0, by circularly shifting the contents of the memory.

The distribution function can be measured in a similar way, by setting S to state 2.

6.7 MEASUREMENT OF SPECTRA

Spectrum analysis means in a wide sense the measurement of a quantity as a function of the frequency (Robinson, 1982). Such functions include amplitude, power, energy density and power density spectra, defined for the different classes of signals (see Subsection 1.3.1). In this section the measurement of each of these functions will be mentioned; however, the attention will be focussed at the most important instruments: spectrum analyzers, which usually measure the power spectral density function (power density spectrum).

As it was already discussed in Subsections 1.2.4 and 6.1.8, spectrum analysis can be based on nonparametric or parametric approaches. Since in commercially available spectrum analyzers usually nonparametric procedures are implemented, in this section we are going to discuss first of all these ones. The determination of the parameters of sine waves, based on nonparametric spectrum measurements, will also be treated. For the parametric approach, we refer to the literature (Haykin, 1983; Candy, 1986; 1988).

Table 6.2 shows the values of the different spectra for the most important signal types. The + signs mark regular functions, the ∞ signs infinite (and not usable) functions, and the δ sign indicates that the spectrum theoretically contains Dirac delta functions. In practical analysis such a Dirac delta means only that measurement gives sharp peaks, since the bandwidth can never reach zero, thus the power (energy) in the band is 'smeared' around the given frequency. This is the reason why spectrum analyzers can be applied in the measurement of mixed signals (periodic component + noise). Another consequence is that because of the sharp peaks the measured spectrum is very sensitive to periodic components, and this sensitivity can be improved by decreasing the bandwidth around a frequency point.

Table 6.2 Values of spectra for different signals. $E(f)$ is the energy spectral density function, $S(f)$ is the power spectral density function, and $P(f)$ is the power spectrum, see Subsection 1.3.1.

		Spectrum		
		$E(f)$	$S(f)$	$P(f)$
Signal	Transient	+	0	0
	Stochastic	∞	(+)	0
	Periodic	∞	(δ)	+

Spectrum analysis

6.7.1 Measurement Principles of Spectra

The obvious way of measuring the spectra is to evaluate the definitions (see Subsection 1.3.1). Fourier transformers (see Subsection 6.7.6) and correlators (Section 6.5) can be applied for this purpose.

However, when considering *functions of frequency*, it seems reasonable to base measurements on frequency selective devices, i. e. bandpass filters. This possibility will be discussed in the following paragraphs.

The use of bandpass filters

The different spectra were defined in Chapter 1 in various ways, but certainly not on the basis of bandpass filters. Thus, it is reasonable to show that bandpass filter based arrangements measure indeed the functions defined in Chapter 1. An ideal bandpass filter (see Fig. 6.55) will be assumed: real filters more or less well approximate this model.

a) Measurement of the power spectrum
The functioning of the arrangement in Fig. 6.56a is clear: if the filter is tuned onto a spectral line, the corresponding power will be passed and measured.

b) Measurement of the energy spectral density function (energy density spectrum)
In the case of the energy density spectrum it is not obvious that the arrangement in Fig. 6.56b measures the same quantity as defined in Subsection 1.3.1 as $E(f) = |X(f)|^2$.

The output signal of the filter is

$$y(t) = h(t) \star x(t) \,. \tag{6.137}$$

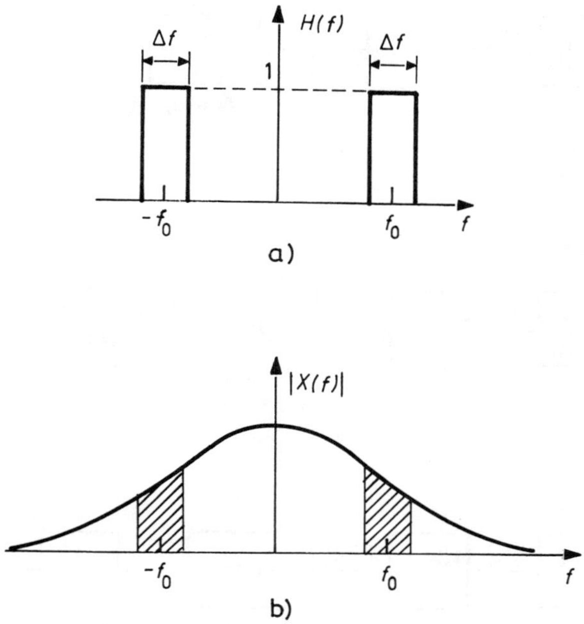

Fig. 6.55 Ideal bandpass filter. *a)* Transfer function *b)* The effect of the filter on the spectrum

For the Fourier transforms it is true that

$$Y(f) = H(f)X(f) \quad \text{and} \quad |Y(f)|^2 = |H(f)|^2|X(f)|^2 . \tag{6.138}$$

The energy can be expressed using Parseval's theorem as

$$W = \int_{-\infty}^{\infty} y^2(t)\,\mathrm{d}t = \int_{-\infty}^{\infty} |Y(f)|^2\,\mathrm{d}f = \int_{-\infty}^{\infty} |H(f)|^2|X(f)|^2\,\mathrm{d}f$$

$$= \int_{-f_0-\frac{\Delta f}{2}}^{-f_0+\frac{\Delta f}{2}} |X(f)|^2\,\mathrm{d}f + \int_{f_0-\frac{\Delta f}{2}}^{f_0+\frac{\Delta f}{2}} |X(f)|^2\,\mathrm{d}f \approx 2\Delta f|X(f_0)|^2 , \tag{6.139}$$

which is just the desired result: dividing by $2\Delta f$, $|X(f_0)|^2 = E(f_0)$ is obtained.

c) Measurement of the power spectral density function (power density spectrum)
The power spectral density function was defined via the Wiener–Khintchine equation:

$$S_{xx}(f) = \mathcal{F}\{R_{xx}(\tau)\} . \tag{6.140}$$

Using the equality $S_{yy}(f) = |H(f)|^2 S_{xx}(f)$ for the arrangement in Fig. 6.56c,

$$P = \psi_x^2 = R_{yy}(0) = \int_{-\infty}^{\infty} S_{yy}(f)\,\mathrm{d}f = \int_{-\infty}^{\infty} |H(f)|^2 S_{xx}(f)\,\mathrm{d}f$$

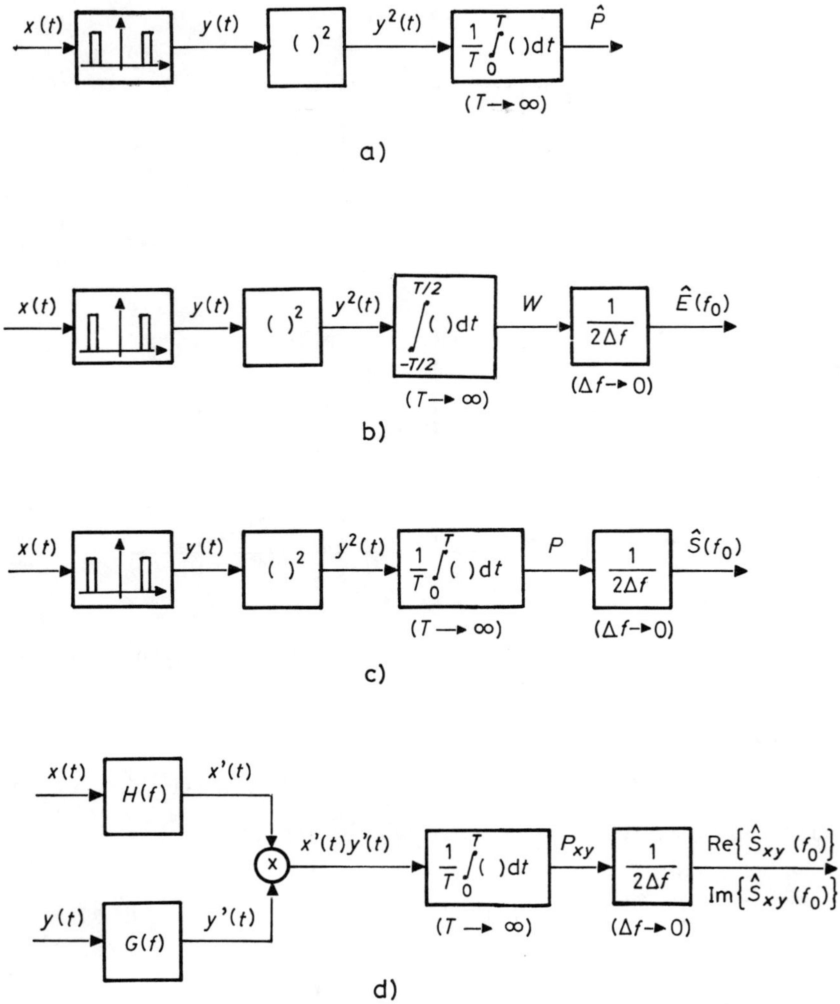

Fig. 6.56 Spectrum measurements using bandpass filter *a)* power spectrum *b)* energy spectral density function *c)* power spectral density function *d)* cross power spectral density function

$$= \int_{-f_0-\frac{\Delta f}{2}}^{-f_0+\frac{\Delta f}{2}} S_{xx}(f)\,\mathrm{d}f + \int_{f_0-\frac{\Delta f}{2}}^{f_0+\frac{\Delta f}{2}} S_{xx}(f)\,\mathrm{d}f \approx 2\Delta f S_{xx}(f_0)\,. \tag{6.141}$$

Indeed, after the division by $2\Delta f$ the estimate of the power spectral density function is obtained.

d) Measurement of the cross power spectral density function

The bandpass filter based measurement of the cross power spectral density function (CPSDF) is very similar to the measurement of the auto spectrum (Fig. 6.56d). The principal difference is a consequence of the fact that the CPSDF is complex, thus the real and imaginary parts have to be measured separately. For the real part, $H(f)$ and $G(f)$ have the same form as shown in Fig. 6.55a. The imaginary part is measured if $G(f)$ is modified: it has a $-90°$ frequency shift for $f>0$, and a $+90°$ frequency shift for $f<0$ (this can be approximately realized in a narrow frequency band).

Errors of bandpass filter based measurements

From the above considerations it is clear that using a bandpass filter, the expected value of the estimate is an expression of this type:

$$\mathcal{E}\{\widehat{S}(f_0)\} = \frac{1}{2\Delta f}\int_{-\infty}^{\infty} |H(f)|^2 S(f)\,\mathrm{d}f\,, \tag{6.142}$$

where $H(f)$ is the transfer function of the (nonideal) bandpass filter. The true bandpass transfer function will be approximated for the error calculations by the ideal filter (Fig. 6.57).

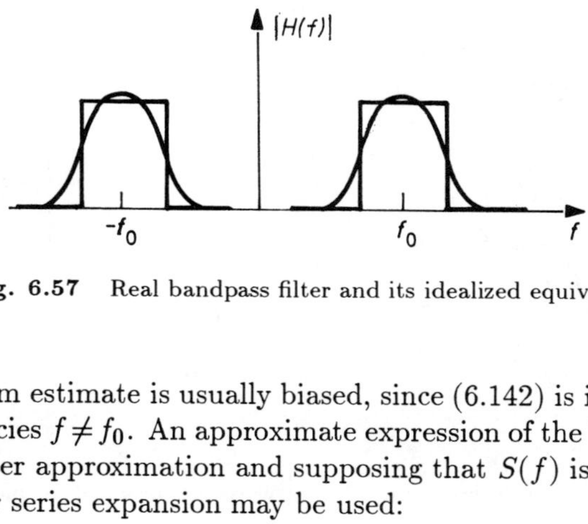

Fig. 6.57 Real bandpass filter and its idealized equivalent

The spectrum estimate is usually biased, since (6.142) is influenced by values of $S(f)$ at frequencies $f \neq f_0$. An approximate expression of the bias can be derived, using the ideal filter approximation and supposing that $S(f)$ is smooth around f_0, that is, the Taylor series expansion may be used:

$$\mathcal{E}\{\widehat{S}(f_0)\} = \frac{1}{\Delta f}\int_{f_0-\frac{\Delta f}{2}}^{f_0+\frac{\Delta f}{2}} \left(S(f_0) + S'(f_0)(f-f_0) + \frac{S''(f_0)}{2}(f-f_0)^2 + \dots\right)\mathrm{d}f$$

$$\approx S(f_0) + \frac{S''(f_0)}{24}\Delta f^2\,. \tag{6.143}$$

The bias is given by the second term; if this is negligible, the higher terms of the Taylor series must be considered.

For signals with periodic components (or for random processes with periodic realizations) the Taylor series expansion cannot be used, even the usual definition of the bias cannot be applied, since a Dirac delta function cannot be well approximated. However, periodic components can be perfectly characterized by the triplet $(U_{\mathrm{p}i}, f_i, \varphi_i)$, and generally it is enough to have good estimates of the first two parameters.

If a sine wave of frequency f_1 is applied to the spectrum analyzer, the mean value of the estimate is

$$\mathcal{E}\{\widehat{S}(f_0)\} = \frac{1}{\Delta f}\int_0^\infty |H_{f_0}(f)|^2 \frac{U_{\mathrm{p}}^2}{4}\delta(f - f_1)\,\mathrm{d}f = \frac{1}{\Delta f}|H_{f_0}(f_1)|^2 \frac{U_{\mathrm{p}}^2}{4}\,, \tag{6.144}$$

where $H_{f_0}(f)$ is the transfer function of the bandpass filter with centre frequency f_0.

If the centre frequency of the bandpass filter is swept along the frequency axis, the shape of the obtained spectrum around f_1 will be

$$|H_{f_1}(2f_1 - f)|^2$$

(see Fig. 6.58), and from the maximum value and the location of the above function, the power $U_{\mathrm{p}}^2/2$ (and from this also the amplitude) of the periodic component can be determined, usually without any bias.

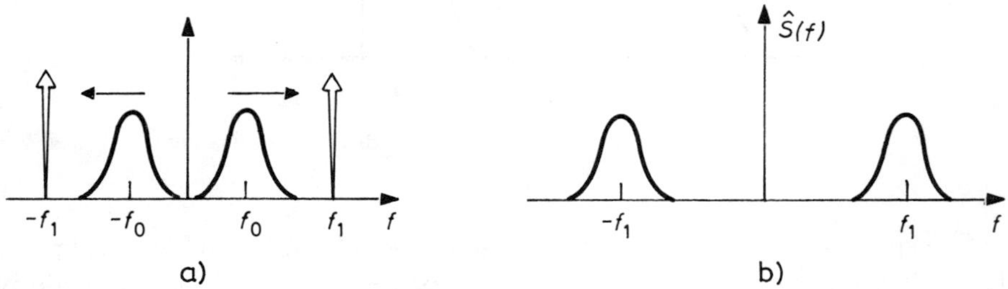

Fig. 6.58 Measurement of the spectrum of a sine wave *a)* tuning of the bandpass filter *b)* measured spectrum

The *variance* of the estimates can be easily determined noticing that after the bandpass filter the mean square value of $y(t)$ is estimated, thus the results of Section 6.4 can be applied.

For *transient* signals as defined in Subsection 1.3.1 the variance is zero, since they are not random.

For stochastic signals with smooth spectra, a similar method can be followed as in Section 6.4, the only difference is that $y(t)$ is a bandpass signal, instead of a lowpass one. The output signal of the filter is a zero mean Gaussian process of bandwidth Δf. Therefore, in the arrangement of Fig. 6.56c, if the measurement time is larger than the width of $C_y(\tau)$ (that is, $T \gg 1/\Delta f$),

$$\mathrm{var}\{\widehat{\psi_y^2}\} \approx \frac{2}{T}\int_{-\infty}^{\infty} C_y^2(\tau)\,\mathrm{d}\tau = \frac{2}{T}\int_{-\infty}^{\infty} S_y^2(\tau)\,\mathrm{d}f \approx \frac{2}{T}S_x^2(f_0)2\Delta f\,, \tag{6.145}$$

and

$$\mathrm{var}\{\widehat{S}_x(f_0)\} = \frac{\mathrm{var}\{\widehat{\psi_y^2}\}}{(2\Delta f)^2} \approx \frac{S_x^2(f_0)}{\Delta f T}\,. \tag{6.146}$$

For random phase sinusoidal signals, using (6.102), the variance is

$$\mathrm{var}\{\widehat{S}_x(f_0)\} = \frac{1}{(2\Delta f)^2}\frac{U_\mathrm{p}^4}{8}\mathrm{sinc}^2(2\pi f_1 T)\,. \tag{6.147}$$

The relative variance is the same as (6.103):

$$\epsilon_\mathrm{v}^2\{\widehat{S}_x(f_0)\} = \frac{\mathrm{var}\{\widehat{S}_x(f_0)\}}{\left(\frac{U_\mathrm{p}^2}{4\Delta f}\right)^2} = \frac{1}{2}\mathrm{sinc}^2(2\pi f_1 T)\,. \tag{6.148}$$

When more than one sine waves are present in the signal, similar expressions can be derived, but special care has to taken of a possible frequency beat. If the frequencies of two sine waves are close to each other, the averaging time must be chosen significantly longer than the reciprocal of the difference of the frequencies.

For the mixed case (6.105) can be used, with $\mu_y = 0$:

$$\mathrm{var}\{\widehat{S}_x(f_0)\} \approx \frac{1}{(2\Delta f)^2}\frac{U_\mathrm{p}^4}{8}\mathrm{sinc}^2(2\pi f_1 T) + \frac{U_\mathrm{p}^2}{\Delta f}\mathrm{sinc}(\pi f_1 T)\frac{S_n(f_0)}{2\Delta f T} + \frac{S_n^2(f_0)}{\Delta f T}\,. \tag{6.149}$$

The periodogram

There is also a third way for the estimation of the power spectral density, the evaluation of the so-called *periodogram*. The idea may be introduced by comparing Figs. 6.56b-c. The only important difference between the two block diagrams is the term $1/T$ before the integral. Thus, it seems to be reasonable to assume that every method, applicable for the measurement of one of the quantities, can be used for the estimation of the other one, if this difference is taken into account.

The energy spectral density function is defined (and may also be measured) by the expression $E(f) = |X(f)|^2$. Why not try to measure the power spectral density by

$$S(f) = \frac{1}{T}|X(f,T)|^2\,, \tag{6.150}$$

where $X(f,T)$ is the Fourier transform of the sample record of $x(t)$ in $(0,T)$?

Indeed, (6.150), called the periodogram, is a very useful estimate of the PSD. Since the Fourier transform can be very effectively performed by the FFT, it is widely used in digital spectrum analyzers. The properties of this estimate will be discussed in the following paragraphs.

The periodogram can be evaluated both in continuous and discrete time and frequency domains. Though important applications exist also in continuous time, the main application is found in digital Fourier analyzers. Therefore, the properties of the discrete realization will be discussed, with some reference to the continuous case (see Priestley, 1981).

For a general treatment, let us consider the *generalized periodogram,* used for the estimation of the cross spectrum, and let us calculate the expected value of the spectral estimate for an arbitrary frequency from sampled input signals. Let $T = N\Delta t$.

$$\widehat{S}(f) = \frac{1}{T}\overline{X(f,T)}Y(f,T)\,. \tag{6.151}$$

The expected value of this quantity can be written as follows:

$$\begin{aligned}\mathcal{E}\{\widehat{S}(f)\} &= \mathcal{E}\Big\{\frac{1}{T}\sum_{i=0}^{N-1} x_i e^{+j2\pi f i\Delta t}\Delta t \sum_{m=0}^{N-1} y_m e^{-j2\pi f m\Delta t}\Delta t\Big\} \\ &= \frac{\Delta t}{T}\sum_{i=0}^{N-1}\sum_{m=0}^{N-1}\mathcal{E}\{x_i y_m\}e^{-j2\pi f(m-i)\Delta t}\Delta t\,. \end{aligned}\tag{6.152}$$

Using $\mathcal{E}\{x_i y_m\} = R_{m-i}$, and introducing a new summation variable $r = m-i$, the following expression is obtained:

$$\mathcal{E}\{\widehat{S}(f)\} = \sum_{r=-(N-1)}^{N-1}\Big(1-\frac{|r|}{N}\Big)R_r e^{-j2\pi f r\Delta t}\Delta t\,. \tag{6.153}$$

This basic result can be interpreted as the discrete Fourier transform of the correlation function, multiplied by a triangle function, called the *Bartlett window.* This multiplication corresponds to convolution in the frequency domain:

$$\mathcal{E}\{\widehat{S}(f)\} = W_{\mathrm{B}}(f) \star S(f)\,, \tag{6.154}$$

where

$$W_{\mathrm{B}}(f) = \sum_{r=-(N-1)}^{N-1}\Big(1-\frac{|r|}{N}\Big)e^{-j2\pi f r\Delta t}\Delta t = \Big(\frac{\sin \pi f T}{\sin \pi f \Delta t}\Big)^2\Delta t\,. \tag{6.155}$$

The effect of the discrete Bartlett window is shown in Fig. 6.59.

Obviously, the Bartlett window appears as a consequence of the finite observation time T. The finite observation time can be modelled by multiplication of

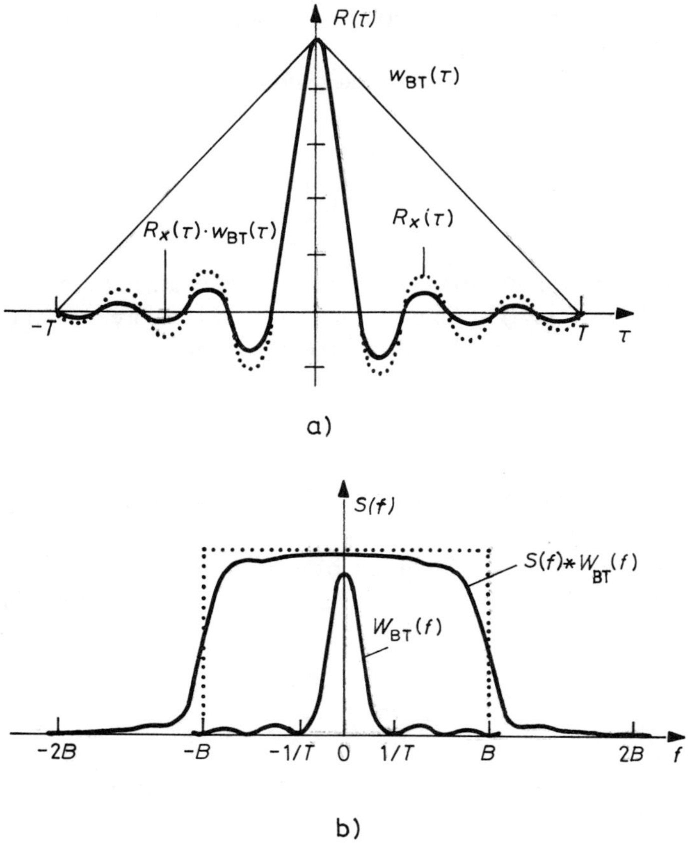

Fig 6.59 The effect of the Bartlett window *a)* in the τ-domain *b)* in the frequency domain

the so-called *rectangular window*. In the frequency domain the rectangular window has the following form:

$$W_{\mathrm{R}}(f) = \frac{\sin \pi f T}{\sin \pi f \Delta t} \Delta t \,. \tag{6.156}$$

The time domain Bartlett window is nothing other than the result of autoconvolution of the rectangular window, while the frequency domain Bartlett window is the square of the rectangular window, multiplied by an appropriate constant. The rectangular window is applied on the signal itself, therefore it is called in this context the *linear window*, while the Bartlett window which appears beside the squared Fourier transform, is called here the *quadratic window*.

In the continuous case (T=const, $\Delta t \to 0$):

$$W_{\mathrm{Rcont}}(f) = T \mathrm{sinc}(\pi f T) \,, \tag{6.157}$$

and

$$W_{\mathrm{Bcont}}(f) = T \mathrm{sinc}^2(\pi f T) \,. \tag{6.158}$$

The discrete and the continuous Bartlett windows do not differ significantly in the frequency domain for $f \ll 1/\Delta t = f_s$.

Eq. (6.154) may be also interpreted as follows:

$$\mathcal{E}\{\widehat{S}(f_1)\} = \int_{-\infty}^{\infty} S(f) W_B(f_1 - f)\, df \tag{6.159}$$

gives the power, appearing at frequency f, *as if* the spectrum was measured using a bandpass filter $W_B(-f) = W_B(f)$, positioned at f_1. Therefore, one may consider the periodogram as a special filter-based spectrum analyzer. The *equivalent bandwidth* of the filter is

$$2B_e = \frac{\int_{-\infty}^{\infty} w_B(f)\, df}{w_B(0)} = \frac{1}{T}\,. \tag{6.160}$$

Eq. (6.154) gives the bias of the periodogram: the edges, peaks etc. will be smeared by convolution with the window (see Fig. 6.59).

The effect of the finite sample record is even more characteristic in the case of periodic components. Since the spectrum of a periodic waveform is

$$S(f) = \sum_{k=-\infty}^{\infty} |c_k|^2 \delta(f - kf_1)\,, \tag{6.161}$$

the expected value of the estimate consists of window functions, repeated at the place of every Dirac delta function:

$$\mathcal{E}\{\widehat{S}(f)\} = S(f) \star W_B(f) = \sum_{k=-\infty}^{\infty} |c_k|^2 W_B(f - kf_1)\,. \tag{6.162}$$

Two important phenomena can be observed: most of the power appears at the appropriate places, but smeared to the width of the main lobe of the window, moreover, a small part of the power 'leaks' to other frequencies.

Here it has to be mentioned that in digital spectrum analysis the spectra are calculated at discrete frequency points only, which causes an additional problem. Since with nonsynchronized measurements the frequency of a sine wave can be anywhere between the calculated spectrum points, only samples of the Bartlett window will be seen (see Fig. 6.60a) and the maximum value will usually significantly differ from the peak value. This phenomenon is called *picket fence effect*. The observable maximum value as a function of the sine frequency is illustrated in Fig. 6.60b.

These undesirable phenomena can be controlled to some extent by appropriate modification of the periodogram (use of modified window functions etc): these methods will be discussed in Subsection 6.7.7.

Beside the bias also the variance has to be examined. This will be done here first for samples of the estimate for random signals with smooth spectra (that

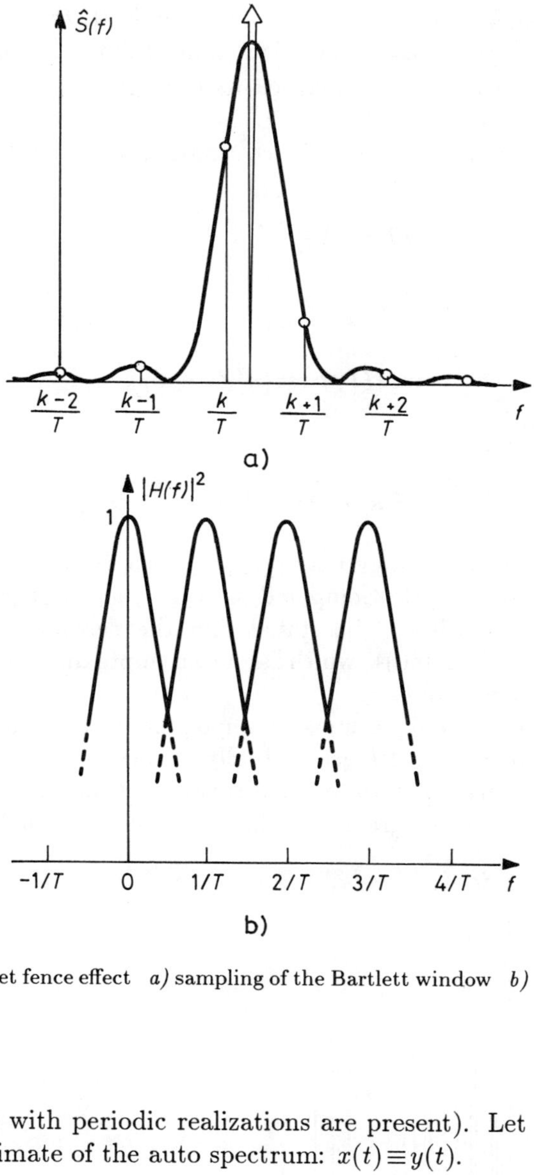

Fig. 6.60 The picket fence effect *a)* sampling of the Bartlett window *b)* the equivalent transfer function

is, no components with periodic realizations are present). Let $\Delta t = 1$, and let us investigate the estimate of the auto spectrum: $x(t) \equiv y(t)$.

$$\widehat{S}_k = \frac{1}{N}\left|\sum_{i=0}^{N-1} x_i e^{-j2\pi\frac{ki}{N}}\right|^2$$

$$= \frac{1}{N}\left(\sum_{i=0}^{N-1} x_i \cos(2\pi\frac{ki}{N})\right)^2 + \left(\sum_{i=0}^{N-1} x_i \sin(2\pi\frac{ki}{N})\right)^2 \quad k = 0, 1 \ldots N-1. \quad (6.163)$$

It is easy to see that for any value of k (except for $k=0$) the two squared quantities are zero mean, approximately Gaussian (central limit theorem[3]), uncorrelated random variables with the same variance. As a consequence, $\widehat{S}_k$ has χ_2^2 (that is, exponential) distribution.

For smooth spectra the above variance can be approximately calculated, noticing that

$$\mathcal{E}\{\widehat{S}_k\} = \mathcal{E}\{\frac{1}{N}(\xi^2 + \eta^2)\} \approx S_k . \tag{6.164}$$

From (6.164) it follows that

$$\mathrm{var}\{\xi\} \approx \mathrm{var}\{\eta\} \approx \frac{N}{2} S_k , \tag{6.165}$$

and

$$\widehat{S}_k \approx \frac{1}{2} S_k \chi_2^2 , \quad \mathrm{var}\{\widehat{S}_k\} \approx S_k^2 . \tag{6.166}$$

The variance is surprisingly large (the relative variance is 1); it will have to be reduced (see Subsection 6.7.7). Compared with (6.174), the two variances are the same if $\Delta f = 1/T$, thus it can be stated that the *equivalent bandwidth* of the periodogram is $1/T$ (cf. (6.160)), which is indeed approximately the width of the main lobe of the Bartlett window.

Using the above technique another important feature can be derived: the estimates for different values of k are mutually uncorrelated, which means that in general the neighbouring spectrum estimates vary virtually independently of each other. Thus, a typical periodogram looks like the one in Fig. 6.61.

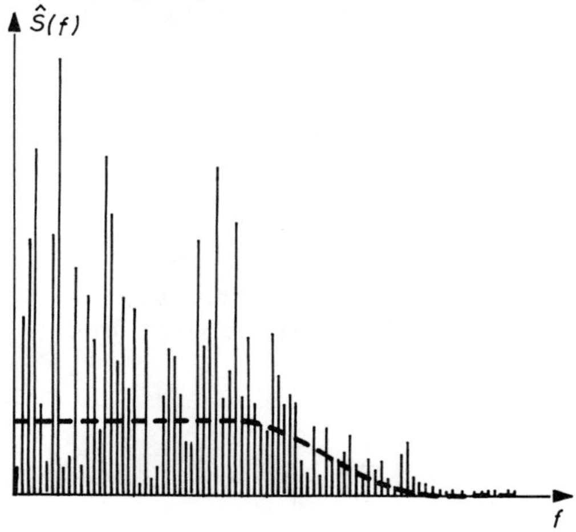

Fig. 6.61 A typical periodogram

[3]The central limit theorem in its usual form refers to the sum of independent, equally distributed random variables. However, the theorem can also be stated for moderately correlated random variables (central asymptotic theorem, see e. g. Loève, 1978, Vol. II, Chapter IX).

Periodic signals behave differently. In Kollár (1984a) it is shown that for random phase sine waves the relative variance is

$$\epsilon_{\mathrm{v}}^2\{\widehat{S}(f)\} = \frac{2W(f-f_1)W(f+f_1)}{\left(W(f-f_1)+W(f+f_1)\right)^2}, \tag{6.167}$$

where $W(f)$ is the applied linear window function.

Since the sidelobes of the windows usually decrease rapidly, for $f_1 \gg 1/T$ the above relative variance can be approximated for $f \approx f_1$ as:

$$\epsilon_{\mathrm{v}}^2\{\widehat{S}(f)\} \approx 2\frac{W(f+f_1)}{W(f-f_1)} \ll 1\,. \tag{6.168}$$

The result is similar for general periodic signals. The peaks of a sine wave in the periodogram have a small relative variance: this is the reason why the periodogram is reliable in the detection of periodic components. A heuristic explanation may be as follows: randomness appears in the phase of the signal and its Fourier transform. In the periodogram the absolute value eliminates the phase information, if no other term is summed up.

However, Eq. (6.168) may not be applied when two main lobes overlap, that is, when two sinusoidal components are very close to each other (*frequency beat*). In such a case the variance is large in the region of overlapping.

The last case that has to be discussed is the sine wave covered by noise. The variance given by Kollár (1984a) is for $f \approx f_1$:

$$\mathrm{var}\{\widehat{S}(f)\} \approx S_{\mathrm{n}}^2(f) + 2S_{\mathrm{n}}(f)S_{\mathrm{p}}(f) + S_{\mathrm{p}}^2(f)2\frac{W(f+f_1)}{W(f-f_1)}, \tag{6.169}$$

where

$$S_{\mathrm{p}}(f) = \frac{U_{\mathrm{p}}^2}{4\Delta f}W^2(f-f_1)\,.$$

In the case of windows with very small sidelobes and not too small an SNR, the second term will dominate, which directly depends on the SNR, and if the peak of the sine wave is large enough, its relative variance will be small.

Example 6.6 Determination of the necessary analysis times (Kollár, 1989)

A coloured noise with bandwidth $B \approx 10\,\mathrm{kHz}$ is investigated. The power spectral density is to be measured in 100 points, with a relative variance of 1%.

a) Using the bandpass filter method how is the bandwidth of the filter to be chosen, and what is the required length of the time record?

b) Using the periodogram method, give the number of periodograms to be averaged, and determine the measurement time.

Solution:

a1) In a heterodyne analyzer the resolution is constant along the frequency axis:

$$\Delta f = \frac{10\,\text{kHz}}{100} = 100\,\text{Hz}.$$

The measurement time must be

$$T = \frac{1}{\epsilon_r^2 \cdot \Delta f} = \frac{100}{\Delta f} = 1\,\text{s}$$

in each band.

The settling time of the filter is in the order of $\frac{1}{\Delta f}$, therefore it can be neglected. The sweeping time over the entire frequency band is: $T_m = 100 \cdot T = 100\,\text{s}$.

a2) Using a tuned filter or a filter bank, $\frac{\Delta f}{f}$ is constant. Measuring in the audio band, the relative frequency resolution, determined by the 100 prescribed bands, is

$$\frac{1}{Q} = \frac{\Delta f}{f} \approx \frac{f_{i+1} - f_i}{f_i} = \frac{f_i + 1}{f_i} - 1 \approx \left(\frac{f_{100} + \frac{\Delta f_{100}}{2}}{f_1 - \frac{\Delta f_1}{2}}\right)^{\frac{1}{100}} - 1$$

$$\approx \left(\frac{20\,\text{kHz}}{20\,\text{Hz}}\right)^{\frac{1}{100}} - 1 \approx 0.0715\,,$$

since the centre frequencies form a geometric series. This resolution is about 3.5 times finer than that of the third octave resolution.

The required measurement time is the largest in the narrowest band Δf_i:

$$T_1 = \frac{1}{\epsilon_r^2 \Delta f_1} = \frac{1}{\epsilon_r^2 \frac{1}{Q} f_1} = \frac{1}{0.01 \cdot 0.0715 \cdot 20\,\text{Hz} \cdot 1.035} \approx 68\,\text{s},$$

since

$$f_1 \approx 20\,\text{Hz}\sqrt{\frac{1}{Q} + 1} \approx 20\,\text{Hz} \cdot 1.035\,.$$

T_1 is also the necessary measurement time for the filter bank.

In the case of a tuneable filter the cumulated measurement time is:

$$T \approx \sum_{i=0}^{99} T_1\left(1 + \frac{1}{Q}\right)^{-i} = T_1 \cdot \frac{1 - \left(1 + \frac{1}{Q}\right)^{-100}}{1 - \left(1 + \frac{1}{Q}\right)^{-1}}$$

$$\approx T_1(Q+1) \approx 15T_1 \approx 1013\,\text{s} = 16'53''.$$

b) Let the record length be $T = \frac{1}{\Delta f}$.

$$T = \frac{1}{100\,\text{Hz}} = 0.01\,\text{s}.$$

The number of the necessary time records is determined by the prescribed relative variance:

$$n = \frac{1}{\epsilon_r^2} = 100,$$

and the total measurement time is $nT = 1$ s.

The number of points to be transformed is $N = 2BT = 200$ in each record.

b1) In an N-point DFT the number of real multiplications and additions (= operations) is $2N^2$, that is, the total number of operations is $M = n \cdot 2N^2 = 8 \cdot 10^6$. The number of operations required for averaging can be neglected.

With $5\,\mu s$ operating time the necessary computation time is $T_c = M \cdot 5\,\mu\text{s} = 40$ s.

Using the FFT algorithm, $M = n \cdot 2N \cdot \log_2 N \approx 3.06 \cdot 10^5$, and $T_c = M \cdot 5\,\mu\text{s} \approx 1.53$ s.

b2) By the Bartlett-method, the entire time record is transformed: $N = 2BT_e = 2 \cdot 10^4$. Using the Discrete Fourier Transform, $M = 2N^2 = 8 \cdot 10^8$, and

$$T_c = M \cdot 5\mu\text{s} = 4 \cdot 10^3\text{s} \approx 1\,\text{h}\ 7\,\text{min}.$$

Using the Fast Fourier Transform, the computation time is considerably reduced:

$$M = 2N \cdot \log_2 N \approx 5.72 \cdot 10^5, \quad T_c \approx 2.86\,\text{s}.$$

Types of spectrum analyzer

Based on the principles presented above, different types of spectrum analyzer can be built (Fig.6.62).

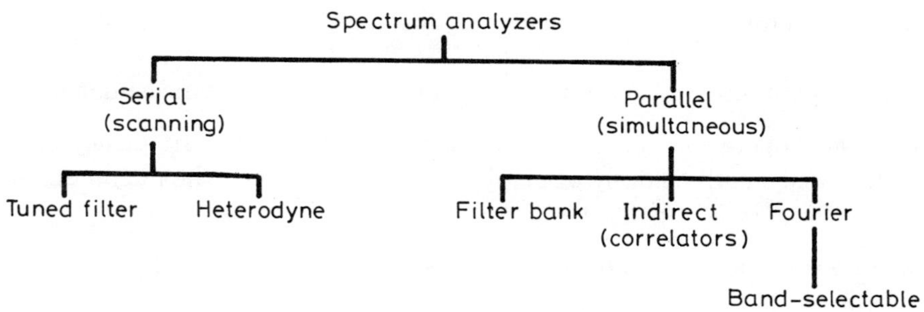

Fig. 6.62 Classification of spectrum analyzers

The first group is scanning type or serial analyzers. These are selective voltmeters (see Subsection 2.5.6), which measure the spectrum at just one frequency point at a time. There are two main types: tuned filter analyzers, and heterodyne analyzers: in the latter type the filter is fixed, and the spectrum is shifted in the frequency domain by appropriate mixing.

The resolution of tuned filter analyzers is limited because of the difficulties associated with tuning, therefore they are rarely applied, usually in the audio domain.

The measurement range of heterodyne analyzers spans from a few Hz to a few 10 GHz. Above the audio band this is the only widely used analyzer type. However, in the audio band the necessary sweeping time may be large, since the settling times of the bandpass filters are long (see Subsection 6.7.3), therefore here parallel analyzers are dominating. Because of logarithmic detection and small-noise stages the dynamics of heterodyne analyzers may be over 100–120 dB. The smallest resolution is in the order of 1 Hz.

Simultaneous (parallel) analyzers measure the spectrum at several frequency points at the same time, thus offering a more effective use of the information in the signal. The filter bank consists of several bandpass filters working at different frequencies; an indirect analyzer is a correlator, followed by a Fourier transformer, while Fourier analyzers are based on the periodogram.

Filter banks are mostly used in the audio band for octave and third octave analysis. The dynamic range is below 70–80 dB.

The frequency range of indirect analyzers spans to 100 MHz. The dynamic range reaches 60 dB, limited by the averager and the Fourier transformer units.

Commercial Fourier analyzers are usually digital ones. The frequency range is somewhat wider than the audio range; the upper frequency limit may reach 100 kHz. Digital Fourier analyzers work from practically 0 Hz, with a resolution down to 1 mHz. The dynamic range is around 70–80 dB. It is interesting to note that Fourier analyzers may also be considered from the point of view of signal processing as special filter banks (see (6.159)).

Tuned filter analyzers and filter banks usually realize constant *relative bandwidth* ($\Delta f/f$), while the others realize constant absolute bandwidth (Δf).

Before discussing the different analyzer types in detail, an often used term has to be introduced: *real-time* analysis. This has different meanings:

- fast operation: the analysis time is not longer than the measurement time,
- parallel operation: the measurement results for different frequency points are obtained simultaneously or nearly simultaneously, and the maximum amount of information is utilized,
- the results are at hand in time to use them in the control of the measured process.

Usually simultaneous analyzers are the ones which meet these definitions.

6.7.2 Tuned Filter Analyzers

The block diagram of tuned filter analyzers is shown in Fig. 6.63.

The conditioned signal $x(t)$ is led to an electronically tuneable bandpass filter. The centre frequency is determined by a sawtooth generator. The squared signal is led to the averager (called video filter), and the averaged signal, proportional to the signal power in the selected band, drives the output unit: an X–Y recorder, or a (maybe storage) CRT.

The sweeping speed (scan rate) is determined by the transient times of the bandpass and the video filters. These transient times are in the order of the reciprocals of the bandwidths. The bandpass filter may not be tuned through a frequency point in less time than the transients.

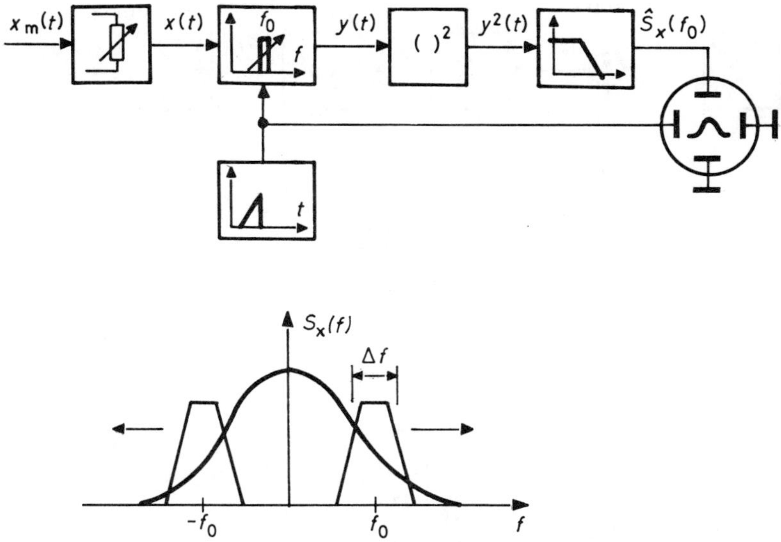

Fig. 6.63 Tuned filter analyzers

Wideband filter tuning is a very difficult task. Electrically tuneable elements (e. g. varicaps) have a relatively narrow tuning range, while mechanical elements (potentiometers, variable capacitors) are difficult to use in automatic instruments, and the tuning range is still not more than two decades. Discrete tuning (switch capacitors on/off) can result in unwanted transients. Moreover, it is difficult to tune higher-order filters keeping the prescribed tolerance scheme. Because of the above facts, tuned filter analyzers have limited resolution.

6.7.3 Heterodyne Analyzers

The main difficulty of tuned filter analyzers is that a selective filter, tuneable in a wide frequency range, is practically not realizable. At a fixed frequency such a filter can be built, but to use this, the spectrum has to be shifted. The technique for frequency shifting is mixing (see Subsection 6.1.7). The block diagram of a heterodyne analyzer, based on mixing, is illustrated in Fig. 6.64.

Choice of the characteristic frequencies

The basic problem of heterodyne analyzers is the choice of the characteristic frequencies. First of all, mixing results in spectrum splitting, and the measured spectrum will be distorted if some power of the other mixed spectrum gets into the measured band (see Fig. 6.65).

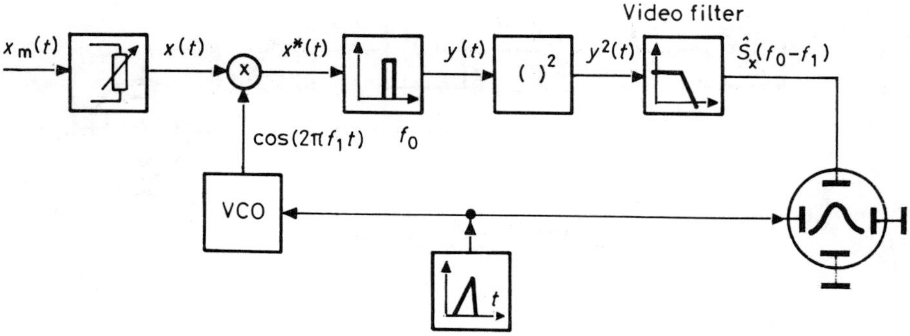

Fig. 6.64 Heterodyne analyzer

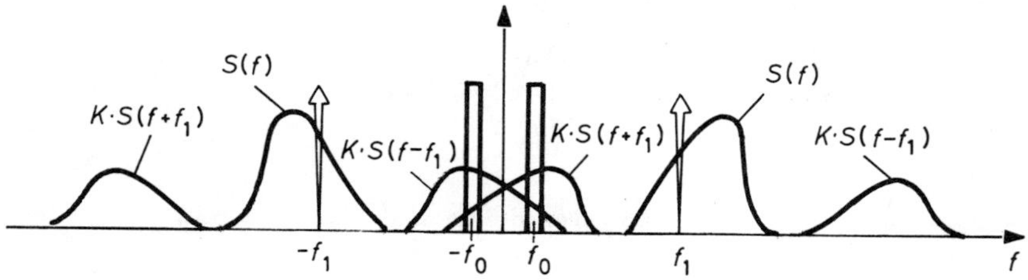

Fig. 6.65 Spectrum distortion caused by improper choice of the characteristic frequencies

There is another problem associated with the realization of mixers. The mixer is often not perfectly balanced, that is, a small part of both input signals appear at the output. Moreover, pure harmonic mixing is very difficult to realize. Usually the higher harmonics of the mixing signal will also appear, especially in the cheapest mixers, which modulate by a square wave. Consequently, the spectrum of the mixed signal is usually

$$S_{x^*}(f) = \sum_{k=-\infty}^{\infty} a_k S(f - kf_1) + b_k \delta(f - kf_1)\,. \tag{6.170}$$

The centre frequency of the bandpass filter, f_0, and the span of the mixing frequency, f_1, have to be chosen in such a way that no distorting component arrives at the bandpass filter (Fig. 6.66).

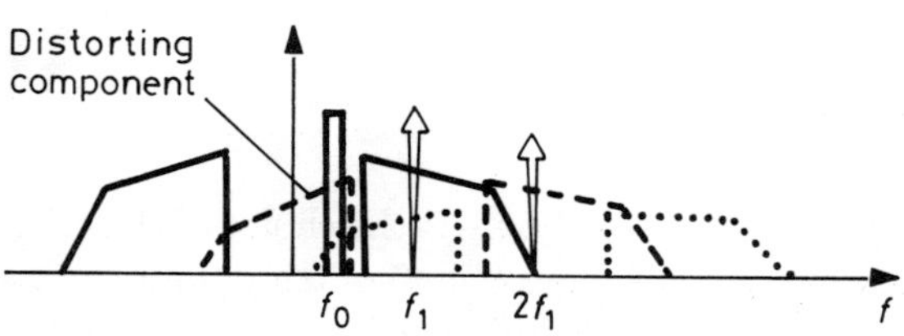

Fig. 6.66 Distortion due to overharmonic mixing

Because of the above problems, f_0 may not be in the measured band. If it is below the measured frequencies, spectral overlapping is difficult to avoid, therefore the usual choice is above the measured frequency band.

In order to avoid the distortions due to harmonic mixing, the span of f_1 is also to be chosen above the measured band. This has another advantage: the frequency span of the band, to be covered by the sweep generator, will also be much smaller. Let us assume e. g. that the band 100 Hz – 1 MHz is to be analyzed. If f_0 is chosen as 50 Hz, the VCO has to sweep in the band (50 Hz – 1 MHz), a frequency ratio of $2 \cdot 10^4$, which is practically not realizable. But if f_0 is chosen as 1 MHz, the desired sweep range is (1 MHz – 2 MHz), with a moderate frequency ratio of 2.

There are still two possibilities for the choice of the frequencies:

$f_m < f_0 < f_1$ (Fig. 6.67a, lower mixing),

$f_m < f_1 < f_0$ (Fig. 6.67b, upper mixing).

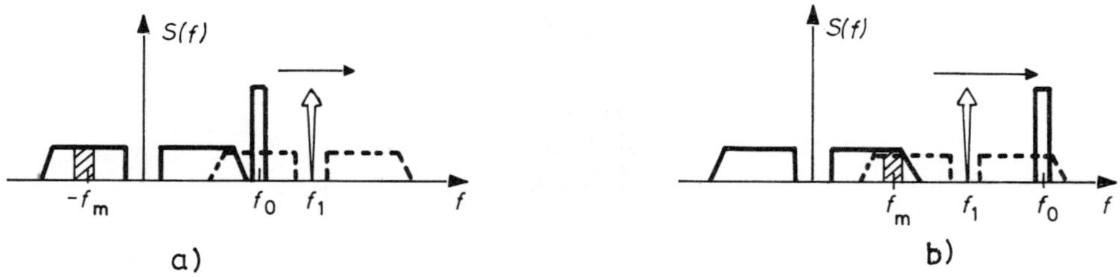

Fig. 6.67 Possibilities for the mixing frequency *a)* lower mixing *b)* upper mixing

The two solutions have similar properties. However, since the signal at frequency f_0 has to be further processed, the first one is usually chosen. A typical solution is illustrated in Fig. 6.68.

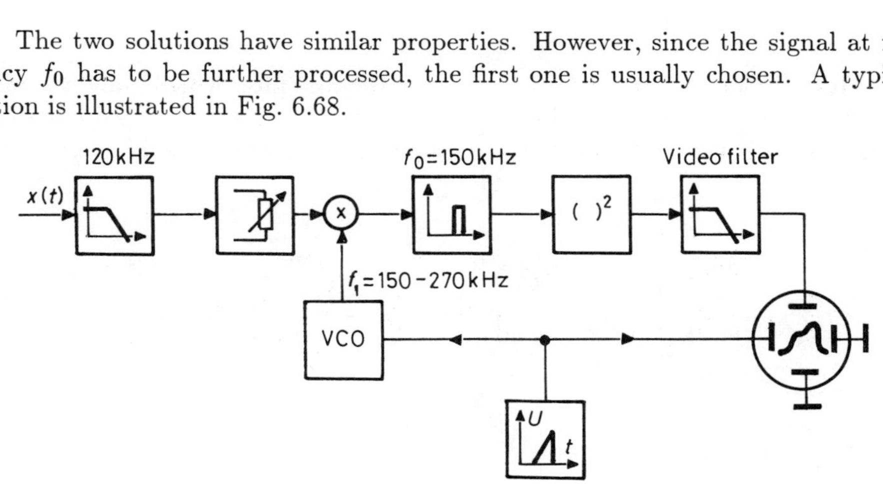

Fig. 6.68 Frequency choice in a heterodyne analyzer

If f_0 is large, it is difficult to realize a narrow enough bandpass filter. Therefore, the output signal of the bandpass filter is usually mixed again to smaller frequencies.

A possibility is to mix to f_{02}=0 Hz. Here any bandwidth can be easily realized; however, because of the flicker noise and the drift of the DC amplifier this solution is usually not chosen.

If $f_{02} \neq 0$, another problem arises. Though the output signal of the filter is bandlimited, by mixing to too small a frequency, overlapping may occur again (Fig. 6.69).

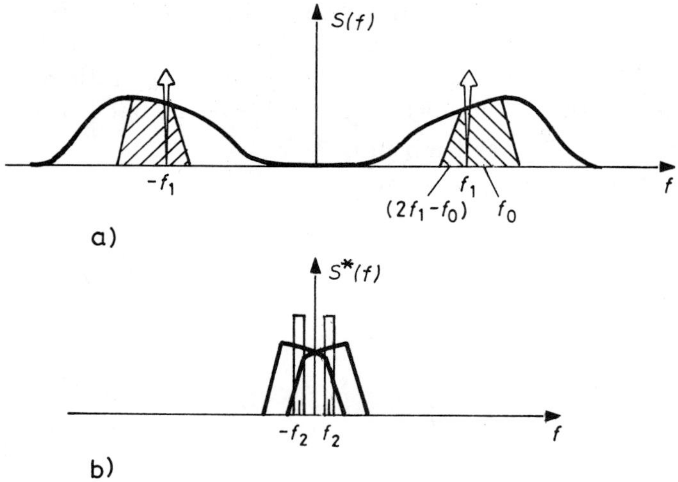

Fig. 6.69 Spectrum overlapping after the second mixing *a)* the result of bandpass filtering with the mixing frequencies *b)* the consequence of too low an intermediate frequency

It is obvious that it is the selectivity of the first filter which limits the achievable frequency ratio f_0/f_{02}, to 10–20 in the case of LC filters. Therefore, frequency transposition is realized in a few steps, applying bandpass filters after each mixing (Fig. 6.70).

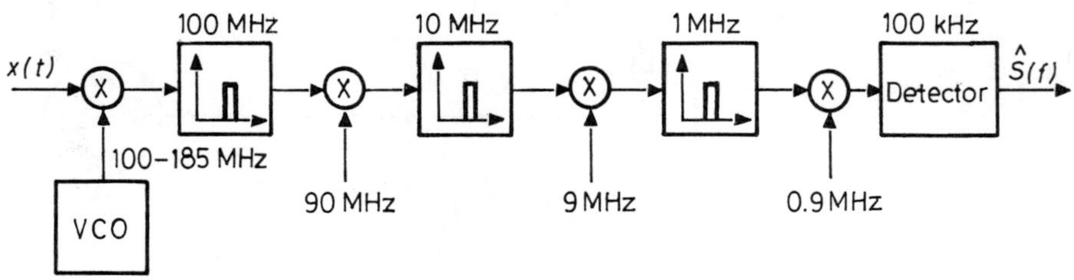

Fig. 6.70 Multistage frequency transposition

Quartz crystal filters have a much better selectivity than LC filters. Using them the frequency ratio can reach 100–200 (Daniels and Atchley, 1975).

The tracking generator

It is an obvious requirement to determine the frequency of spectrum peaks with maximum accuracy. However, to measure the instantaneous value of a changing frequency is difficult and inaccurate. Therefore, it is desirable to generate a signal of this very frequency, and use a frequency meter. The so-called tracking generator (Page, 1971) is built into spectrum analyzers partly for this purpose (Fig. 6.71).

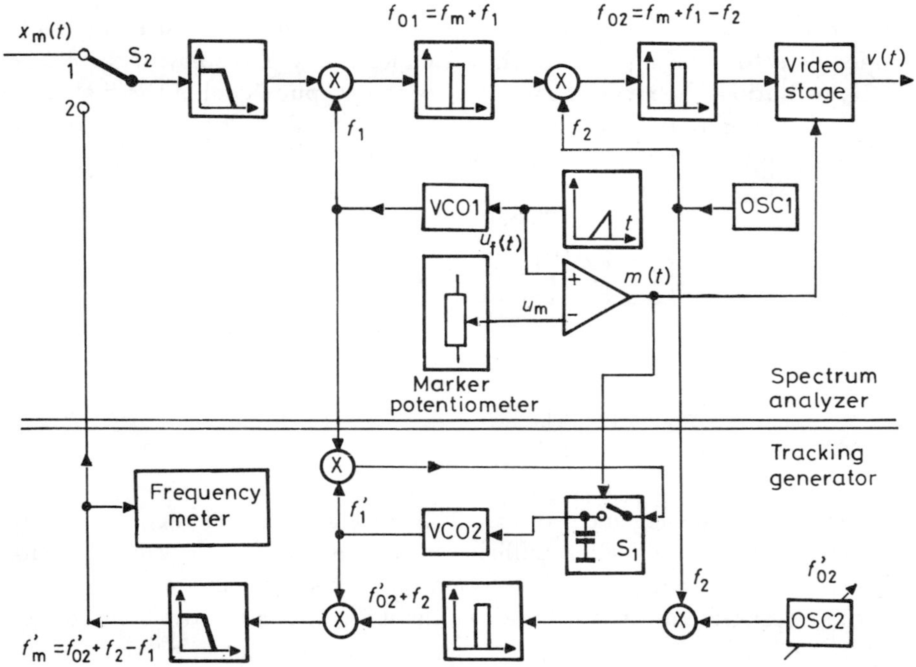

Fig. 6.71 Tracking generator

The aim is to produce a frequency, always equal to the centre frequency of the actually measured band, and to 'freeze' this frequency when it coincides with the desired one. The intermediate frequency is generated in the spectrum analyzer as $f_{02} = f_m + f_1 - f_2$. The tracking generator does the inverse: it synthesizes the measured frequency from the internal signals of the spectrum analyzer: $f_m' = f_{02}' + f_2 - f_1'$. The mixing frequency f_1' is synchronized to f_1 by the PLL, consisting of VCO2, the phase detector, and the sample and hold circuit, containing the switch S_1. The intermediate frequency is not available, therefore it is generated by OSC2. The frequency of OSC2 is adjusted to the right value before the measurement, by seeking for the maximum value on the analyzer in position 2 of S_2.

During measurement, the desired spectrum peak is selected by setting the display marker, indicated on the screen using the $m(t)$ marker signal. When this is set and $m(t)$ goes high, S_1 opens, and the frequency of the tracking generator will be equal to the desired one, and it can be measured by the frequency meter.

The tracking generator can be used also for transfer function measurement. Since its frequency is always equal to the actually analyzed frequency, leading this (with constant amplitude) to the input of a linear system, the measured output spectrum will be equal to $|H(f)|^2$.

The video detector

An important part of heterodyne analyzers is the unit after the last bandpass filter: the video detector, which produces the signal for the X–Y display of the analyzer.

By definition, the mean square value of the output signal of the filter is to be measured. Averaging is done via the so-called video filter. Because of continuous sweeping, forgetting averaging is to be applied (Fig. 6.72).

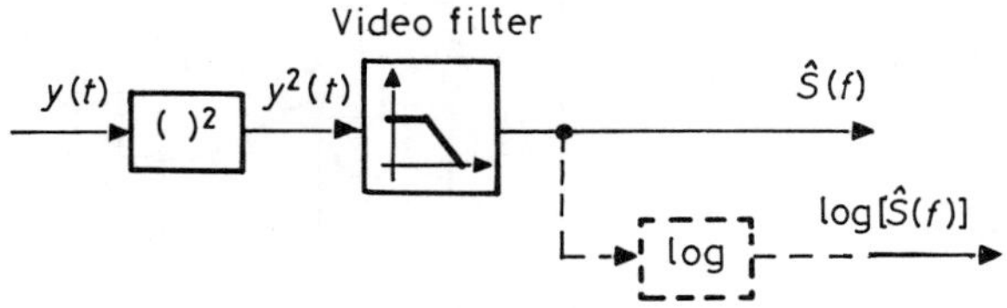

Fig. 6.72 Theoretically correct video detector

However, the arrangement of Fig. 6.72 has a serious difficulty. Although the logarithmic display would allow sufficiently high dynamics, the squaring circuit, which cannot be realized with high accuracy, limits the dynamic range to 40–50 dB. Thus, this arrangement must not be used as it is: either the dynamic range, expected of the squaring circuit, is to be decreased, or the squaring circuit has to be eliminated.

The necessary dynamic range at the squaring circuit can be decreased by using a controlled amplifier (Fig. 6.73). The gain of the quasilinear amplifier depends on the actual value of the video signal v. Since this changes very slowly,

$$A(v) = \frac{\sqrt{K_2 v}}{e^{\frac{v}{2K_2}}} \tag{6.171}$$

can be realized with sufficient accuracy.

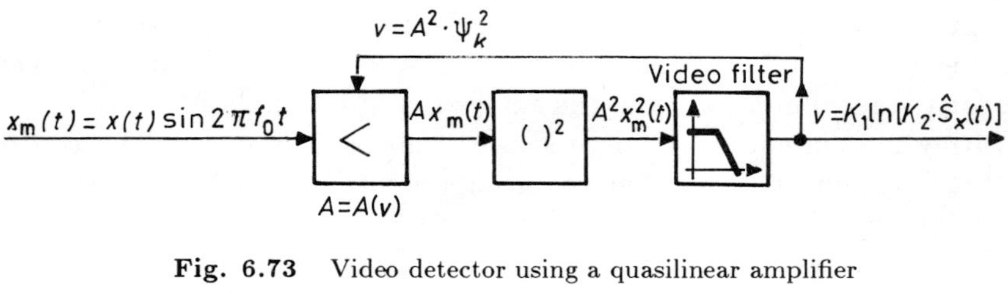

Fig. 6.73 Video detector using a quasilinear amplifier

The logarithm of the instantaneous value of the intermediate frequency signal (or rather of its rectified value) can also be taken (Fig. 6.74). Demodulation before the logarithmation is not advisable because of noise and drift problems.

It can be seen that the result is theoretically not correct, since the mean value of the logarithm is generally not equal to the logarithm of the mean square value. However, if the distribution of the envelope is known (for continuous spectra it has Rayleigh or χ_2 distribution, while for pure sine it has the distribution of a sine wave), their relation can be determined. Analyzers of this type are usually calibrated using sine waves, and correction diagrams can be used if the signal has continuous spectrum, or it is a mixed one in the analyzed band.

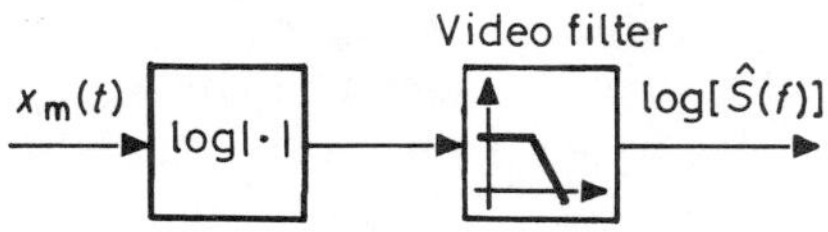

Fig. 6.74 Video detector based on logarithm of the instantaneous value of the intermediate frequency

It is often desirable to have a static display of the measured spectrum, even with slow sweep speeds, and to further process the measured function. This can be done by a digital video processor, which samples the spectrum and displays the stored values. To do this properly, the sampling theorem must be met for the sampling of the spectrum. If there are sharp peaks in it, the necessary point number may be high.

In heterodyne analyzers the point number of the video processor is usually limited, and special solutions help to avoid the loss of important information (the sampled value is the maximum of the spectrum in the last interval Δf, or alternately the minimum and the maximum in successive intervals, etc (Holdaway et al, 1978).

Sweeping speed (scan rate)

An important parameter to be set by the user is the sweeping speed of the frequency f_1. If this speed is too high, the measured spectrum will be distorted, and its variance may be too large.

The time for which the analyzer has to work in a given frequency band of width Δf depends on the following factors:

- the total length of the transients of the narrowest (the last) bandpass filter, which are inversely proportional to the bandwidth,
- the length of the transients of the so-called video filter (the averaging lowpass filter), which are inversely proportional to the bandlimit,

- the averaging time, necessary for the desired decrease of the variance — this can be set by the bandlimit of the video filter, and is again inversely proportional to this bandlimit.

An interesting consequence of the above statements is that since the scan rate is inversely proportional to the prescribed bandwidth Δf, the necessary measurement time to scan a given frequency band is inversely proportional to the *square* of Δf. Therefore, too small a Δf value may result in a very long measurement time.

6.7.4 Filter Banks

The block diagram of a filter bank is presented in Fig. 6.75. The filters should somewhat overlap in order not to leave some peaks undetected (Fig. 6.75b-c).

The disadvantage of the filter bank is the limited resolution. In the audio band (20 Hz–20 kHz) with third octave resolution about 30 filters are necessary — the reasonable number of bands in a filter bank is of this order. The advantages are the straightforward realization, the constant relative frequency resolution, and the standardized filter shapes, used for audio applications.

In the audio band both analog and digital filters can be used.

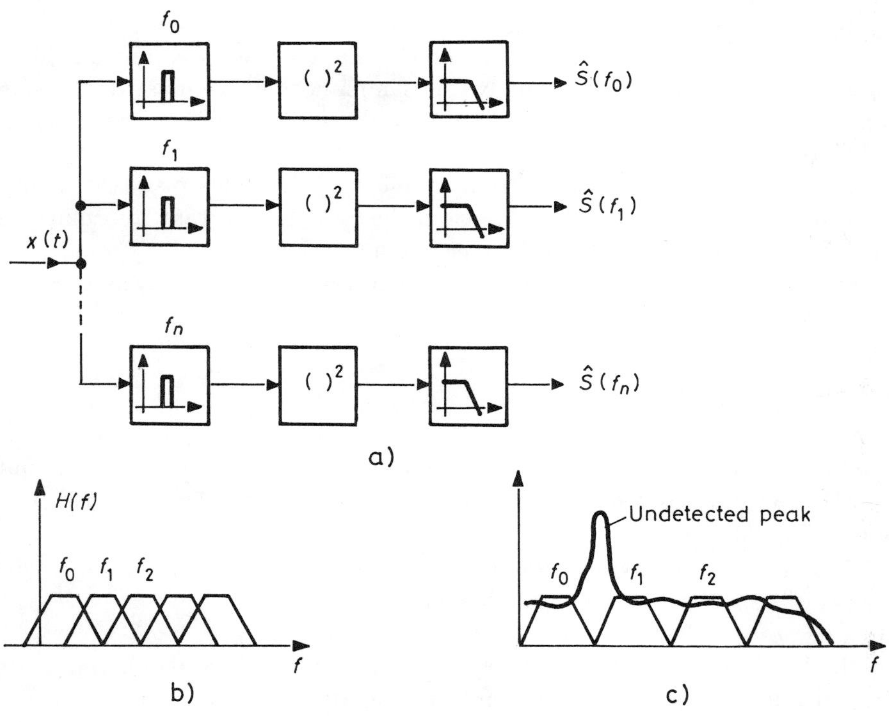

Fig. 6.75 Filter bank *a)* block diagram *b)*proper frequency setting *c)* improper frequency setting

6.7.5 Indirect Spectrum Analyzers

These analyzers are based on correlators (see Section 6.4) and on Fourier transformers (see Subsection 6.7.6), thus they will not be treated here in detail.

6.7.6 Fourier Analyzers

Fourier analyzers are based on the evaluation of the periodogram (see Subsection 6.7.1). Therefore, in the first part of this subsection the possibilities of the realization of the Fourier transform will be discussed.

Tools of the Fourier transform

There are several possibilities for the realization of the Fourier transform. Here only some characteristic solutions will be presented:

- hardware realizations of the FFT,
- digital comb filter as a Fourier transformer,
- charge-coupled device (CCD),
- surface acoustic wave device (SAW),
- optical Fourier transform.

Hardware realizations of the FFT

The basics of the FFT algorithm have been presented in Subsection 6.1.5 (see Fig. 6.24). This algorithm can be implemented in different ways.

In a *serial FFT* the butterflies are executed one by one by a single arithmetic unit, and the whole algorithm is controlled by a general purpose (micro)computer. This is the method used in computer programs. The transform speed is about $T_{\mathrm{b}}\frac{N}{2}\log_2 N$, where T_{b} is the execution time of the butterfly.

The FFT, performed by a general purpose microcomputer, is cheap and simple; however, the calculation of the data addresses and the software multiplication may be too slow. Therefore, these tasks are often executed by additional hardware elements (Fig. 6.76).

The *cascade FFT* means that a butterfly unit is assigned to each stage. The organization of the data flow is not trivial, therefore this less important solution will not be treated in detail.

The *parallel iterative processor* contains $N/2$ butterfly units, which execute one stage of Fig. 6.24 at a time (Fig. 6.77). The data record is stored in the memory,

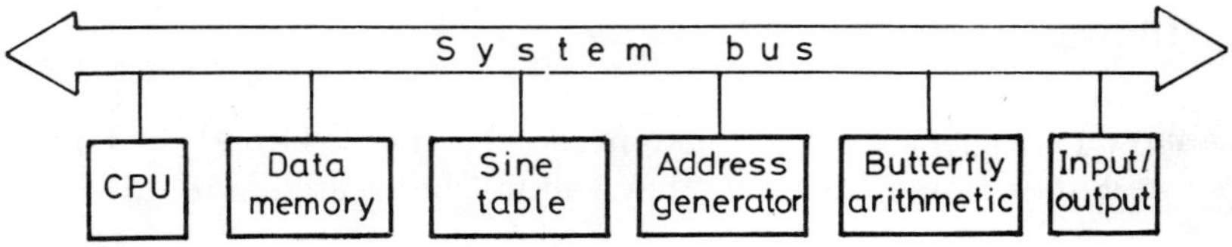

Fig. 6.76 Serial FFT, accelerated by additional hardware

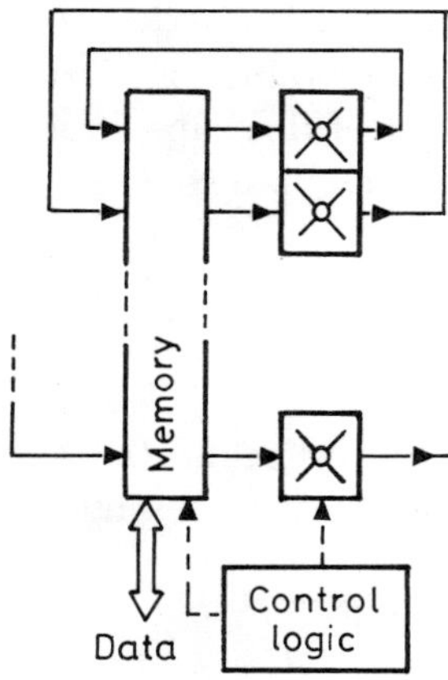

Fig. 6.77 Parallel iterative processor

and the result of each butterfly stage is stored into the memory to prepare the next one. The transform time is $T_b \log_2 N$.

In an *array FFT* processor a butterfly unit is assigned to each butterfly of the flow diagram (see Fig. 6.78). This is the quickest (and most expensive) FFT realization, it can be a combination circuit or a synchronous one.

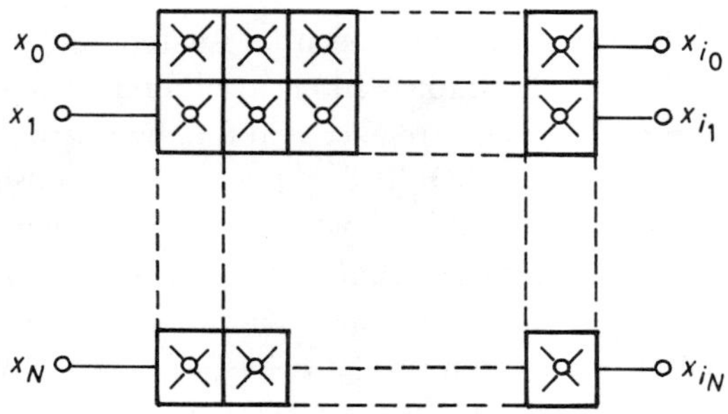

Fig. 6.78 Array FFT

An often used variant of the array FFT is the pipeline structure (Fig. 6.79). The stages work in a synchronous way, and pass the results to the next stage always at the same instant. This makes it possible for the different stages to process different data records at the same time. One data record is transferred in time $T_b \log_2 N$ through the FFT unit, but the hardware produces a transformed series and reads a new record every time period T_b.

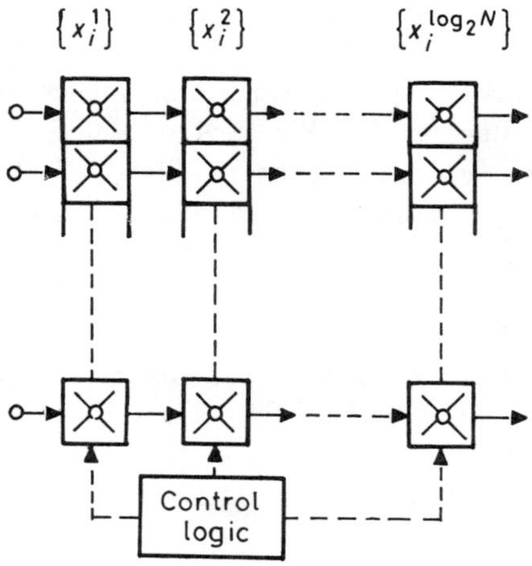

Fig. 6.79 Pipeline FFT

Digital comb filter as DFT processor

The DFT of a sampled series can also be obtained using digital IIR filters (Péceli, 1985). The arrangement in Fig. 6.80, the so-called spectral observer, recursively calculates the DFT of each sample record $\{x_{0+k}, x_{1+k}, \ldots, x_{N-1+k}\}$, for $k=0, 1\ldots$

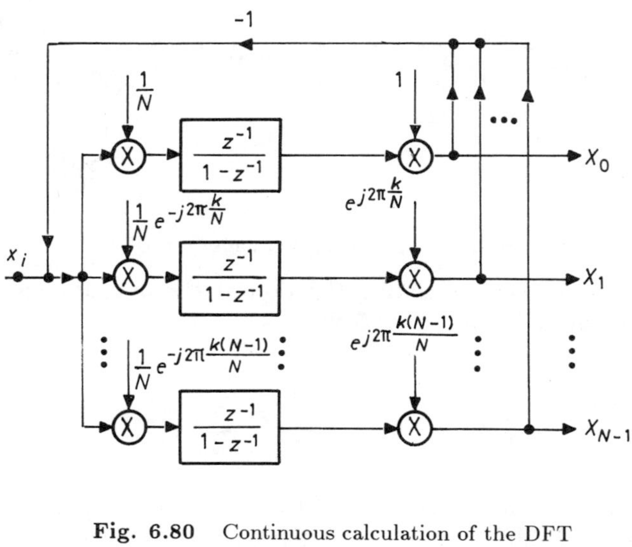

Fig. 6.80 Continuous calculation of the DFT

Charge coupled device (CCD)

Charge coupled devices are analog shift registers, where the samples can be weighted and summed: this is FIR filtering or discrete convolution. Consequently, if the DFT can be re-written into a form which contains discrete convolution with a fixed function, multiplications by constants, and summations, CCD-s can be applied.

Such a formulation is the chirp z-transform (Subsection 6.1.5, Eq. 6.78). The CCD implementation is illustrated in Fig. 6.81.

Eq. (6.78) contains complex numbers. The CCD Fourier transformer can handle real quantities only, therefore the real and imaginary parts are treated separately: altogether four CCD convolvers are used.

The CCD Fourier transformer in Fig. 6.81 can calculate the DFT of the input series in $2N$ steps, if the N-point input series is *repeated* at the input (it is easy to see that in the first $N-1$ steps, when not all the necessary input data are in the transformer, the output is not yet the DFT). Thus, this Fourier transformer works discontinuously. However, if the signal $x(t)$ is applied continuously to the input, an interesting observation can be made (Fig. 6.82).

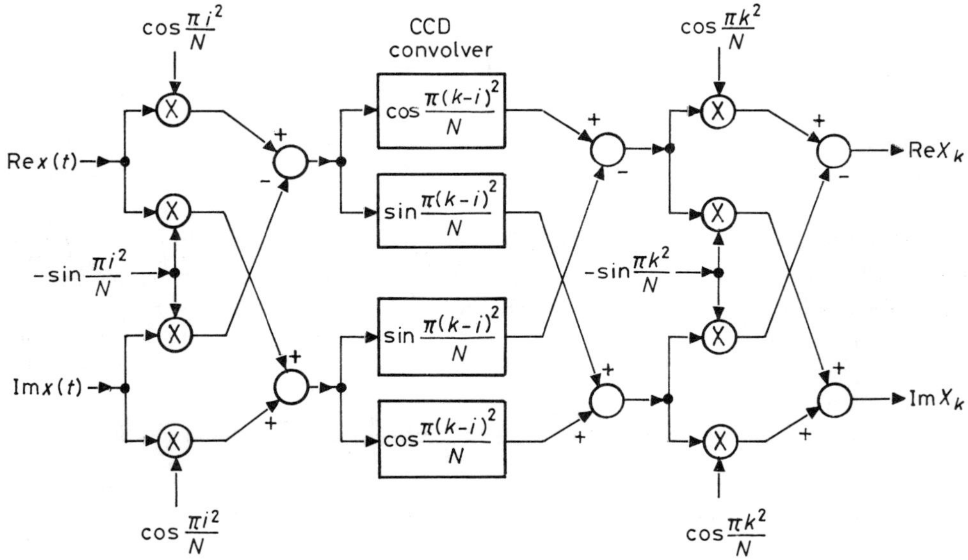

Fig. 6.81 CCD implementation of the DFT

Let us conjugate the modulating function at the output, and examine the output quantity in the kth step.

$$X'_k = e^{+j\pi\frac{k^2}{N}} \sum_{i=0}^{N-1} x_{i+k} e^{-j\pi\frac{(i+k)^2}{N}} e^{j\pi\frac{i^2}{N}} = \sum_{i=0}^{N-1} x_{i+k} e^{-j2\pi\frac{ki}{N}} . \tag{6.172}$$

X'_k is just the kth element of the DFT of the series $x_k, x_{k+1}, \ldots, x_{k+N-1}$. The successive transformed values, produced by the CCD transformer, do not belong to

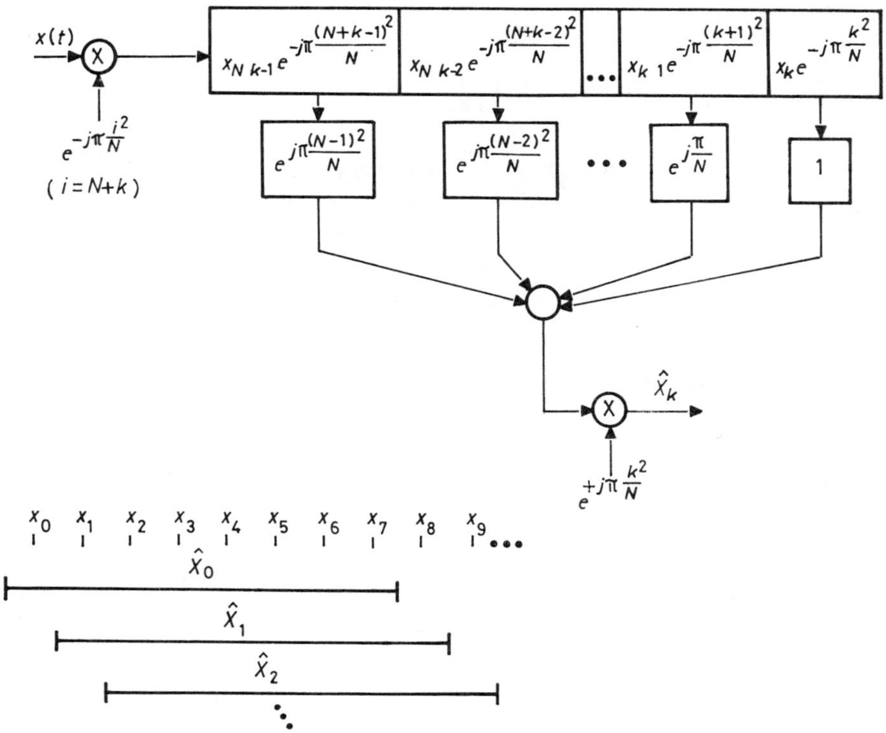

Fig. 6.82 Continuous spectrum measurement with a CCD Fourier transformer

the same record, but to records shifted in time. In spectrum analysis this is usually no problem, since the spectrum estimates are usable even if the different spectrum points do not originate from exactly the same data record.

CCD-s work with sampling frequencies from 1 kHz to a few hundred MHz, the achievable SNR is about 60 dB.

Surface acoustic wave (SAW) device

As it was illustrated in Subsection 6.5.1, SAW devices are analog delay units. Since differently delayed signals may be available by using several sensor electrodes, the chirp-z transform can be applied. However, the convolution can also be executed in a more effective way, illustrated in Fig. 6.83 (Ingebrigtsen, 1976).

The two signals to be convolved, $f(t)$ and $g(t)$ modulate the carrier of frequency ω. The resulting signals $m_1(t)$ and $m_2(t)$ are led to the two transducers on the piezo-electric crystal. The generated mechanical waves travel with velocity v towards each other. The electromagnetic field, travelling together with the mechanical potential, penetrates into the silicon crystal above.

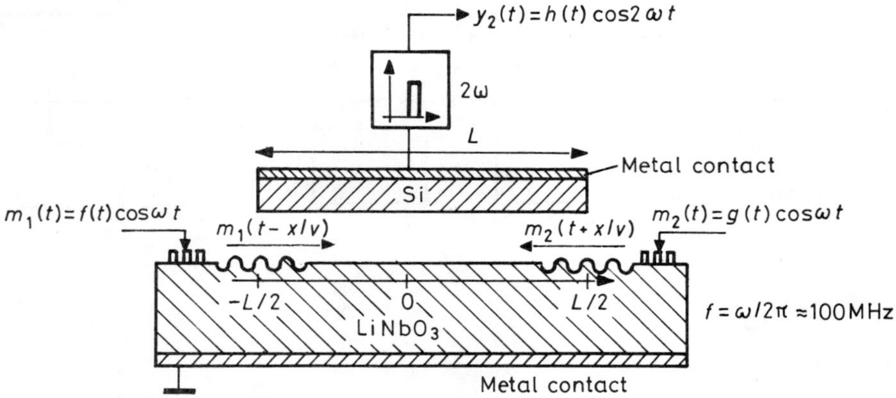

Fig. 6.83 SAW convolver

The potential at distance x is

$$U(x,t) = k_1 V\left(m_1\left(t-\frac{x}{v}\right)+m_2\left(t+\frac{x}{v}\right)\right), \tag{6.173}$$

where V is a nonlinear function.

The metal layer averages this potential on the length L:

$$y(t) = k_2 \int_{-\frac{L}{2}}^{\frac{L}{2}} V\left(m_1\left(t-\frac{x}{v}\right)+m_2\left(t+\frac{x}{v}\right)\right) \mathrm{d}x\,. \tag{6.174}$$

The signal which appears at the output of the bandpass filter at angular frequency 2ω can be calculated from the Taylor series expansion of V:

$$V(m) = V(0) + V'(0)m + \frac{V''(0)}{2}m^2 + \frac{V'''(0)}{6}m^3 + \ldots\,. \tag{6.175}$$

It can be shown that in any term of the series, the component with frequency 2ω contains an x/v term in the argument of the sine, thus these terms will be cancelled by the averaging. The only exception is the component

$$y_2(t) = k_3 \int_{-\frac{L}{2}}^{\frac{L}{2}} V\left(f\left(t-\frac{x}{v}\right)+g\left(t+\frac{x}{v}\right)\right) \mathrm{d}x \cos 2\omega t\,, \tag{6.176}$$

originating from the second-order term of the Taylor series. Consequently, the output signal with frequency 2ω contains the convolution integral of the two signals in the time interval L/v. With $L = 10$ cm and v=1000 m/s this time is 1 ms: SAW devices are to be applied for relatively high frequency signals.

Optical Fourier transform

Optical signal processing is a special means of spectrum analysis (see Papoulis, 1968). There is no space in this book to discuss every interesting detail, here the basic principles will be illustrated instead, on simple examples. It will be shown that if in the focal plane of a thin lens there is a wave front

$$f(x)e^{j\omega t} = a(x)e^{j\varphi(x)}e^{j\omega t}, \tag{6.177}$$

in the other focal plane the front $F(ky)e^{j\omega t}$ will appear, where k is a constant, y is the perpendicular distance from the optical axis, and $F(ky) = \mathcal{F}\{f(x)\}$.

The simplest examples are well known, though not in this context: rays parallel with the optical axis will be concentrated into the focal point by the lens (the Fourier transform of a constant function is a Dirac delta at zero), and vice versa (see Fig. 6.84). It is also known from elementary optics that parallel rays, falling to the lens at an angle with the optical axis, will also be focussed to a point in the focal plane, different from the focus.

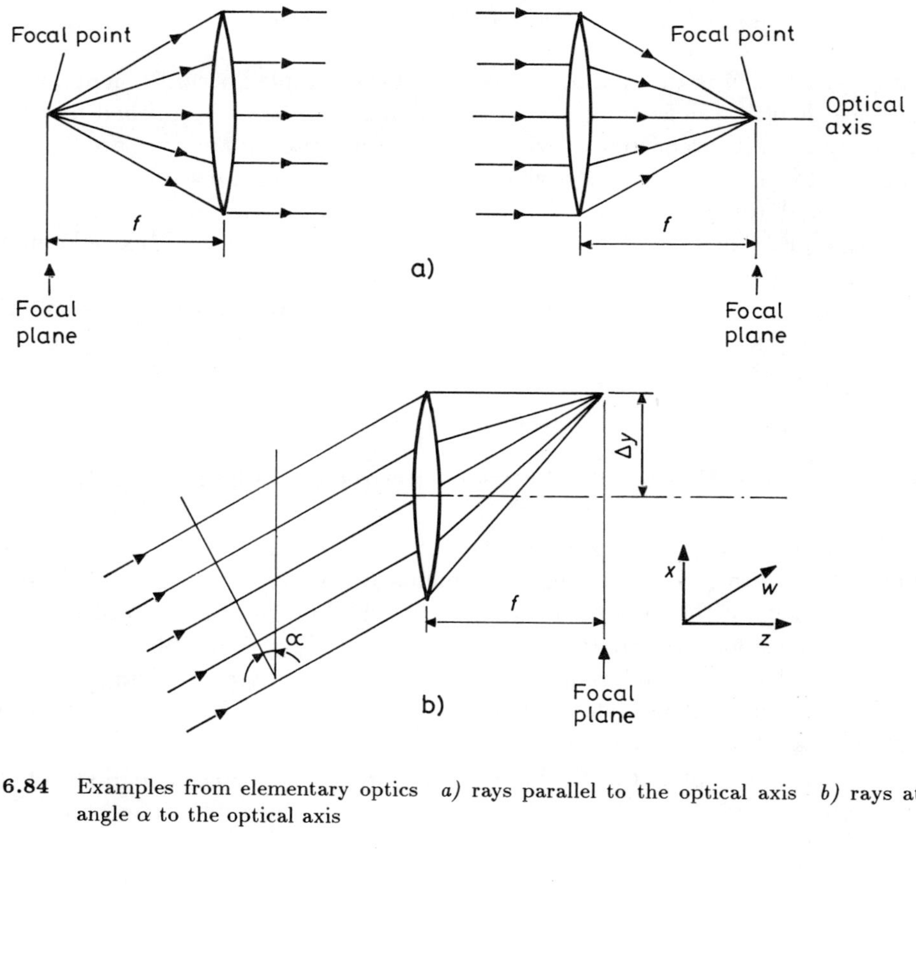

Fig. 6.84 Examples from elementary optics *a)* rays parallel to the optical axis *b)* rays at angle α to the optical axis

Let us examine this latter case in more detail. The equation of parallel rays (as a scalar wave), using the coordinate w, in the direction of the rays, is:

$$V(w,t) = k_1 e^{j\omega(t-\frac{w}{c})}\,. \tag{6.178}$$

V is any component of **H** or **B**, orthogonal to the travelling direction. In the focal plain, the phase delay as a function of x is

$$\Delta\varphi = \frac{x\sin\alpha}{c}\omega\,. \tag{6.179}$$

Substituting this into (6.178),

$$V(x,t) = k_1 e^{j\omega(t-\frac{x\sin\alpha}{c})} = k_1 e^{-j\omega\frac{x\sin\alpha}{c}} e^{j\omega t}\,. \tag{6.180}$$

(6.180) shows that the equation of a slantwise ray is a complex periodic function in the focal plain. The position of the concentration point from the ray, crossing the middle of the lens, is:

$$\Delta y = f\tan\alpha \approx f\frac{c}{\omega}f_x\,, \tag{6.181}$$

where $f_x = \omega\frac{\sin\alpha}{c}$ is the spatial 'frequency' of the complex periodic signal, and f is the focal length. Thus, complex periodic signals will be distinguished by the thin lens: the one with frequency f_x will produce strong intensity at distance $\Delta y \approx f\frac{c}{\omega}f_x$ from the focus. This is exactly what the Fourier transform does.

The above considerations do not give a proof, they are rather illustrations of the principle. The exact treatment can be given by solving the Maxwell equations (see Papoulis, 1968).

The Fourier transformed wave can be modified by optical means (filtering), and applying a second transform, complex data (image) processing equipment can be constructed.

6.7.7 Digital Fourier Analyzers — Errors and Their Reduction

In the low frequency domain (up to 100 kHz) digital Fourier analyzers are the basic tools. These are rather complex instruments, with a lot of built-in facilities that should be set by the user. Therefore, understanding the basic problems of such instruments cannot be avoided for the proper and effective use of them.

Digital Fourier analyzers are based on the periodogram, calculated from the DFT of the input signal. The basic problems were already briefly mentioned, they will be treated here in more detail.

First let us summarize the basic difficulties and error sources of digital Fourier analyzers:

1. Consequences of the finite record length and the use of the periodogram
   a) bias of the spectrum (convolution with the window function):
      - limited frequency resolution
      - spectral leakage which limits the dynamic range
      - picket fence effect (biased amplitude measurement of sine waves)
   b) large variance of the spectrum estimate of signals with smooth power spectral density
2. Consequences of the discrete algorithm
      - frequency band, limited to half of the sampling frequency
3. Consequences of the digital implementation
   a) bias and variance, originating from the finite bit numbers in
      - the ideal uniform quantizer
      - the coefficients or the samples of the window function
      - the FFT
      - averaging
      - the complex demodulator (see later)
   b) speed limits:
      - conversion rate of the A/D converter
      - FFT processing time
      - speed of the complex demodulator
   c) nonuniform relative frequency resolution (due to the FFT)
4. Nonideal behaviour of signal processing units:
   a) errors associated with sampling
      - finite rolloff of the anti-aliasing filter, nonlinear distortion and passband ripples
      - timing uncertainties (jitter), frequency inaccuracy, droop etc.
      - erroneous samples
   b) nonidealities of the A/D converter
      - nonideal uniform characteristics (integral and differential nonlinearity)
      - limited amplitude range
      - statistical uncertainties of quantization levels.

There is no space here to analyze each of the above error sources in detail. However, the ones important in the usage of digital Fourier analyzers will be discussed. But before dealing with these error sources, an important concept, the *dynamic range* has to be defined.

There may be several definitions of the dynamic range. The one used in this book is the ratio of the power of the maximum and the minimum sine wave that can be analyzed simultaneously by the analyzer. This is limited by several factors: internal noises, spectral leakage etc. One of the basic aims of analyzer design is to achieve the largest possible dynamic range.

Effect of an ideal uniform quantizer at the input

As was shown in Subsection 6.1.3, if the signal amplitude is large enough (and the larger sine meets this requirement) the effect of an ideal quantizer can be described by the noise model. The variance of the quantization noise is $q^2/12$, and this power is uniformly distributed in the frequency band $(0, f_s)$. In the Fourier transformed signal the noise will be approximately normally distributed because of the central limit theorem. In the variance analysis of the spectrum estimate, the results of Subsection 6.7.1 can be applied.

Example 6.7 Quantization noise and the dynamic range (Kollár, 1989)

Let us calculate the dynamic range (that is, the power ratio of the maximal sine wave and the minimal detectable sine wave) for the Fourier analyzer, based on an N=512-point FFT, if no averaging is applied, and the amplitude of the minimal detectable sine wave is determined by the input quantization noise, which is produced by the B=10-bit A/D converter.

Let detectability mean that the peak belonging to the sine wave is at least twice the maximal peak of the calculated quantization noise spectrum, with 95% probability.

a) Let us assume that the sampling frequency is an integer multiple of the sine wave (that is, the picket fence effect does not play a role);

b) Let us take into account the picket fence effect, with no special window applied.

Solution: Let the quantum size be denoted by q. The peak-to-peak value of the maximal analysable sine wave is

$$U_{\mathrm{pp}} = 2^B q\,,$$

and thus the peak value is

$$U_{\mathrm{p}} = 2^{B-1} q\,.$$

In the case of sine waves the peak value in the periodogram is (neglecting the effect of the noise)

$$\frac{U_{\mathrm{p}}^2}{4\Delta f} K_{\mathrm{P}}\,,$$

where $\Delta f = 1/T$, $T = N\Delta t$ is the observation time and K_{P} is the factor introduced by the picket fence effect: if the sampling frequency is an integer multiple of the frequency of the sine wave, $K_\mathrm{p}=1$; otherwise K_p is less than 1, its minimum value for the rectangular window is

$$K_\mathrm{p} = \left(\frac{2}{\pi}\right)^2 \approx 0.41 .$$

The quantization noise is white. Since its variance is $q^2/12$, the mean value of any spectrum line is

$$S_q = \frac{q^2}{12} \cdot \frac{1}{N\Delta f} .$$

The real and imaginary parts of the transformed time record are approximately independent and of Gaussian distribution. Because of this fact, at nonzero frequencies the periodogram contains random variables of χ_2^2 distribution and, as a consequence of symmetry, from these only $N/2$ variables are independent.

This χ_2^2 distribution is nothing other than an exponential distribution with parameter $1/S_q$, thus the probability of the event that every line is smaller than the value S_l is

$$P = \left(1 - e^{-\frac{S_\mathrm{l}}{S_q}}\right)^{\frac{N}{2}} .$$

Solving this for S_l the following expression is obtained:

$$S_\mathrm{l} = -S_q \ln\left(1 - \sqrt[N/2]{P}\right) ,$$

which can be well approximated using $P \approx 1$:

$$S_\mathrm{l} \approx S_q \ln\left(\frac{N}{2(1-P)}\right) .$$

From this expression the inequality for the minimal detectable sine wave with factor K_p can be obtained:

$$\frac{U_\mathrm{pd}^2}{4\Delta f} K_\mathrm{p} > 2S_q \ln\left(\frac{N}{2(1-P)}\right) = 2\frac{q^2}{12}\frac{1}{N\Delta f} \ln\left(\frac{N}{2(1-P)}\right) ,$$

and

$$U_\mathrm{pd} > \sqrt{\frac{2q^2}{3K_\mathrm{p}N} \ln\left(\frac{N}{2(1-P)}\right)} .$$

From this the dynamic range is

$$d = 20\lg\left(\frac{U_\mathrm{p}}{U_\mathrm{pdmin}}\right) = 20\lg \frac{2^B}{\sqrt{\frac{8}{3K_\mathrm{p}N} \ln\left(\frac{N}{2(1-P)}\right)}} ,$$

and, substituting the given data and using $K_\mathrm{p}=1$,

$$d = 73.7\,\mathrm{dB} .$$

The worst case is when the picket fence effect reduces the peak by a factor of $K_\mathrm{p}=(2/\pi)^2 = 0.41$, and thus the dynamic range is smaller:

$$d = 69.8\,\mathrm{dB} .$$

It should be mentioned here that when the picket fence effect is considerable, the sidelobes may seriously modify the distribution of the spectral lines around the frequency of the sine wave. This effect has been neglected here.

Finite record length and window functions

It was shown in Subsection 6.7.1 that the distortion of the spectrum can be described by the convolution of the true one with the quadratic spectral window (the Bartlett window in that case), which is the square of the linear one (the rectangular window). Both the spectral leakage and the picket fence effect are caused by this phenomenon.

However, it should be noted here that there are cases when there is no spectral distortion.

For periodic signals, when an integer number of periods is measured, neither leakage, nor picket fence effect will appear in the result of the DFT. Thus, when the sampling rate can be synchronized to the measured signal, no special consideration is necessary. This fact can be effectively utilized when transfer functions are measured with artificially generated excitation signals.

When a *transient signal* is measured with appropriately long observation time, the finite value of T will not cause any distortion.

In the above cases no special techniques have to be used, moreover, the application of a window function will even cause extra problems. Therefore, additional windows are to be switched off.

With the exception of these cases, in the calculated spectrum some distortion will generally be present, thus it must be reduced in some way. Since this distortion depends on the window shape, the idea arises to modify the window. Indeed, windows can be designed with small sidelobes (Rife and Vincent, 1970; Harris, 1978; Nuttall, 1981; Nagy, 1991), and, if desired, a flat-top main lobe (Pendergrass and Farnbach, 1978; Nagy, 1991). In the literature several window functions are described, here some typical and often used windows will be presented.

The most popular family of windows is *harmonic windows*, which are defined in $(-T/2, T/2)$ as

$$w(t) = \sum_{n=-N_{\mathrm{w}}}^{N_{\mathrm{w}}} A_n \cos 2\pi \frac{n}{T} t\,. \tag{6.182}$$

The frequency domain equivalent of these windows is

$$W(f) = \sum_{n=-N_{\mathrm{w}}}^{N_{\mathrm{w}}} A_n W_{\mathrm{rect}}\left(f - \frac{n}{T}\right). \tag{6.183}$$

If a sine wave of frequency k/T is measured, it is easy to see that the height of the corresponding peak will be

$$\frac{U_{\mathrm{P}}^2}{4} A_0^2 \frac{1}{T}\,, \tag{6.184}$$

see (6.162). Therefore, these windows are usually normalized to $A_0 = 1$.

In Table 6.3 the most important properties of some windows are summarized (see also Fig. 6.85). The simplest window, the Hann or Hanning window is defined with $A_1 = 0.5$. For the Hamming, Blackman, and Blackman-Harris windows the maximum sidelobe is minimized; for the Flat Top window also the flatness of the

Table 6.3 Properties of window functions

Window	Coeff. (A_0=1)	Sidelobe	Rolloff	Noise BW	Ampl. error
Rect		-13.3 dB	-20 dB/dec.	1	-36.3%
Hann (Hanning)	A_1=0.5	-31.5 dB	-60 dB/dec.	1.5	-15.1%
Hamming	A_1=0.429	-43.1 dB	-20 dB/dec.	1.37	-18.1%
Blackman	A_1=0.58602 A_2=0.09228	-71.5 dB	0 dB/dec.	1.70	-12.2%
Blackman-Harris (4-term)	A_1=0.672742 A_2=0.187879 A_2=0.014639	-98.1 dB	0 dB/dec.	1.98	-9.32%
Blackman-Harris (5-term)	A_1=0.7294503 A_2=0.2716618 A_3=0.0441349 A_4=0.0019569	-125.3 dB	0 dB/dec.	2.22	-7.53%
Flat top (5-term)	A_1=0.96441 A_2=0.64446 A_3=0.19801 A_4=0.01717	-95.3 dB	0 dB/dec.	3.77	-0.19%

main lobe is adjusted in order to minimize the picket fence effect. If both the main lobe flatness and the sidelobes are optimized, the optimization goal is not definite; thus the given window represents just one possibility.

Windowing can be applied at three different stages of spectrum estimation: to the time record (linear windowing in the time domain), to the result of the DFT (linear windowing in the frequency domain, spectral windowing), and to the periodogram (quadratic windowing). The quadratic window may also be applied to the correlation estimate (lag windowing). The bias of the spectral estimate will be determined by all the applied windows. However, while the expected value of the leaked power lines can be modified by the quadratic window, it can be seen that a large part of the associated variance of the lines will remain, since by squaring a zero mean random term is produced, which cannot be eliminated any more by quadratic windowing. Therefore, spectral leakage must be decreased by the use of linear windowing: it is 'too late' to try to eliminate the leakage effects when already having calculated the periodogram.

The problem of spectral leakage is solved using appropriate windows. The power (and from the power also the amplitude) of the sine can be determined either by direct read-off using a window with moderate picket fence effect, or by summing the power of the appropriate spectral lines.

For direct read-off of sine levels, the window functions need to be normalized as above. However, by this the total signal power is changed: the spectral values of smooth-spectrum signals (or, in general, the power calculated by the integration of the measured spectrum in a band) have to be corrected, see the noise bandwidth figures in Table 6.3.

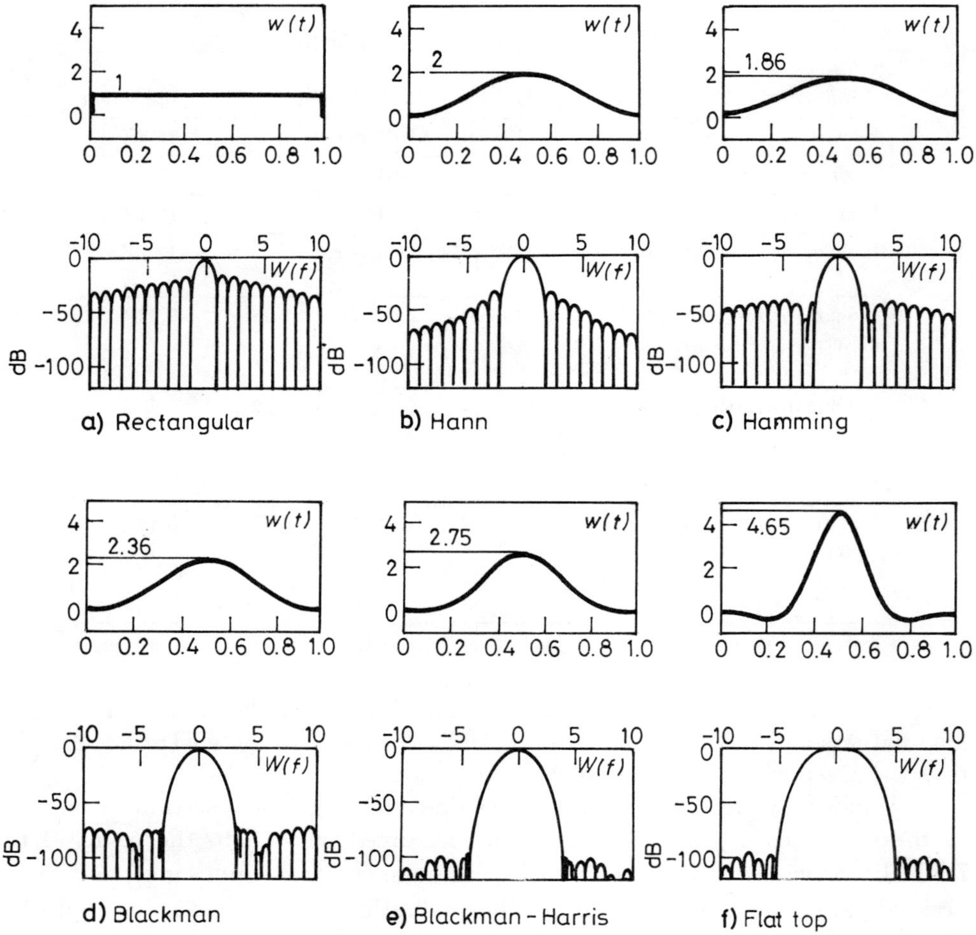

Fig. 6.85 Window functions in time and frequency domains *a)* rectangular *b)* Hann *c)* Hamming *d)* Blackman *e)* Blackman-Harris, 4-term *f)* flat top, 5-term

Frequency resolution

The frequency resolution of the periodogram is of the order of $1/T$. T is usually limited by the FFT length and the minimum sampling frequency.

There is a straightforward solution for the improvement of the accuracy of frequency measurement with DFT (FFT) in the case of sinusoidal signals: the so-called interpolated FFT (Rife and Vincent, 1970; Renders et al, 1984). The basic idea is that the window function can be fitted to the spectral lines, and the position of the fitted window gives the sine frequency with much higher accuracy than $1/T$. This method is very effective if the SNR is high. The variance of the estimated parameters was investigated e. g. by Schoukens et al (1991), and Offelli and Petri (1991).

A general possibility of the increase of T without the need of a longer FFT is complex demodulation or band-selectable Fourier analysis (McKinney, 1975). This method selects a narrow band of interest, and analyzes it with high resolution. The idea is very simple: the band of interest is to be transformed to zero frequency, the other frequencies are to be filtered out, and what remains can be sampled at a very low rate, allowing a large value of T.

However, the transform to zero frequency has to be done with care. Conventional mixing by a sinusoidal signal causes spectral splitting, and the two spectrum parts added around zero cannot be separated any more (Fig. 6.86).

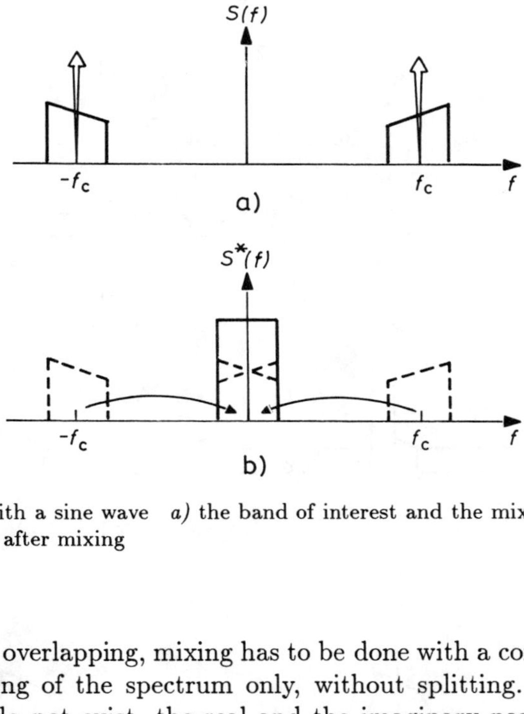

Fig. 6.86 Mixing with a sine wave *a)* the band of interest and the mixing frequency *b)* the spectrum after mixing

To avoid this overlapping, mixing has to be done with a complex exponential, which causes shifting of the spectrum only, without splitting. Though complex signals physically do not exist, the real and the imaginary parts can be handled separately, in two channels. This is realized in complex demodulation (Fig. 6.87).

The highest frequency, present in the input signal, is limited by the anti-aliasing filter, then the signal is sampled above the Nyquist rate. The spectrum is shown in Fig. 6.87b; it is to be measured around f_c, thus the signal is mixed with $e^{j2\pi f_c t}$. This complex modulation cannot be directly realized, the real and imaginary parts $\mathrm{Re}x_i*$ and $\mathrm{Im}x_i*$ are calculated instead. The spectrum in Fig. 6.87c is that of the complex signal.

The desired frequency band can be cut out by an appropriate lowpass filter pair. Since the bandlimit is reduced, the signal can be resampled at a much lower rate (Fig. 6.87e), or the filtered signal has to be calculated for these time instants only. The goal is achieved: the complex sampled signal $\{x_j\}$ has the desired spectrum, and is sampled at the optimum rate to have maximum frequency resolution; the FFT can be performed on the complex samples.

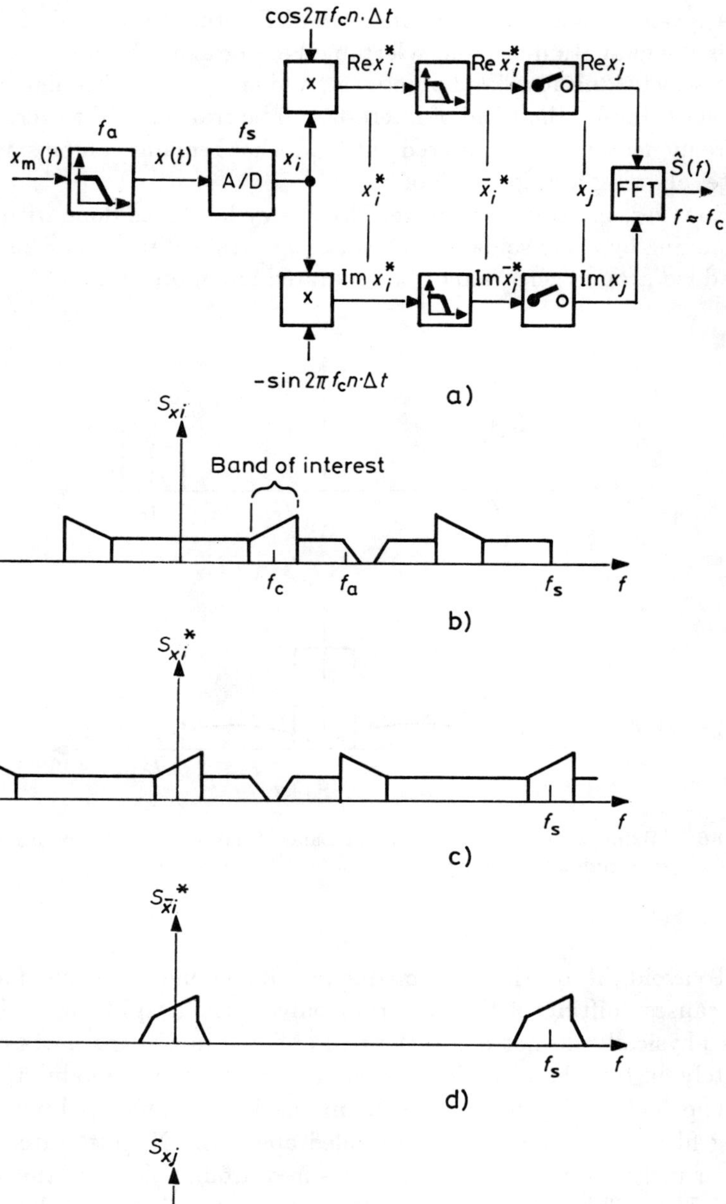

Fig. 6.87 Complex demodulation or band-selectable Fourier analysis *a)* block diagram *b)* selection of the band of interest *c)* result of complex mixing *d)* result of lowpass filtering *e)* result of decimation (Reprinted by permission from McKinney, 1975; © 1975 Hewlett-Packard Co.)

There is one crucial point in the above complex signal processing: the two channels must perform identical processing steps. Since even modulation is done digitally, this requirement can be exactly fulfilled.

Variance reduction

The relative variance of the periodogram of stochastic signals with smooth spectra is 100%, and does not depend on the observation length: the periodogram is an inconsistent estimate of the spectrum. Consequently, it has to be modified to reduce the variance.

The basic method of variance reduction is straightforward: the periodogram is calculated M times from M independent sample records, and the results are averaged (see Fig. 6.88a). The variance is reduced by a factor of $1/M$.

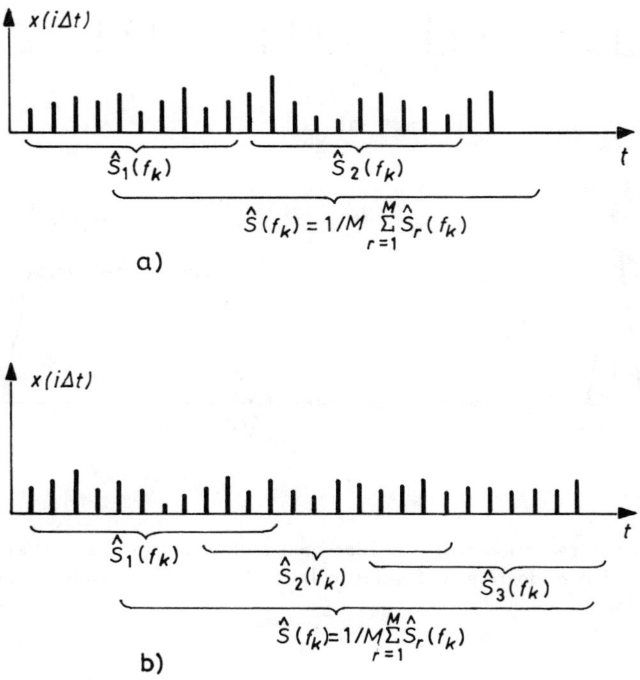

Fig. 6.88 Variance reduction by averaging of periodograms *a)* usual method *b)* Welch's method, based on overlapped segments

The efficiency can be increased if the periodograms are calculated from overlapped sample records (Welch, 1967, see Fig. 6.88b). It can be shown (Kollár and Nagy, 1982) that usually even 50% overlapping will not significantly deteriorate the factor $1/M$, especially if windowing is applied (weighted overlapped segment averaging, WOSA): the individual periodograms remain practically independent. This method is important for the analysis of low frequency signals, since for these the measurement time may be very long.

The Blackman and Tukey (or Bartlett's) method (Fig. 6.89) follows another approach. The periodogram of a long sample record is calculated, and the neighbouring M spectral estimates are averaged. Since these are uncorrelated, the variance is decreased again by a factor of $1/M$.

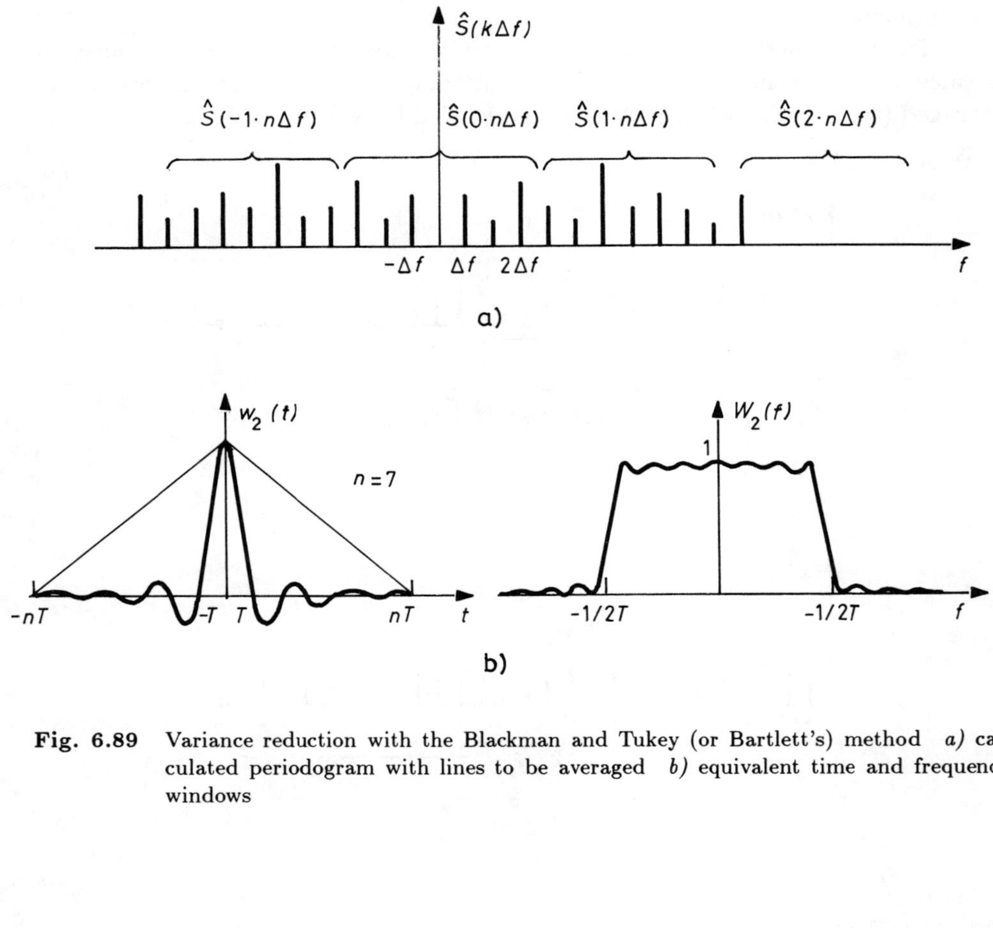

Fig. 6.89 Variance reduction with the Blackman and Tukey (or Bartlett's) method *a)* calculated periodogram with lines to be averaged *b)* equivalent time and frequency windows

This latter method has a side effect: since spectrum estimates belonging to slightly different frequencies are averaged, the shape of the window function is changed (Fig. 6.89b). Bartlett's method can be considered as a quadratic windowing (convolution) with a window of rectangular shape (equal weights) in the frequency domain. The spectral lines can also be averaged with different weights, thus the window shape can be modified during averaging, or changed afterwards, using another quadratic window.

Roundoff errors

Digital instruments always suffer from roundoff errors. From this aspect, the most critical part of FFT analyzers is the FFT processor, since to improve calculation speed, usually block-float number representation is applied.

The roundoff error depends on the actual circumstances: the type and properties of the signal, the number representation, the structure and length of the FFT, the rounding algorithm. In the literature there are different formulas given for the mean square error of the FFT result. For example, Welch (1969) gave the following formula: for two's complement number representation and white random noise of uniform distribution,

$$\frac{\psi\{Z\}}{\psi\{X\}} < h = \frac{2^{\frac{M+3}{2}} 2^{-B} 0.9}{\psi\{x\}} = \frac{\sqrt{N} 2^{-B} 1.2}{\psi\{x\}}, \tag{6.185}$$

where $M = \log_2 N$, B is the bit number (without the sign), ψ denotes the rms value, x denotes the input series, X the output series and Z the error of X.

Assuming that the roundoff error is white and normally distributed, this formula can be used for the estimation of the power of the roundoff noise. However, this assumption is not valid for large sinusoids, since the amplitude increases in the FFT from stage to stage (from 1 to $N/2$), and since the whole block is rescaled regularly to avoid overflow, the different roundoff error sources have different weight, and the power is not uniformly distributed in the frequency band. Anyway, a typical result is that 16-bit number representation (15 bits + sign) limits the dynamic range to approximately 75 dB.

6.8 MEASUREMENT OF LINEAR SYSTEMS

The analysis of linear systems is an often arising task in engineering: examples are the checking of the transfer function of a filter, or the identification of the multiple-element models of impedances. Most of the basic principles of the measurement of system description quantities are common with signal analysis methods, this is why measurement of systems is treated in this chapter.

The measurement of systems — *system identification* — can be based on two basic approaches (see also Subsection 6.1.8). In the *nonparametric approach* the desired functions (impulse response, transfer function etc.) are measured as functions without a priori information about the structure of the system. Since no extensive information is required, this approach is be taken in most of the commercially available instruments, thus mainly these methods will be treated in this section.

The *parametric approach* means that a certain structure is assigned to the system under test (rational form of the transfer function with or without delay; the physical structure of the model, e. g. a parallel or serial RLC circuit etc.), and the aim of the measurement is to determine the actual values of the parameters from the measurements. Modern impedance analyzers already offer such services.

System identification is a stand-alone subject; only some of the basic ideas can be presented here. More details can be found in the literature (Bendat and Piersol, 1971, 1980; Eykhoff, 1974; Goodwin and Payne, 1977; Sorenson, 1980; Norton, 1986; Ljung, 1987; Söderström and Stoica, 1989; Schoukens and Pintelon, 1991).

Similarly to signal analysis, system identification can be performed in time or in frequency domain. Time domain methods were first elaborated, and the above mentioned books deal mostly with these, with the exception of the work of Schoukens and Pintelon (1991), which is devoted to frequency domain system identification.

Parametric or nonparametric, time domain or frequency domain, system identification is a rather complicated task, and requires the careful design of the whole procedure, from the optimum choice of excitation signals through the choice of the object function of the optimization, and the numerical solutions, to the verification of the results. Sometimes systematic methods can solve only a part of the problem, and much is based on intuition.

In the following section some of the basic procedures will be presented, with a focus on the basics of commercially available instruments. No attention will be paid to the physical dimensions of the measured quantities, signal and system theoretical aspects will be emphasized instead.

6.8.1 Measurement of the Impulse Response and of the Step Response

Theoretically, the impulse response is the output signal when the system is excited by a Dirac delta function. Consequently, the direct measurement of the impulse response is only approximately possible. The main difficulties encountered:

- the Dirac delta function has to be approximated by a large amplitude, narrow impulse,
- the signal-to-noise ratio may be quite bad, the amplitude usually has to be large to provide an acceptable SNR,
- the power cannot be concentrated to the 'interesting' frequency bands,
- the system may behave nonlinearly when excited by a large amplitude,
- recording may be difficult because of the possible large dynamic range and sometimes long duration of the response.

Because of the above difficulties, the impulse response is rarely measured in a direct way which has the most important applications in the situations where impulse-like excitation can only be applied (e. g. explosion tests in geophysics).

It often happens that the excitation impulse is a bad approximation of the Dirac delta function (its duration is not significantly less than the smallest time constant of the system). In these cases both the input and the output signals can be measured, and the following deconvolution problem has to be solved: determine $h(t)$ if $x(t)$ and $y(t)$ are measured in

$$y(t) = x(t) \star h(t) . \tag{6.186}$$

The difficulties of deconvolution were briefly discussed in Subsection 6.1.6.

Sometimes the step response is much easier to measure, first of all because of the required moderate dynamic range, and because the excitation signal is easier to generate.

Theoretically, the impulse response can be calculated from the measured step response by numerical derivation. However, this often gives useless results, because derivation amplifies the usually wide-band measurement noises.

The impulse response is nothing other than the inverse Fourier transform of the frequency response (or transfer function). Thus, another possibility is the measurement of the frequency response followed by inverse Fourier transform.

The impulse response can also be measured applying stochastic excitation signals. A possibility is to apply an additive wide-band white noise at the input: the cross correlation between the output and the noise will give the impulse response (Fig. 6.90):

$$R_{ny}(\tau) = h(t) \star R_{nn}(\tau) \approx h(\tau) . \tag{6.187}$$

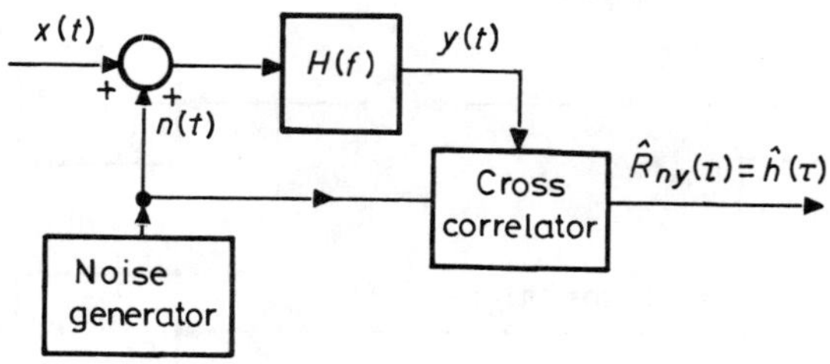

Fig 6.90 Measurement of the impulse response on a system in operation with external stochastic excitation

In Fig. 6.90 $x(t)$ is the operational input signal of the system.

Sometimes the measurement of cross correlation may give important information on the impulse response of the system, even if the input noise is not strictly white in the desired band. A typical application is the study of acoustic noise propagation paths in mechanical systems: the maxima of the cross correlation function give the delay of the individual propagation paths, enabling the designer to reduce annoying noise e. g. in vehicles.

6.8.2 Measurement of the Frequency Response

The simplest method of the measurement of the frequency response is to measure response of the system to a given sine wave in the stationary state. Since the measurement of sinusoidal voltages and of the phase between the input and output sine waves is required, the methods discussed in Chapters 2 and 5 can be applied (e. g. phase sensitive selective voltmeters can be used). During the measurement two important aspects are to be considered:

- the measurement is valid only if the system transients are over,
- the distortion components and the noise must not spoil the measurement, usually it is advisable to use a selective voltmeter (Chapter 2).

Selectivity can be significantly increased by using a so-called orthogonal correlator (Fig. 6.91). The system under test is excited by the usual input signal, and in addition by the sinusoidal output signal of the generator, which also generates a sine wave with 90° phase shift. The output signal is cross correlated with both signals, thus the real and imaginary parts of the response are measured. When the measurement time is long enough, the result is practically not disturbed by the input $x(t)$. It should be mentioned that from a signal processing point of view, the orthogonal correlator is essentially a very selective, phase sensitive voltmeter.

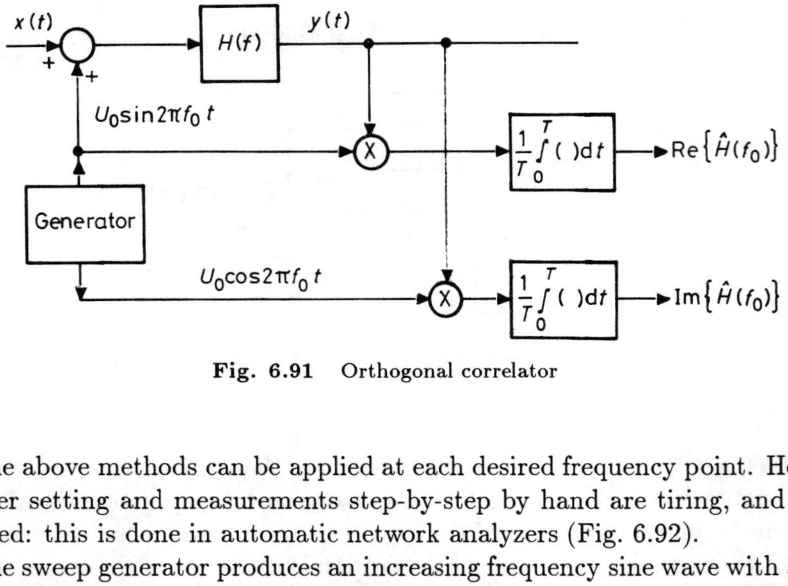

Fig. 6.91 Orthogonal correlator

The above methods can be applied at each desired frequency point. However, parameter setting and measurements step-by-step by hand are tiring, and can be automated: this is done in automatic network analyzers (Fig. 6.92).

The sweep generator produces an increasing frequency sine wave with a stabilized amplitude level, and the parameters of the output sine are measured by some kind of classical means. The sweep speed must be small enough to measure always in (quasi-)steady-state.

The phase usually can be measured with an accuracy better than 1°, e. g. by the measurement of appropriate fill-in factors (Chapter 5). At frequencies over 10–100 MHz it is often enough and also easier to measure rather the group delay instead of the phase.

In the following paragraphs some typical solutions used in analog network analyzers will be briefly described.

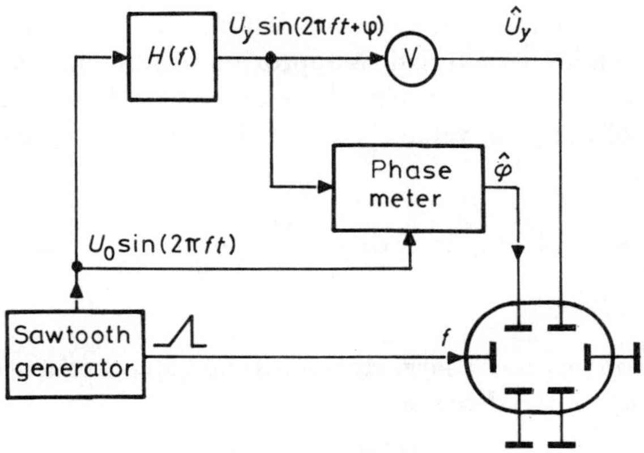

Fig. 6.92 Automatic network analyzer

In the analyzer of Fig. 6.93 (Nelson et al, 1974) the output signal is mixed down to frequency f_0. The mixing signal is correct in phase, because it is synchronized to the excitation signal by the PLL. The phase response can only be measured correctly when the total phase shift of the bandpass and the lowpass filters at f_0 is zero, or at least known.

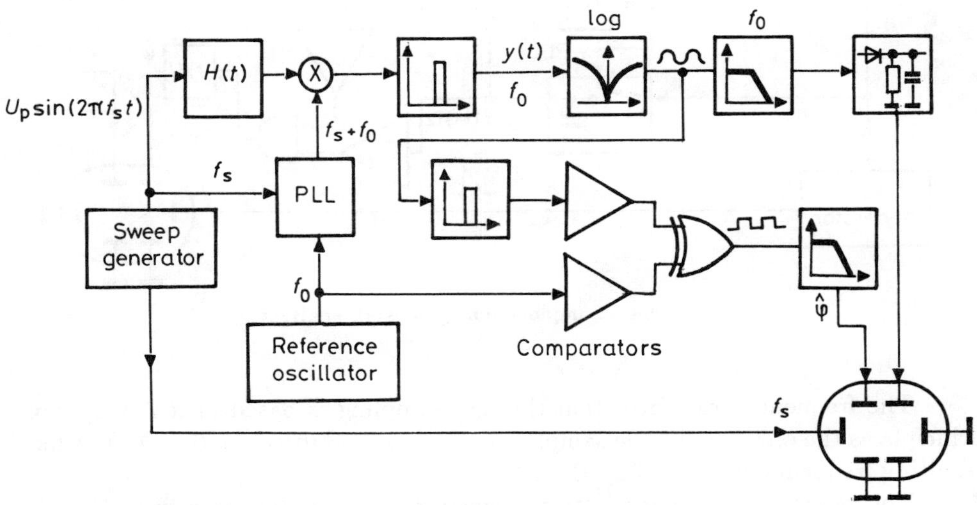

Fig. 6.93 Analog network analyzer with mixing

The signal at the intermediate frequency is processed in two channels:

- the coincidence gate converts the phase difference between the reference signal and the measured signal to a fill-in factor, which is further converted to DC voltage by the lowpass filter.

- the intermediate frequency signal is led directly to the logarithmic amplifier. The logarithmic characteristic is approximate only, since for $U_\mathrm{p} \to 0$ the output would tend to $-\infty$. The base harmonic is in direct relationship with the logarithm of the peak value U_p of the IF frequency signal:

$$U_1 = \frac{1}{\pi}\int_0^\pi \sin(x)\ln(U_\mathrm{p}\sin x)\,\mathrm{d}x = \frac{\ln U_\mathrm{p}}{\pi} + \frac{2}{\pi}(\ln 2 - 1) = \frac{1}{\pi}\ln\big(U_\mathrm{p}(2\ln 2 - 2)\big)\,. \tag{6.188}$$

  This solution provides a large dynamic range, a consequence of the logarithmic amplifier *before* the detector.

The network analyzer in Fig. 6.94 (Zellers, 1979) is similar to the above one, with some interesting solutions, like the mixing realized via sampling, or the amplitude detector, based on a feedback controlled amplifier.

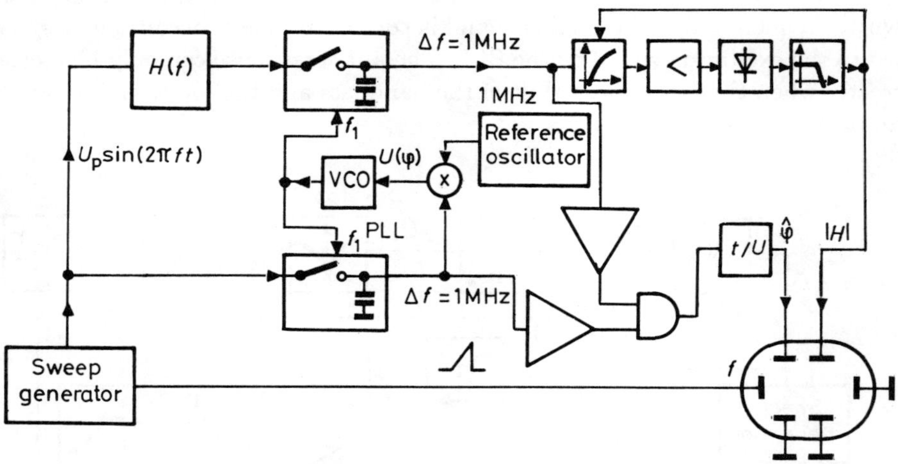

Fig. 6.94 Sampling analog network analyzer

The frequency transformation through sampling is based on the very same principle as the one used by the sampling oscilloscope (Subsection 6.2.1). The fixed resulting frequency is assured by the PLL.

The reference frequency is transformed in exactly the same way, thus the relative phase difference of the two signals remains the same, and can be measured at the lower intermediate frequency, as the fill-in factor at the output of the AND gate.

The amplitude response can also be measured by an analog spectrum analyzer, using a tracking generator (see Fig. 6.95 and also Subsection 6.7.3). However, the spectrum analyzer cannot measure the phase. If necessary, a minimum phase response can be calculated from the amplitude response, though the result is usually rather inaccurate.

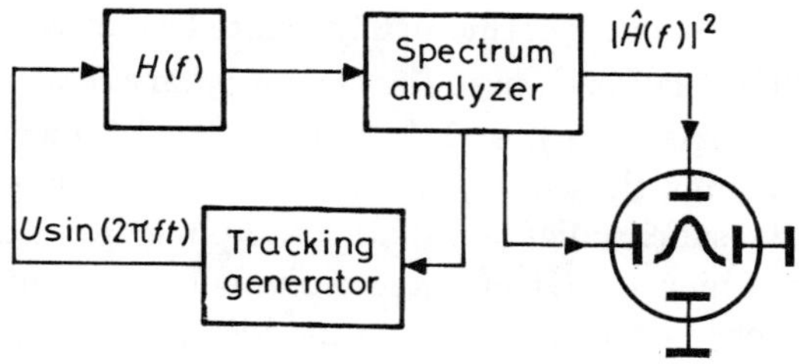

Fig. 6.95 Network analysis with tracking generator

Digital network analyzers

At lower frequencies sweeping network analyzers need a long measurement time, since they measure at just one frequency at a time. An improvement can be achieved by instruments which work at several frequencies simultaneously. These are digital correlators and digital network analyzers. These apply one of the following equations, valid for linear systems:

$$Y(f) = H(f)X(f) \tag{6.189}$$

$$R_{xy}(\tau) = h(\tau) \star R_{xx}(\tau), \quad R_{yy}(\tau) = h(\tau) \star R_{yx}(\tau) \tag{6.190}$$

$$S_{xy}(f) = H(f)S_{xx}(f), \quad S_{yy}(f) = H(f)S_{yx}(f). \tag{6.191}$$

$S_{yy}(f) = |H(f)|^2 S_{xx}(f)$ can also be used (e. g. in spectrum analyzers), but the phase information is not measured, though theoretically it can be reconstructed from $|H(f)|^2$ for minimum phase systems.

Further considerations depend on the actual measurement task itself:

- whether the excitation signal can be measured noise-free,
- whether the system is internally noise-free,
- whether the system has to be measured under operation conditions (with excitation, undesirable for the measurement, at the input),
- what kinds of excitation signals may be applied,
- etc.

These questions are discussed in the literature of system identification, here some basic principles can only be given.

The most straightforward method is to measure the input and output signals, calculate the Fourier transforms, and calculate $H(f)$ from (6.189). This can be done with both transient and periodic excitations. However, in order to have a high signal-to-noise ratio, periodic signals are the most advantageous. There are several forms of periodic excitations (see Schoukens and Pintelon, 1991); the most important ones are multisines (harmonically related sine waves with phases adjusted to have small crest factors), pseudo-random binary sequences (PRBS, binary signals having a white spectrum), and discrete interval binary sequences (binary signals with zero crossings chosen appropriately to approximate the desired excitation spectrum). By synchronization of the sampling frequency to the periodic excitation, it can be achieved that an integer number of periods is measured, thus no leakage or picket fence effect will disturb the identification procedure.

$H(f)$ used to be calculated as $Y(f)/X(f)$. However, if the input and output signals are both measured with observation noise, the result will be distorted: in this case synchronous averaging has to be performed, and/or special frequency domain identification techniques have to be applied (Schoukens and Pintelon, 1991). Since both $Y(f)$ and $X(f)$ are complex, and the observation noise is also complex, the ratio $Y(f)/X(f)$ has uncommon properties. If the results of several measurements are to be averaged, the complex *geometric mean* is desirable instead of the commonly used arithmetic mean (Schoukens and Pintelon, 1990). If both $X(f)$ and $Y(f)$ contain practically observation noise only, the expected value of the ratio will be around σ_y/σ_x (since $|Y(f)|^2/|X(f)|^2$ has F-distribution), thus special attention has to be paid for proper excitation at every frequency point of interest, otherwise misleading results will be obtained.

In a noisy environment, when synchronous averaging of measured records cannot be performed, the measurement of the cross spectrum may also help (see Subsection 6.191). This can be estimated by the generalized periodogram:

$$\widehat{S}_{xy}(f) = \frac{1}{T}\overline{X(f,T)}Y(f,T)\,, \tag{6.192}$$

which is a generalization of the regular periodogram (see Subsection 6.7.1), and can be evaluated by two-channel Fourier analyzers. The treatment of the generalized periodogram (window functions, averaging etc.) is very similar to that of the regular one, with one important difference: the relative variance for stochastic signals with smooth spectrum is not 100%, but

$$\frac{\operatorname{var}\{\widehat{S}_{xy}(f)\}}{|S_{xy}(f)|^2} \approx \frac{S_{xx}(f)S_{yy}(f)}{|S_{xy}(f)|^2} = \frac{1}{|\gamma(f)|^2}\,, \tag{6.193}$$

where $|\gamma(f)|^2$ is the so-called magnitude-squared coherence function. If the input and the output are more or less independent of each other, the relative variance may be large: the estimate may be useless.

There is an additional problem of the correlation (or power spectral) method that must also be mentioned. As seen from (6.190) and (6.191), for the determination of $H(f)$ the measurement of the input quantities ($R_{xx}(t)$, $S_{xx}(f)$) or the output quantities ($R_{yy}(t)$, $S_{yy}(f)$) is also necessary. If input and/or output observation noise is present, the relevant quantities can be measured with bias only.

6.8.3 Measurement of the Coherence Function

In the measurement of linear networks it is usually assumed that the system may contain no internal signal or noise sources, and the system works in linear mode. These assumptions have to be checked: a tool is the measurement of the magnitude-squared coherence function:

$$|\gamma(f)|^2 = \frac{|S_{xy}(f)|^2}{S_{xx}(f)S_{yy}(f)}, \tag{6.194}$$

which can be easily measured by a digital Fourier analyzer.

The coherence function is a measure of the portion of the output power at a given frequency, resulting by linear transformation from the same frequency in the input signal. If the system is linear, and no extra noise sources are present,

$$|\gamma(f)|^2 = \frac{|S_{xy}(f)|^2}{S_{xx}(f)S_{yy}(f)} = \frac{|H(f)S_{xx}(f)|^2}{S_{xx}(f)|H(f)|^2S_{xx}(f)} = 1\,. \tag{6.195}$$

It can also be proved that in general $|\gamma(f)|^2 \leq 1$, and if e. g. an independent noise source is present in the system, it is definitely smaller than 1.

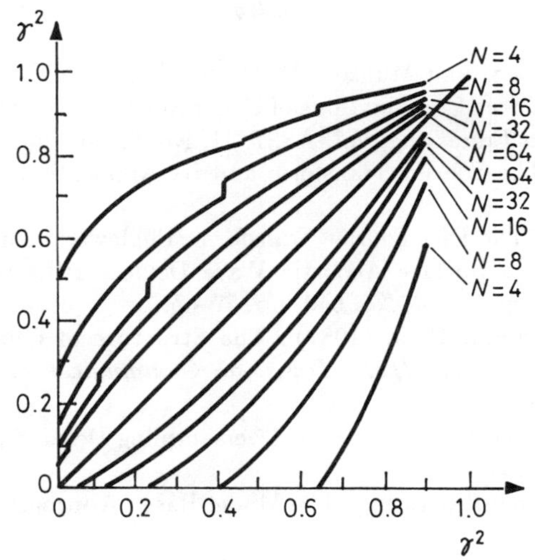

Fig. 6.96 Confidence intervals of the coherence function (Reproduced by permission from the paper of Beningus (1969); © 1969, IEEE)

From (6.194) it can be seen that the measurement of the coherence function is based on the measurement of spectra. This has been extensively discussed in the previous chapters. However, at the coherence function there is an additional difficulty, namely because of the division the variance is large and difficult to calculate.

The coherence function needs to be evaluated using averaged spectrum estimates (Pendergrass, 1978). Beningus (1969) gave the confidence intervals for coherence estimates of stochastic signals with smooth spectra if N (generalized) periodograms were averaged in order to obtain the spectral estimates (see Fig. 6.96).

References

Agarwal, R. C. and Cooley, J. W. (1977): New Algorithms for Digital Convolution. *IEEE Trans. on Acoustics, Speech and Signal Processing*, Vol. 25, No. 5, pp. 392–410, June 1977.

Balakrishnan, A. V. (1962): On the Problem of Time Jitter in Sampling. *IRE Trans. on Information Theory*, Vol. 8, No. 3, pp. 226–36, April 1962.

Bendat, J. S. and Piersol, A. G. (1971): Random Data: Analysis and Measurement Procedures. New York, John Wiley & Sons.

Bendat, J. S. and Piersol, A. G. (1980): Engineering Applications of Correlation and Spectral Analysis. New York, John Wiley & Sons.

Beningus, V. A. (1969): Estimation of the Coherence Spectrum and Its Confidence Interval Using the Fast Fourier Transform. *IEEE Trans. on Audio and ELectroacoustics*, Vol. 17, No. 2, pp. 145–150, June 1969.

Bennett, W. R. (1948): Spectra of Quantized Signals. *Bell System Technical Journal*, Vol. 27, No. 4, pp. 446-472.

Bergland, G. D. (1969): Fast Fourier Transform Hardware Implementations — An Overview. *IEEE Trans. on Audio and Electroacoustics*, Vol. AU-17, No. 2, pp. 104–8, June 1969.

Bovik, A. C., Huang, T. S. and Munson, D. C. Jr. (1983): A Generalization of Median Filtering Using Linear Combinations of Order Statistics. *IEEE Trans. on Acoustics, Speech and Signal Processing*, Vol. ASSP-31, No. 6, pp. 1342–50, Dec. 1983.

Bracewell, R. (1965): The Fourier Transform and Its Applications. New York, McGraw-Hill.

Brigham, E. O. (1974): The Fast Fourier Transform. Englewood Cliffs, NJ, Prentice-Hall.

Butler, M. B. N. (1980): Surface Acoustic Wave Devices I-II. *Electronic Engineering*, Vol. 52, No. 639, pp. 37–44; No. 640, pp. 57–65.

Candy, J. C. and Benjamin, O. J. (1981): The Structure of Quantization Noise from Sigma–Delta Modulation. *IEEE Trans. on Communications,* Vol. 29, No. 9, pp. 1316–23, Sept. 1989.

Candy, J. C. and Temes, G. C., eds. (1991): Oversampling Delta–Sigma Data Converters. New York, IEEE Press.

Candy, J. V. (1986): Signal Processing: The Model-Based Approach. New York, McGraw-Hill.

Candy, J. V. (1988): Signal Processing: The Modern Approach. New York, McGraw-Hill.

Claasen, T. A. C. M. and Jongepier, A. (1981): Model for the Power Spectral Density of Quantization Noise. *IEEE Trans. on Acoustics, Speech and Signal Processing*, Vol. ASSP-29, No. 4, pp. 914–17, Aug. 1981.

Cooley, J. W. and Tukey, J. W. (1965): An Algorithm for the Machine Calculation of Complex Fourier Series. *Mathematics of Computation*, Vol. 19, No. 90, pp. 297–301.

Cox, R. G. (1978): Window Functions for Spectrum Analysis. *Hewlett-Packard Journal*, Vol. 29, No. 13, pp. 10–11, Sept. 1978.

Crochiere, R. E. and Rabiner, L. E. (1983): Multirate Signal Processing. New York, Prentice-Hall.

Daniels, J. W. and Atchley, R. L. (1975): A Precision Spectrum Analyzer for the 10 Hz-to-13 MHz Range. *Hewlett-Packard Journal*, Vol. 26, No. 9, pp. 18–24, May 1975.

De Lotto, I. and Paglia, G. E. (1986): Dithering Improves A/D Converter Nonlinearity. *IEEE Trans. on Instrumentation and Measurement*, Vol. IM-35, No. 2, pp. 170–177.

Dobrowiecki, T. (1984): Modelling a Quantizer — Models and Possible Approaches. *Periodica Polytechnica Ser. Electrical Engineering*, Vol. 28, Nos. 2-3, pp. 159–172.

Eykhoff, P. (1974): System Identification. London, John Wiley & Sons.

Frye, G. J. and Nahman, N. S. (1964): Random Sampling Oscillography. *IEEE Trans. on Instrumentation and Measurement*, Vol. IM-13, No. 1, pp. 8–13. March 1964.

Gold, B. and Rader, C. (1969): Digital Processing of Signals. New York, McGraw-Hill.

Goodwin, G. C. and Payne, R. L. (1977): Dynamic System Identification: Experiment Design and Data Analysis. New York, Academic Press.

Hagen, J. B. and Farley, D. T. (1973): Digital-Correlation Techniques in Radio Science. *Radio Science*, Vol. 8, Nos. 8-9, pp. 275–284, Aug.–Sept. 1973.

Hamming, R. W. (1977): Digital Filters. Englewood Cliffs, NJ, Prentice-Hall.

Harris, F. J. (1978): On the Use of Windows for Harmonic Analysis with the Discrete Fourier Transform. *Proc. IEEE*, Vol. 66, No. 1, pp. 51–83, Jan. 1978.

Haykin, S. (1983): Nonlinear Methods of Spectral Analysis. Berlin, Springer-Verlag.

Heute, U. (1981): Results of a Deterministic Analysis of FFT Coefficient Errors. *Signal Processing*, Vol. 3, No. 4, pp. 321–31, Oct. 1981.

Holdaway, S. N., Molinari, D. H., Linkwitz, S. H. and Neering, M. J. (1978): Signal Processing in the Model 8568A Spectrum Analyzer. *Hewlett-Packard Journal*, Vol. 29, No. 10, pp. 9–16, June 1978.

Ingebrigtsen, K. A. (1976): The Schottky Diode Acoustoelectric Memory and Correlator — A Novel Programmable Signal Processor. *Proc. IEEE*, Vol. 64, No. 5, pp. 764–9, May 1976.

Inose, H., Yasuda, Y. and Murakami, J. (1962): A Telemetering System by Code Modulation — Δ-Σ Modulation. *IRE Trans. on Space Electronics Telemetry*, Vol. 8, No. 3, pp. 204-9, Sept. 1962.

Jazwinski, A. H. (1970): Stochastic Processes and Filtering Theory. New York, Academic Press.

Jerry, A. J. (1977): The Shannon Sampling Theorem — Its Various Extensions and Applications: A Tutorial Review. *Proc. IEEE*, Vol. 65, No. 11, pp. 1565–1596.

Katzenelson, J. (1962): On Errors Introduced by Combined Sampling and Quantization. *IRE Trans. on Automatic Control*, Vol. AC-7, pp. 58–68.

Kay, S. M. and Marple, S. L. jr. (1981): Spectrum Analysis — A Modern Perspective. *Proc. IEEE*, Vol. 69, No. 11, pp. 1380–1419, Nov. 1981.

Kolba, D. P. and Parks, T. W. (1977): A Prime Factor FFT Algorithm Using High-Speed Convolution. *IEEE Trans. on Acoustics, Speech and Signal Processing*, Vol. 25, No. 4, pp. 281–94, August 1977.

Kollár, I. and Nagy, F. (1982): On the Design and Use of FFT-Based Spectrum Analyzers. *Periodica Polytechnica Ser. Electrical Engineering*, Vol. 26, Nos 3-4, pp. 295–315.

Kollár, I. (1984a): The Variance of the Power Density Spectrum of Periodic Signals. *Measurement*, Vol. 12, No. 14, pp. 1203-1206.

Kollár, I. (1984b): Statistical Theory of Quantization: Results and Limits. *Periodica Polytechnica Ser. Electrical Engineering*, Vol. 28, Nos 2-3, pp. 173–190.

Kollár, I. (1986): The Noise Model of Quantization. *Proc. 1st IMEKO TC4 Symposium 'Noise in Electrical Measurements'*, Como (Italy), June 19-21, 1986. Bp. OMIKK-Technoinform, 1987. pp. 125–129.

Kollár, I. (1988): Understanding Quantization by Simple Models. *11th IMEKO World Congress,* Houston, Texas, Oct. 16-21, 1988. *Proceedings: Acta IMEKO 1988.* ISA, Research Triangle Park, NC, 1988. Vol. 5, pp. 487-494.

Kollár, I. (1989): Méréstechnika II. Jelanalízis Feladatgyűjtemény (Technology of Measurement II. Signal Analysis. Problems and Solutions). Textbook of the Technical University of Budapest. Budapest, Tankönyvkiadó, 1989. J5-1441. In Hungarian. (Parts available also in English.)

Kollár, I., Pintelon, R., Rolain, Y. and Schoukens, J. (1990): Equalization of Data Acquisition Channels Using Digital Filters. *Periodica Polytechnica Ser. Electrical Engineering,* Vol. 34, No. 3, pp. 165–182.

Korn, G. A. (1966): Random-Process Simulation and Measurements. New York, McGraw-Hill.

Kunt, M. (1986): Digital Signal Processing. Norwood, MA, Artech House.

Lee, Y. M. and Kassam, S. A. (1985): Generalized Median Filtering and Related Nonlinear Filtering Strategies. *IEEE Trans. on Acoustics, Speech and Signal Processing,* Vol. ASSP-33, pp. 672–83, June 1985.

Ljung, L. (1987): System Identification: Theory for the User. Englewood Cliffs, Prentice-Hall.

Loève, M. (1978): Probability Theory I.–II. New York, NY, Springer Verlag.

Mayer, G. J. (1975): The Chirp z-Transform — A CCD Implementation. *RCA Review,* Vol. 36, No. 4, pp. 759–73, Dec. 1975.

McKinney, H. W. (1975): Band-Selectable Fourier Analysis. *Hewlett-Packard Journal,* Vol. 26, No. 8, pp. 20–24, April 1975.

Mendel, J. M. (1983): Optimal Seismic Deconvolution. New York, Academic Press.

Mulvey, J. (1970): Sampling Oscilloscope circuits. Beaverton, Tektronix Inc.

Nagy, F. (1991): A Novel Method for the Design of Optimal Window Functions. Unpublished paper.

Nahman, N. S. (1986): Software correction of Measured Pulse Data. In: Thompson, J. E. and Luessen, L. H.: Fast Electrical and Optical Measurements, Vol. I. pp. 351–417. Dordrecht, The Netherlands, Martinus Nijhoff Publishers.

Nelson, G., Thomas, P. and Atchley, R. (1972): Faster Gain-Phase Measurements with New Automatic 50 Hz-to-13 MHz Network Analyzers. *Hewlett-Packard Journal,* Vol. 24, No. 2, pp. 12–19, Oct. 1972.

Norsworthy, S. R., Post, I. G. and Fetterman, H. S. (1989): A 14-bit 80-kHz Sigma–Delta A/D Converter: Modeling, Design, and Performance Evaluation. *IEEE Journal of Solid-State Circuits,* Vol. 24, No. 2, pp. 256-66, April 1989.

Norton, J. P. (1986): An Introduction to Identification. London, Academic Press.

Nussbaumer, H. J. (1982): Fast Fourier Transform and Convolution Algorithms. 2nd corrected and updated edition. Berlin, Heidelberg, New York, Springer-Verlag.

Nuttall, A. H. (1981): Some Windows with Very Good Sidelobe Behavior. *IEEE Trans. on Acoustics, Speech and Signal Processing,* Vol. 29, No. 1, pp. 84–89, Feb. 1981.

Nyquist, H. (1928): Certain Topics in Telegraph Transmission Theory. *AIEE Transactions,* Vol. 47, pp. 617–644.

Offelli, C. and Petri, D. (1991): Effect of Noise on Real-Time Parameter Estimation of Multifrequency Signals. *Conference Record of the IEEE Instrumentation and Measurement Technology Conference,* May 14–16, 1991, Atlanta, GA, pp. 217–221. 91CH2940-5.

Oppenheim, A. V. and Schafer, R. W. (1975): Digital Signal Processing. Englewood Cliffs, NJ, Prentice-Hall.

Oppenheim, A. V., Willsky, A. S. and Young, I. T. (1983): Signals and Systems. Englewood Cliffs, NJ, Prentice-Hall.

Page, J. (1971): Tracking Generators. *Hewlett-Packard Journal*, Vol. 23, No. 1, pp. 10–11, Sept. 1971.

Papoulis, A. (1962): The Fourier Integral and Its Applications. New York, McGraw-Hill.

Papoulis, A. (1968): Systems and Transforms with Applications in Optics. New York, McGraw-Hill.

Papoulis, A. (1977): Signal analysis. New York, McGraw-Hill.

Péceli, G. (1986): A Common Structure for Recursive Discrete Transforms. *IEEE Trans. on Circuits and Systems,* Vol. 33, No. 10, pp. 1035-36, Oct. 1986.

Peebles, P. Z. (1976): Communication System Principles. Reading, MA, Addison-Wesley.

Peled, A. and Liu, B. (1976): Digital Signal Processing. New York, John Wiley & Sons.

Pendergrass, N. A. (1978): Coherence Function and Averaging Boost Confidence in Spectrum Measurements. *Electronics*, Vol. 51, No. 19, pp. 132–37, Sept. 1978.

Pendergrass, N. A. and Farnbach, J. S. (1978): A High-Resolution, Low-Frequency Spectrum Analyzer. *Hewlett-Packard Journal*, Vol. 29, No. 13, pp. 2–13, Sept. 1978.

Priestley, M. B. (1981): Spectral Analysis and Time Series. New York, Academic Press.

Rabiner, L. R. and Allen, J. B. (1979): Short-Time Fourier Analysis Techniques for FIR Systems Identification and Power Spectrum Estimation. *IEEE Trans. on Acoustics, Speech and Signal Processing*, Vol. 27, No. 2, pp. 182–192.

Rabiner, L. R. and Gold, B. (1975): Theory and Application of Digital Signal Processing. Englewood Cliffs, NJ, Prentice-Hall.

Rabiner, L. R. and Schafer, R. W. (1969): The Chirp z-Transform Algorithm. *IEEE Trans. on Audio and Electroacoustics*, Vol. 17, No. 2, pp. 86–92, June 1969.

Renders, H., Schoukens, J. and Vilain, G. (1984): High-Accuracy Spectrum Analysis of Sampled Discrete Frequency Signals by Analytical Leakage Compensation. *IEEE Trans. on Instrumentation and Measurement,* Vol. 33, No. 4, pp. 287–292, Dec. 1984.

Rife, D. C. and Vincent, G. A. (1970): The Use of the Discrete Fourier Transform in the Measurement of Frequencies and Bands of Tones. *Bell System Technical Journal,* Vol. 49, No. 2, pp. 197–228.

Robinson, E. A. (1982): A Historical Perspective of Spectrum Estimation. *Proc. IEEE,* Vol. 70, No. 9, pp. 885–907. Sept. 1982.

Robinson, E. A. and Silvia, M. T. (1986): Digital Signal Processing and Time Series Analysis. San Francisco, CA, Holden-Day.

Schoukens, J. and Pintelon, R. (1990): Measurement of Frequency Response Functions in Noisy Environment. *IEEE Trans. on Instrumentation and Measurement,* Vol. 39, No. 6, pp. 905–909, Dec. 1990.

Schoukens, J. and Pintelon, R. (1991): Identification of Linear Systems: a Practical Guideline for Accurate Modelling. London, Pergamon Press.

Schoukens, J., Pintelon, R. and Van hamme, H. (1991): The Interpolated Fast Fourier Transform: A Comparative Study. *Conference Record of the IEEE Instrumentation and Measurement Technology Conference,* May 14–16, 1991, Atlanta, GA, pp. 358–64. 91CH2940-5.

Shannon, C. E. (1949): Communication in the Presence of Noise. *Proc. IRE*, Vol. 37, pp. 10–21.

Sheingold, D. H., ed. (1986): Analog/Digital Conversion Handbook. Englewood Cliffs, NJ, Prentice-Hall.

Silverman, H. F. (1977): An Introduction to Programming the Winograd Fourier Transform Algorithm (WFTA). *IEEE Trans. on Acoustics, Speech and Signal Processing,*

Vol. 25, No. 2, pp. 152–65, April 1977. Addendum and Correction: Vol. 26, No. 3, p. 268, June 1978.

Söderström, T. and Stoica, P. (1989): System Identification. Englewood Cliffs, Prentice-Hall.

Sorenson, H. W. (1980): Parameter Estimation: Principles and Problems. La Jolla, Marcel Dekker.

Souders, T. M., Flach, D. R., Hagwood, C. and Yang, G. L. (1990): The Effects of Timing Jitter in Sampling Systems. *IEEE Trans. on Instrumentation and Measurement,* Vol. 39, No. 1, pp. 80–85, Feb. 1990.

Sripad, A. B. and Snyder, D. L. (1977): A Necessary and Sufficient Condition for Quantization Errors to Be Uniform and White. *IEEE Trans. on Acoustics, Speech and Signal Processing*, Vol. ASSP-25, No. 5, pp. 442–448.

Tikhonov, A. V. and Arsenin, V. Y. (1972): Solutions to Ill-Posed Problems. Washington, John Wiley & Sons.

Thomson, D. J. (1982): Spectrum Estimation and Harmonic Analysis. *Proc. IEEE,* Vol. 70, No. 9, pp. 1055–96. Sept. 1982.

Tumfart, S. (1976): Stochastic-Ergodic Measurement Techniques and Their Contribution towards the Solution of Metrological Problems in Science and Technology. *IMEKO VII. World Congress,* London, 10–14 May, 1976. *Proceedings: Acta IMEKO 1976,* pp. 1263-276.

Urkowitz, H. (1983): Signal Theory and Random Processes. Norwood, MA, Artech House.

Wehrmann, W. (1973): Einführung in die stochastisch-ergodische Impulstechnik. Vienna, R. Oldenburg Verlag. In German.

Weinstein, C. J. (1969): Roundoff Noise in Floating Point Fast Fourier Transform Computation. *IEEE Trans. on Audio and Electroacoustics*, Vol. 17, No. 3, pp. 209–215, June 1967.

Welch, P. D. (1967): The Use of Fast Fourier Transform for the Estimation of Power Spectra: A Method Based on Time Averaging over Short, Modified Periodograms. *IEEE Trans. on Audio and Electroacoustics*, Vol. 15, No. 2, pp. 70–73, June 1967.

Welch, P. D. (1969): A Fixed-Point Fast Fourier Transform Error Analysis. *IEEE Trans. on Audio and Electroacoustics*, Vol. 17, No. 2, pp. 151–57, June 1969.

Widrow, B. (1956a): A Study of Rough Amplitude Quantization by Means of Nyquist Sampling Theory. Sc. D. Dissertation, M.I.T. Elec. Eng. Dept., Electronics Systems Lab., Cambridge, Mass.

Widrow, B. (1956b): A Study of Rough Amplitude Quantization by Means of Nyquist Sampling Theory. *IRE Trans. on Circuit Theory*, Vol. CT-3, Dec. 1956, pp. 266–276.

Widrow, B. (1961): Statistical Analysis of Amplitude-Quantized Sampled-Data Systems. *Trans. AIEE*, Vol. 79, Part II. Appl. and Ind., No. 52, pp. 555–568.

Widrow, B. and Stearns, S. D. (1985): Adaptive Signal Processing. Chapter 10: Inverse Adaptive Modeling, Equalization, and Deconvolution. pp. 231-269. Englewood Cliffs, Prentice-Hall.

Winograd, S. (1976): On Computing the Discrete Fourier Transform. *Proc. Nat. Acad. Sci. USA*, Vol. 73, No. 4, pp. 1005–6, April 1976.

Zellers, J. R. (1979): An Economical Network Analyzer for the 4-to-1300 MHz Frequency range. *Hewlett-Packard Journal*, Vol. 30, No. 10, pp. 9-17, Oct. 1979.

7

Electrical Units

L. Schnell

Editorial introduction

The base units of the SI system have to be realized with high accuracy and long term stability in order to meet strict international requirements on reproducibility and worldwide agreement of measurements. In the following, the presently used base units (primarily electrical units) will be discussed, together with their main properties in measurement applications.

7.1 BASE UNITS OF THE SI SYSTEM

Let us investigate how far base units are determined by relations between physical quantities, and to what extent an arbitrary choice of these base units is possible.

Assume a set of m equations with n unknown quantities ($n > m$). It is known that the degree of freedom is then given by

$$n_0 = n - m\,,$$

allowing an arbitrary choice of n_0 unknown quantities. In specific fields of natural sciences, the degree of freedom means the number of base units which can be arbitrarily chosen (see Table 7.1).

At present, the internationally accepted International System of Units (in French: Système International d'Unités, abbreviated SI) is based on the following seven base units:

- the metre, unit of length,
- the kilogram, unit of mass,
- the second, unit of time,

Technology of Electrical Measurements, Edited by L. Schnell
© 1993 John Wiley & Sons Ltd.

- the ampere, unit of electrical current,
- the kelvin, unit of thermodynamic temperature,
- the candela, unit of luminous intensity,
- the mole, unit of amount of substance.

Table 7.1 Number of base units which can be arbitrarily chosen

	The field of natural science	n_0
I.	Mechanics	3
II.	I + Electricity	4
III.	II + Thermology	5
IV.	III + Optics	6
V.	Physics and Chemistry	7

From these seven base units, all physical and chemical units can be derived by using the equations relating these units to the base units. The base units are defined in Table 7.2.

The physical realization of the base units given in Table 7.2 provides the international standards which are kept in the International Bureau of Weights and Measures in Sèvres (in French: Bureau International des Poids et Mésures, abbreviated BIPM). This Bureau has set up the International Committee of Weights and Measures (Comité International des Poids et Mésures: CIPM) which has the responsibility to deal with problems associated with the base units and international standards. The committee related to electrical units is designated the Consultative Committee on Electricity (CCE).

According to Table 7.1, the field of electricity is covered by four arbitrarily chosen base units, the metre, the kilogram, the second and the ampere. From these base units, all other electrical units can be derived. To produce a new unit not included among the base units, the physical relationship must be found that connects the known and unknown quantities, then it must be realized by an experiment. In this way, all the electrical units can be produced step-by-step (Taylor, 1987).

The important electrical and magnetic units are now summarized.

- The volt (V) is the difference of electric potential between two points of a conducting wire carrying a constant current of 1 A, when the power dissipated between these points is equal to 1 W.
- The ohm (Ω) is the electric resistance between two points of a conductor when a constant difference of potential of 1 V, applied between these two points, produces in this conductor a current of 1 A, this conductor not being the source of any electromotive force (e.m.f.).
- The coulomb (C) is the quantity of electricity transported in 1 s by a current of 1 A.

Table 7.2 Summary of SI base units

Unit designation	Definition	Year of international acceptance
m	The metre is the length of the path, travelled by light in vacuum during a time interval of 1/299 792 458 of a second.	1983
kg	The kilogram is the unit of mass: it is equal to the mass of the international prototype of the kilogram.	1899
s	The second is the duration of 9 192 631 770 periods of the radiation corresponding to the transition between the two hyperfine levels of the ground state of the caesium-133 atom.	1967
A	The ampere is that constant current which, if maintained in two straight parallel conductors of infinite length, of negligible circular cross section, and placed 1 metre apart in vacuum, would produce between these conductors a force equal to $2 \cdot 10^{-7}$ newton per metre of length.	1946
K	The kelvin, unit of thermodynamic temperature, is the fraction 1/273.16 of the thermodynamic temperature of the triple point of water.	1967
cd	The candela is the luminous intensity, in a given direction, of a source that emits monochromatic radiation of frequency $540 \cdot 10^{12}$ hertz and that has a radiant intensity in that direction of (1/683) watt per steradian.	1983
mol	The mole is the amount of substance of a system which contains as many elementary entities as there are atoms in 0.012 kilogram of carbon 12.	1967

- The farad (F) is the capacitance of a capacitor between the plates of which there appears a difference of potential of 1 V when it is charged by a quantity of electricity equal to 1 C.
- The henry (H) is the inductance of a closed circuit in which an e.m.f. of 1 V is produced when the electric current in the circuit varies uniformly at a rate of 1 A/s.
- The weber (Wb) is the magnetic flux which, linking a circuit of one turn, produces in it an e.m.f. of 1 V as the flux is reduced to zero at a uniform rate in 1 second.

- The tesla (T) is a flux density of $1\,\mathrm{Wb/m^2}$.

The BIPM is responsible for the uniformity of electrical units as it is for other units. This provides a central basis for a consistent system of physical measurements that is coordinated with those of other nations.

7.2 REALIZATION OF ELECTRICAL UNITS

The international standards of electrical units should have the following properties:

- extremely high accuracy,
- long term stability,
- portability.

However, there is a discrepancy between the finite accuracy of the realization process resulting in the international standard and the ideal circumstances assumed for the unit definition. This applies mainly to the ampere base unit from which the voltage unit could be derived by using the relation $P = UI$, and the resistance unit could be derived by using the relation $P = I^2R$, by measuring the thermal energy produced by the electrical power P. (The watt does not belong to the base units, it can be derived from mass, length and time.) However, the uncertainties of the ampere standard, to be discussed in Subsection 7.2.1, would then be further extremely increased by the uncertainty of the thermal energy measurement which is much higher than that of electrical measurements. Therefore, from the direct realization of the unit of the current intensity, the other electrical units cannot be derived with sufficient accuracy.

As a result of the technical development, it has become possible to realize the units of voltage and resistance based on the Josephson and the quantum Hall effect, respectively (described in Subsections 7.2.2 and 7.2.3), independently from the current intensity unit. These units are called derived units. These derived units are then subsequently applied for calculating the ampere by using Ohm's law. However, in spite of these practical realization considerations, the electrical base unit of the SI system remains the ampere.

7.2.1 Current Standard by the Current Balance Technique

According to Table 7.2, the ampere is defined as that constant current which, if maintained in two straight parallel conductors of infinite length of negligible circular cross section, and placed 1 metre apart in vacuum, would produce between these conductors a force equal to $2 \cdot 10^{-7}$ newton per metre of length. This base unit can be realized by a physical model based on a relation comprising, in addition to the current unit, only other base units such as the metre, the kilogram and the second.

This relation, given by

$$F = \frac{\mu_0}{2\pi} I^2 \frac{l}{d} \tag{7.1}$$

presents the force F appearing between two parallel conductors, l and d being the distances defined in Fig. 7.1a, μ_0 the permeability of vacuum, and I the current flowing through the two conductors.

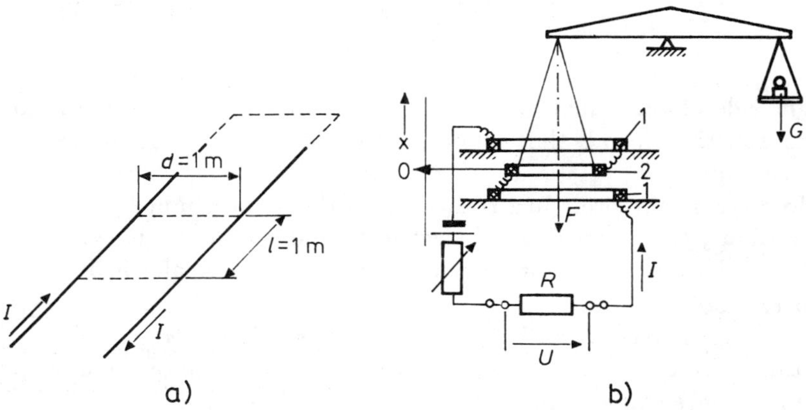

Fig. 7.1 Current balance technique *a)* principle *b)* realization

However, in Eq. (7.1), two unknown quantities are comprised, μ_0 and I. In order to solve Eq. (7.1) for the current I, μ_0 is given the (arbitrary) value of

$$\mu_0 = 4\pi \cdot 10^{-7} \, \frac{\text{Vs}}{\text{Am}} \, .$$

In the case of $d = 1\,\text{m}$ and $I = 1\,\text{A}$, the force appearing between the two parallel conductors along a conductor length of $l = 1\,\text{m}$ is given by

$$F = 2 \cdot 10^{-7}\,\text{N} \, . \tag{7.2}$$

However, this very small force cannot be measured with sufficient accuracy, so a current balance (a device for multiplying this force) has to be used.

Referring to the arrangement shown in Fig. 7.1b, the fixed coil 1–1 and the coil 2, fastened to the balance arm, are connected in series so that the connecting leads do not produce additional moments. By suitably choosing the direction of the current flow, the generated force F will produce a downwards motion of coil 2, introducing a displacement of the balance arm by x. A weight G is then chosen to counteract this displacement, resulting in $x = 0$. The following relation will then hold:

$$F = G = I^2 \left. \frac{\mathrm{d}M_{12}}{\mathrm{d}x} \right|_{x=0} \tag{7.3}$$

Here I is the current flowing through the two coils, and $M_{12} = M_{12}(x)$ gives the change of mutual inductance between coil 1 and coil 2 as a function of x.

Selecting a suitable geometrical layout, this function can be calculated with high accuracy, and can also be measured by applying the Thompson-Lampard capacitor (see Sec. 7.2.2). From Eq. (7.3), we have

$$I = \sqrt{\frac{G}{\left.\frac{\mathrm{d}M_{12}}{\mathrm{d}x}\right|_{x=0}}} \tag{7.4}$$

The right side of this equation comprises μ_0 in M_{12}, and further the metre, the kilogram and the second, thus allowing the calculation of the ampere unit from these base units.

However, in the preceding realization of the ampere unit by the current balance technique, the comparison of electrically induced forces is complicated, and the associated uncertainties are relatively large (Olsen et al, 1985; Weiyand and Willenberg, 1985).

The current balance technique has been used in different forms. However, these forms have no importance any more in the realization of the electrical units, because nowadays simpler and more accurate solutions exist.

7.2.2 Resistance Standard by the Thompson-Lampard Calculable Capacitor

On the basis of Lampard's theoretical work (Lampard, 1957) and Thompson's proposal (Thompson, 1959, 1968), it is possible to build a calculable capacitor standard suitable for realizing a resistance standard.

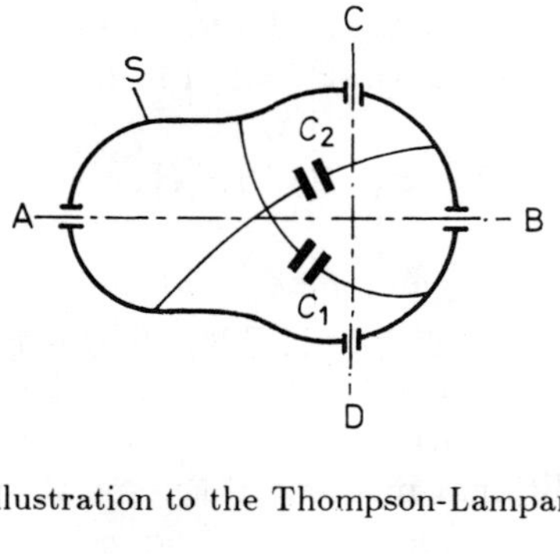

Fig. 7.2 Illustration to the Thompson-Lampard capacitor

Let the closed curve S (see Fig. 7.2) be the right cross section of a conducting cylindrical shell, the cross section having one axis of symmetry AB, but being otherwise arbitrary. Further, let this shell be divided into four parts by two planes at right angles, the line of intersection of the planes being parallel to the generators of the cylinder, and one of the planes containing the line AB. The C_1 and C_2 capacitors marked on Fig. 7.2 have the following properties:

- $C_1 = C_2$,
- their capacitance value is independent from the shape of the cross section, and depends only slightly on the finite width of the cutting,
- the capacitance value depends only on the length of the cylinder as given by

$$C_1 = C_2 = C_0 = \frac{\varepsilon_0}{\pi} \ln 2 = 1.953\,549\,02\,\mathrm{pF/m}, \tag{7.5}$$

where ε_0 is the permittivity of the vacuum, calculated to be

$$\varepsilon_0 = \frac{1}{c_0^2 \mu_0} = \frac{1}{9 \cdot 4\pi} \cdot 10^{-9} \frac{\mathrm{As}}{\mathrm{Vm}}, \tag{7.6}$$

c_0 being the velocity of light in the vacuum, and μ_0 the permeability of the vacuum.

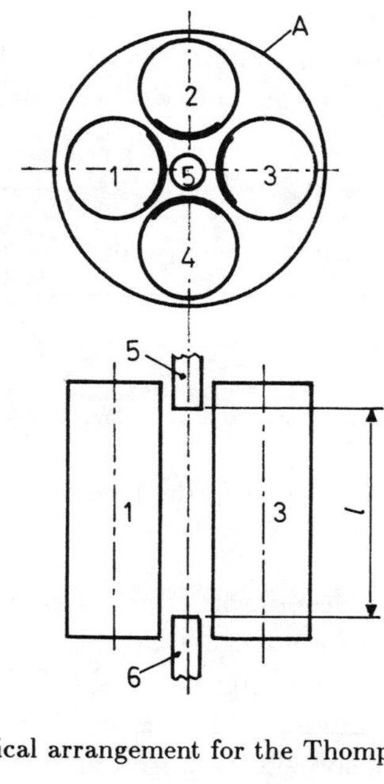

Fig. 7.3 Practical arrangement for the Thompson-Lampard capacitor

The practically implemented variant, as shown in Fig. 7.3, was proposed by Thompson (1959; 1968). Cylinders 1, 2, 3 and 4 are electrically insulated from each other, and symmetrically placed within the shielding cylinder A. In the figure, the surfaces realizing the capacitor C_{13} between cylinders 1 and 3, and the capacitor C_{24}

between cylinders 2 and 4, are shown by heavy lines so the statements referring to Fig. 7.2 are applicable here, too. It can be shown that the distance between the cylinders and a small asymmetry of the arrangement have only negligible effects on the average capacitance, given by

$$C_0 = \frac{C_{13} + C_{24}}{2} . \tag{7.7}$$

Cylinders 5 and 6 are electrically connected to cylinder A, and by their exact positioning by interferometric distance measurement, the length l can be precisely adjusted. This method results in the realization of a small capacitance ($C_0 \approx 0.5\,\text{pF}$) with a length of $l = 0.25\,\text{m}$, at an uncertainty of 10^{-8}.

The next step is the establishment of the resistance standard from this capacitance standard. This is realized by first generating a higher capacitance by a high precision ratio transformer, and subsequently deriving from its reactance $1/\omega C$ — again by a ratio transformer — the international resistance standard. Utilizing present measurement methods, this standard may have an uncertainty of less than 10^{-7}.

7.2.3 Voltage Standard by the Josephson Effect

The Josephson effect (Josephson, 1964), has a high significance from the measurement aspect since it provides a proportionality between a dc voltage and a frequency, the proportionality factor being the ratio of two basic natural constants.

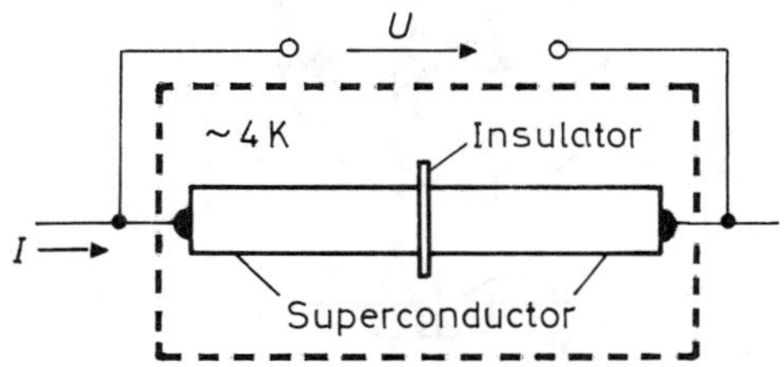

Fig. 7.4 Illustration of the Josephson effect

The arrangement producing the Josephson effect is shown in Fig. 7.4. According to this arrangement, two parts of a superconducting material (e. g. niobium) are separated by an extremely thin insulating material (of the order of 10^{-10} m). The part within the dashed frame is immersed into liquid helium. In the following, only a short description of the Josephson effect is presented, the theoretical background is to be found in Josephson (1964).

[1]▷When such a Josephson junction is irradiated with microwave radiation of frequency f, its current versus voltage curve exhibits current steps at highly precise

[1]In the next pages the text between triangles (▷ and ◁) is cited from Taylor, B. N., New International Representations of the Voltage and Ohm Effective January 1, 1990, *IEEE Trans. on*

quantized Josephson voltage U_J. The voltage of the nth step $U_J(n)$, where n is an integer, is related to the frequency of the radiation by

$$U_J(n) = \frac{nf}{K_J}, \tag{7.8}$$

where K_J, the Josephson constant, is a universal quantity assumed to be equal to the invariant quotient of fundamental constants $2e/h$, where e is the elementary charge and h is the Planck constant. The quantized Josephson voltages depend only upon a readily measured frequency and an invariant quotient of fundamental constants of nature (Melchert, 1979; Taylor, 1985).

To ensure the consistency of national representations of the volt based on the Josephson effect, the CCE also suggested at its October 1972 meeting that national standards laboratories use an 'agreed upon' or conventional value of K_J, namely

$$K_{J-72} = 483\,594\,\text{GHz/V} \tag{7.9}$$

exactly.◁

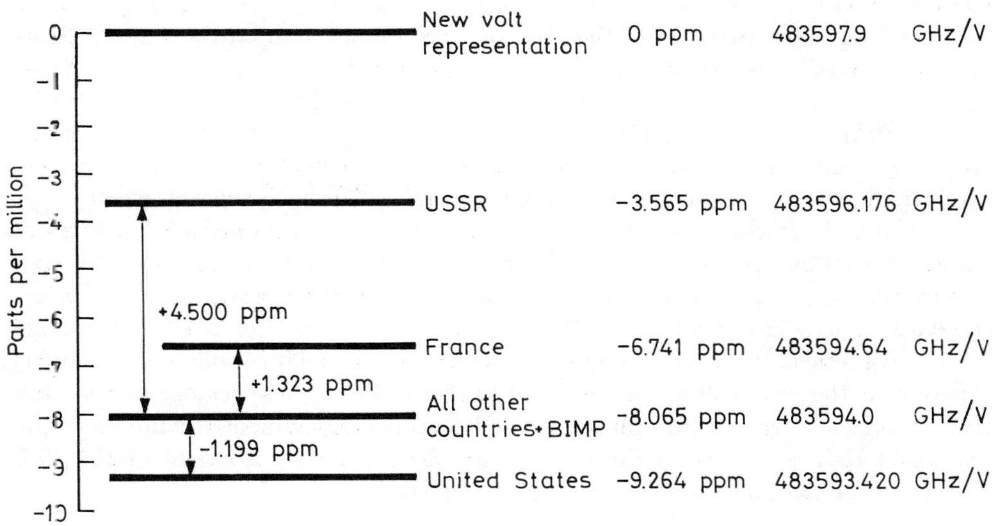

Fig. 7.5 Values of national voltage standards before the correction introduced 1st January, 1990

While most national laboratories adopted this value, three countries voted for the following Josephson constants:

USA:	483 593.420 GHz/V,
FRANCE:	483 594.64 GHz/V,
USSR:	483 596.176 GHz/V.

Instrumentation and Measurement, Vol. 39, No. 1, pp. 2–5, 1990, with the permission of the Publisher.

However, novel experimental results after 1972 induced the CCE to recommend the new conventional value of

$$K_{J-90} = 483\,597.9 \pm 0.2\,\mathrm{GHz/V}\,. \tag{7.10}$$

This value of the Josephson constant was finally accepted by all national laboratories, and came into force in 1990, as shown by the subscript J-90. This adoption necessitated a change in the voltage values of Weston voltage standards kept in national laboratories, as illustrated in Fig. 7.5. The e.m.f., defined by the Josephson constant K_{J-90}, has an estimated uncertainty of $4 \cdot 10^{-7}$, and a much better reproducibility (Taylor, 1987, 1990).

7.2.4 Resistance Standard by the Quantum Hall Effect

▷The resistance standard, based on the Thompson-Lampard capacitor, is an apparatus difficult to use routinely, most national standards laboratories have continued to base their representation of the ohm on wire-wound standard resistors. However, such artifact standards age, and the various national representations of the ohm now differ significantly from each other. Therefore, von Klitzing's 1980 discovery of the quantum Hall effect (QHE, von Klitzing et al, 1980) was welcomed, since it promised to provide a method for basing a representation of the ohm on invariant fundamental constants, in direct analogy with the Josephson effect.

The QHE is characteristic of certain high mobility semiconductor devices of standard Hall-bar geometry when placed in a large applied magnetic field, and cooled to a temperature of about 1 K. For a fixed current I through a QHE device, there are regions in the curve of Hall voltage versus gate voltage, or of Hall voltage versus magnetic field, depending upon the device, where the Hall voltage U_{H} remains constant as the gate voltage or magnetic field is varied. These regions of constant Hall voltage are termed Hall plateaus. Under proper experimental conditions, the quantized Hall resistance of the ith plateau, defined as the quotient of the Hall voltage of the ith plateau to the current I, is given by

$$R_{\mathrm{H}}(i) = \frac{U_{\mathrm{H}}(i)}{I} = R_{\mathrm{K}}/i \tag{7.11}$$

where i is an integer and R_{K}, the von Klitzing constant, is a universal quantity assumed to be equal to the invariant quotient of fundamental constants h/e^2.

At its September 1988 meeting, the CCE also recommended that the national standards laboratories base their national representation of the ohm on the QHE in order to eliminate the problems associated with artifact resistance standards. All national laboratories that choose to use the QHE for this purpose use the same conventional value of the von Klitzing constant, namely

$$R_{\mathrm{K}-90} = 25\,812.807 \pm 0.005\,\Omega \tag{7.12}$$

exactly, where the subscript 90 again indicates that the new ohm representation came into effect on January 1, 1990.◁

The resistance value, defined by this von Klitzing constant, has an estimated uncertainty of $2 \cdot 10^{-7}$ (Taylor, 1990), and a much better reproducibility than the previous standard (Taylor and Witt, 1989).

7.2.5 Artifact Standards

Artifact standards such as Weston elements (e.m.f. $\approx 1.018\,\mathrm{V}$), solid-state voltage standards based on Zener diodes or other semiconductor reference devices, and standard wire-wound resistors, are intended for the long term preservation of externally presented certified voltage or resistance values. In national laboratories, these voltage and resistance values are occasionally generated by performing Josephson and QHE experiments, and by 'copying' these values onto the artifact standards. These are then used to represent the standard values in the periods between these experiments, but should be re-calibrated with suitable frequency to counteract aging and temperature variations. This re-calibration is carried out with relative ease because only small-sized chips are used in the hardware needed for generating the Josephson voltage and the QHE resistance. These are straightforward procedures as compared with the experiments based on the current balance technique or on the Thompson-Lampard calculable capacitor.

7.3 UNCERTAINTY OF SI BASE UNITS AND DERIVED UNITS

The uncertainty of international standards is due to

- the uncertainty of the experiment carried out to realize the unit according to the SI definition, and
- the uncertainty in transferring the standard value from the SI standard to the artifact standard which provides the unit value.

The uncertainty of the above transfer is generally, but not necessarily, one order of magnitude lower than the realization uncertainty.

The uncertainties of the first four base units given in Table 7.2, and further those of the volt and ohm derived units, are presented in Table 7.3. All uncertainties given in this Table correspond to one standard deviation (1σ) estimates (see Subsection 1.6.2).

7.4 CHANGES IN ELECTRICAL STANDARD VALUES

It has been mentioned in Subsection 7.2.3 that, effective from January 1, 1990, the value of the volt has been changed, affecting a change in other units too. Further,

Table 7.3 Unit uncertainties

	Fundamental constant on which realization is based	Uncertainty of realization experiment	Note	Artifact standard description
SI base units:				
metre	$c_0 = \frac{1}{\sqrt{\mu_0 \varepsilon_0}}$ (by experimental determination of light velocity)	$4 \cdot 10^{-9}$	–	Pt rod preserved at BIPM
kg	–	10^{-8}	Due to the uncertainty of weighting	Prototype kg
s	transition between hyperfine levels of the caesium-133 atom	10^{-13}	–	atomic clock
ampere	$\mu_0 = 4\pi \cdot 10^{-7}$	$3 \cdot 10^{-7}$	complicated experiment	–
SI derived units:				
ohm	$\varepsilon_0 = \frac{1}{9.4\pi} \cdot 10^{-9}$ (Thompson-Lampard calculable capacitor)	10^{-7}	complicated realization	wire-wound resistor
ohm	$R_K = h/e^2$ (von Klitzing constant)	$2 \cdot 10^{-7}$		standard
volt	$K_J = 2e/h$ (Josephson constant)	$4 \cdot 10^{-7}$		Weston element Zener diode voltage standard

various efforts have been made to reduce the uncertainties in derived units volt and ohm.

The first substantial result was the discovery of the Thompson-Lampard capacitor in the early sixties which allowed the realization of the ohm unit with satisfactory accuracy. However, in spite of the correct measurement, experimental difficulties prevented a widespread use of this method.

In the middle of the sixties, a decisive breakthrough was achieved by the Josephson effect voltage standard. In 1972, the CCE advised national laboratories to apply the Josephson standard for the definition of the voltage unit. A further substantial result was the discovery of the quantum Hall effect in 1980, capable of relating the ohm unit to a natural constant, similarly to the Josephson effect which allowed the same to happen with the volt unit.

Since 1972, Josephson voltage standard experiments have been analyzed, and the CCE meeting in 1988 decided to introduce a correction, effective from January 1990. Figure 7.5 shows the deviations of earlier voltage standard values from the values introduced in 1990. This meant that any previously measured voltage became smaller by this amount. This figure also shows that in most of the countries, and also in the voltage standard represented by BIPM, this correction amounts to 8.065 ppm.

Following the recommendations of CCE and CIMP, all other countries, as shown in Fig. 7.5, have adopted the ohm value corresponding to the quantum Hall effect, by adopting a von Klitzing constant of $R_K = 25\,812.807$ ohm.

The introduction of these new representations has thus changed the earlier volt and ohm values, and affected also the ampere and watt values. By these changes, the differences between standard values in different national laboratories have been removed. For example, in the United States there were increases of 9.264, 1.69, 7.57, and 16.84 parts per million for the volt, ohm, ampere, and watt representations, respectively.

What these changes mean in a practical sense is that, effective from the 1st of January, 1990, any standard or instrument having an uncertainty approximately equal to or less than 10 times the changes (rounded up) had to be considered for adjustment. For example, in the United States, this encompassed standards or instruments with uncertainty levels of 100, 20, 80, and 170 ppm or better for voltage, resistance, current, and power measurements, respectively. If the application was not especially critical, a factor of five times could probably be safely followed (Taylor, 1989).

NIST Technical Note 1263 gives detailed technical instructions on how to bring standards of voltage and resistance and related instrumentation into conformity with the new representations.

References

Josephson, B. D. (1964): Coupled Superconductors. *Review of Modern Physics,*, Vol. 36, No. 1, pp. 216–220, Jan. 1964.

Lampard, D. G. (1957): A New Theorem in Electrostatics and Its Application to Calculable Standards of Capacitance. *Proc. IEE, Part C,* Vol. 104, No. 13, pp. 271–280. Monogr. 216M.

Melchert, F. (1979): Darstellung der Spannungseinheit mit Hilfe des Josephson-Effektes (Representation of the Voltage Unit Using the Josephson Effect). *Technisches Messen,* Vol. 46, No. 2, pp. 59–64. (In German)

NIST (1989): Technical Note 1263: Guidelines for implementing the new representations of the volt and ohm effective January 1, 1990. By Belecki, N. B., Dziuba, R. F., Field, B. F. and Taylor, B. N.

Olsen, P. T., Bower, V. E., Phillips, W. D., Williams, E. R. and Jones, G. R. jr. (1985): The NBS Absolute Ampere Experiment. *IEEE Trans. on Instrumentation and Measurement,* Vol. 34, No. 2, pp. 175–181, June 1985.

Taylor, B. N. (1985): Electrical Units, Fundamental Constants. *IEEE Trans. on Instrumentation and Measurement,* Vol. 34, No. 2, pp. 155–162, June 1985.

Taylor, B. N. (1987): History of the Present Value of $2e/h$ Commonly Used for Defining National Units of Voltage and Possible Changes in National Units of Voltage and Resistance. *IEEE Trans. on Instrumentation and Measurement,* Vol. 36, No. 3, pp. 659–664, June 1987.

Taylor, B. N. (1989): Preparing for the New Volt and Ohm. *IEEE Spectrum,* Vol. 26, No. 7, pp. 20–23, July 1989.

Taylor, B. N. and Witt, T. J. (1989): New International Electrical Reference Standards Based on the Josephson and Quantum Hall Effects. *Metrologia,* Vol. 26, No. 1, pp. 47–62.

Taylor, B. N. (1990): New International Representations of the Voltage and Ohm Effective January 1, 1990. *IEEE Trans. on Instrumentation and Measurement,* Vol. 39, No. 1, pp. 1–5, Feb. 1990.

Thompson, A. M. (1959): The Cylindrical Cross-Capacitor as a Calculable Standard. *Proc. IEE, Part B,* Vol. 106, No. 27, pp. 307–310, May 1959. Paper 2887M.

Thompson, A. M. (1968): An Absolute Determination of Resistance Based on a Calculable Standard of Capacitance. *Metrologia,* Vol. 4, pp. 1–7.

von Klitzing, K., Dorda, G. and Pepper, M. (1980): New Method for High Accuracy Determination of the Fine-Structure Constant Based on Quantized Hall Resistance. *Phys. Rev. Lett.,* Vol. 45, No. 6, pp. 494–497, Aug. 1980.

Weiyand, K. and Willenberg, G. D. (1985): The Proposed Ampere Experiment at the PTB. *IEEE Trans. on Instrumentation and Measurement,* Vol. 34, No. 2, pp. 192–198, June 1990.

Author Index

Subject Index